Airline Transport and IATRA

Ground School Course

Michael J. Culhane

Accelerated Aviation Training
Richmond BC

AcceleratedAviation.com

INTRODUCTION

Congratulations on your decision to obtain your ATPL (or Individual Aircraft Type Rating)!

This course has been completed to assist you in preparing for the Transport Canada Airline Transport written examinations for Aeroplanes - SARON (Aviation Regulations and Air Traffic Procedures, Aeroplane Operations and Navigation General) and SAMRA (Meteorology, Radio Aids to Navigation and Flight Planning). This course can also be used to prepare for the Individual Aircraft Type Rating written - IATRA (since all IATRA topics derive from ATPL subjects, study as if you were preparing for the ATPL). The course is divided into the following sections:

1.00   Air Law and Procedures
2.00   Airframes, Engines, Systems
3.00   Instruments
4.00   Flight Operations
5.00   Navigation General
6.00   Meteorology
7.00   Radio Communications and Aids to Navigation

Each subject in this course has been assigned a numbered subject reference code. This subject numbering is the basis on which we classify the relevant subjects in this course so that you can quickly cross reference a desired topic. As part of completing this course and to ensure that you are fully prepared to write the TC writtens, you should also obtain and complete the sample exams provided in our **Airline Transport and IATRA Written Test Book, 2007 Edition**.

By completing this course manual and thereafter completing our sample exams, and by referencing and re-learning any weak subjects identified by the sample exams, you will be assured of strong performance on the actual TC writtens. To study for the ATPL writtens on a test-by-test basis, study Sections 1, 2, 3 and 4 for the SARON, followed by Sections 5, 6, 7 and a review of Section 1 (flight planning topics only) for the SAMRA.

With respect to flight computers, a basic E6B, CR circular type or electronic flight computer will all suffice for the purposes of the ATPL/IATRA writtens. Regardless of the type of computer used, be sure that you are competent at completing basic, standard flight calculations such as wind drift - heading - ground speed, time - speed - distance etc. A pocket calculator with a square root function is very useful for quickly adding up sets of distances, weights etc. and the square root function is needed to complete various calculations.

Other than our supplemental written test book, you will not need any other secondary publications to prepare for the ATPL/IATRA writtens: we have carefully selected and summarized all the subjects you will need to know so that you can quickly learn the required material without having to independently source and study from any outside references.

This training manual has been written with the assumption that you are CURRENT to the level of the Canadian Commercial Pilot Licence. If your basic CPL knowledge is weak or a little out-of-date, or if you are a foreign pilot converting to Canadian requirements for the first time, it is highly recommended that you first obtain and complete our **Commercial Pilot Ground School Course, 2007 Edition** before you commence this course.

Given the rapid rate of change to the field of aeronautics, this publication is subject to ongoing revisions. Any amendments that we make to this text during its currency period will be posted to our website. We welcome any feedback you may wish to provide that could improve the content, accuracy or effectiveness of this course.

This 2007 revised edition is the most current edition of this text and supersedes all previous editions. This edition is not intended to remain valid indefinitely and contains time sensitive information. We normally revise and re-issue this text on an annual, ongoing basis. Upon our publication of a revised edition to this text, we consider this text to have become obsolete and we ask that users not use or rely upon this edition of this text should it have fallen out of date.

Those wishing to supplement their study for the ATPL/IATRA can use our ATPL/IATRA Online Ground School. Our internet course consists of approximately 1,000 programmed questions and answers, inclusive of detailed instructions for complex calculations and problem chains, multiple sets of review quizzes and sample exams, all of which have been cross referenced to this training text. Please visit our website for further details on our online courses.

We take this opportunity to thank you for purchasing this course and wish you every success on your ATPL/IATRA writtens!

Michael J. Culhane
ATPL BSc BA LLB

September, 2006
Revised December, 2006

Accelerated Aviation Training
Unit 130-5980 Miller Road
YVR International Airport
Richmond BC V7B 1K2 CANADA
Tel. 604-279-0179   FAX 604-279-0161

www.acceleratedaviation.com

TABLE OF CONTENTS

ABBREVIATIONS

AC	Alternating Current
ACA	Arctic Control Area
ACAS	Airborne Collision Avoidance System
ACATS	Arctic Control Area Track Structure
ACC	Area Control Centre
AD	Airworthiness Directive
ADF	Automatic Direction Finder
ADIZ	Air Defence Identification Zone
AFTN	Aeronautical Fixed Telecommunication Network
AGL	Above Ground Level
AI	Attitude Indicator
AIC	Aeronautical Information Circular
AIM	Aeronautical Information Manual
AIP	Aeronautical Information Publication
AM	Amplitude Modulation
AME	Aircraft Maintenance Engineer
ANAL	Analysis
AOC	Air Operator Certificate
ASL	Above Sea Level
ASR	Airport Surveillance Radar
AST	Atlantic Standard Time
ATC	Air Traffic Control
ATPL	Airline Transport Pilot License
ATS	Air Traffic Services
BFS	Bearing from Station
BOW	Basic Operating Weight
BTS	Bearing to Station
cA	Continental Arctic
CADC	Central Air Data Computer
CAP	Canada Air Pilot
CARs	Canadian Aviation Regulations
CARS	Community Aerodrome Radio Station
CAS	Calibrated Airspeed
CAT	Clear Air Turbulence
CAVOK	Ceiling and Visibility OK
Cb	Cumulonimbus
CDA	Canadian Domestic Airspace
CDI	Course Deviation Indicator
CFS	Canada Flight Supplement
CLC	Course Line Computer
CMC	Canada Meteorological Center
CMNPS	Canadian Minimum Navigation Performance Specifications Airspace
COM	Company Operations Manual
CO2	Carbon Dioxide
C of G	Centre of Gravity
CPC	Canadian Pilotage Chart
CPL	Commercial Pilot License
CRFI	Canadian Runway Friction Index
CST	Central Standard Time
CP	Critical Point
CTA	Control Area
CVR	Cockpit Voice Recorder
DALR	Dry Adiabatic Lapse Rate
DND	Department of National Defence

DOT	Department of Transport
DME	Distance Measuring Equipment
DT	Daylight Savings Time
DVFR	Defence VFR
EADI	Electronic Attitude Director Indicator
EAS	Equivalent Airspeed
ECU	Electronic Control Unit
EHSI	Electronic Horizontal Situation Indicator
EOW	Empty Operating Weight
EPR	Exhaust Pressure Ratio
ESCAT	Emergency Security Control of Air Traffic Rules
EST	Eastern Standard Time
ETOPS	Extended Range Twin Engine Operations
FCU	Fuel Control Unit
FD	Upper Winds Forecast
FDR	Flight Data Recorder
FIR	Flight Information Region
FL	Flight Level
FM	Frequency Modulation
FMS	Flight Management System
FSS	Flight Service Station
GFA	Graphic Area Forecast
GHA	Greenwich Hour Angle
GPS	Global Positioning System
GPWS	Ground Proximity Warning System
GRIV	Grivation
HF	High Frequency
Hg	Mercury
Hz	Hertz
IAS	Indicated Airspeed
ICAO	International Civil Aviation Organization
ILS	Instrument Landing System
INS	Inertial Navigation System
IRS	Inertial Reference System
ISS	Integrated Satellite System
KHz	Kilohertz
KPH	Kilometers per Hour
LE	Enroute Low Altitude
LEMAC	Leading Edge Mean Aerodynamic Chord
LF	Low Frequency
LGW	Landing Gross Weight
LHA	Local Hour Angle
LORAN	Long Range Navigation
LSB	Lower Sideband
LRC	Long Range Cruise
MASPS	Minimum Aircraft System Performance Specification
mA	Maritime Arctic
MAC	Mean Aerodynamic Chord

Mb	Millibar
MCTOW	Maximum Certified Take-Off Weight
MEA	Minimum Enroute Altitude
MEL	Minimum Equipment List
METAR	Aviation Routine Weather Report
MF	Mandatory Frequency
MFA	Military Flying Area
MFD	Multi Function Display
MH	Magnetic Heading
MHz	Megahertz
MOT	Ministry of Transport
mP	Maritime Polar
MSC	Maximum Speed Cruise
MSL	Mean Sea Level
MST	Mountain Standard Time
MT	Maritime Tropical
NAM	Nautical Air Miles
NARP	North American Route Program
NAT MNPS	North Atlantic Minimum Navigation Performance Specifications Airspace
NCA	Northern Control Area
NRP	National Route Program
NTS	Northern Track System
OAT	Outside Air Temperature
OCA	Oceanic Control Area
OM	Outer Marker
ONS	Omega Navigation System
OTS	Organized Track System
PAR	Precision Approach Radar
PFD	Primary Flight Display
PIC	Pilot in Command
PIREP	Pilot Report
PNR	Point of No Return
PPC	Pilot Proficiency Check
PRM	Preferred Route Message
PST	Pacific Standard Time
PTC	Pitch Trim Compensator
PTS	Polar Track Structure
RA	Resolution Advisory
RADAR	Radio Detection and Ranging
RAFC	Regional Area Forecast Center
RAREP	Radar Report
RB	Relative Bearing
RCO	Remote Communications Outlet
RMI	Radio Magnetic Indicator
RNAV	Area Navigation
RNPC	Required Navigation Performance Capability Airspace
RVR	Runway Visual Range
RVSM	Reduced Vertical Separation Minimum
S	Speed of Sound
SALR	Saturated Adiabatic Lapse Rate
SAR	Search and Rescue
SAR	Specific Air Range
SCA	Southern Control Area
SELCAL	Selective Calling System
SGR	Specific Ground Range
SHA	Sidereal Hour Angle

SIGMET	Significant Meteorology
SSB	Single Sideband
SSR	Secondary Surveillance Radar
STAR	Standard Terminal Arrival Routing
SVFR	Special VFR
TA	Traffic Alert
TACAN	Tactical Air Navigation
TAF	Aerodrome Forecast
TAS	True Airspeed
T/B	Track Bar
TC	Transport Canada
TC	True Course
TC AIM	Transport Canada Aeronautical Information Manual
TCAS	Traffic Alert and Collision Avoidance System
TEMAC	Trailing Edge Mean Aerodynamic Chord
TFS	Track from Station
TODA	Take-Off Distance Available
TORA	Take-Off Run Available
TROWAL	Trough of Warm Air Aloft
TSB	Transportation Safety Board of Canada
TRU	Transformer Rectifier Unit
TSR	Terminal Surveillance Radar
TTS	Track to Station
TOC	Time of Useful Consciousness
UHF	Ultra High Frequency
USB	Upper Sideband
UTC	Universal Time Coordinated
VHF	Very High Frequency
VLF	Very Low Frequency
VMC	Visual Meteorological Conditions
VNC	Visual Navigation Chart
VOR	VHF Omnirange
VOT	Omnitest
WAC	World Aeronautical Chart
WCA	Wind Correction Angle
Z	Universal Time Coordinated
ZFW	Zero Fuel Weight

ACKNOWLEDGEMENTS

I wish to thank those pilots I have trained over the years ... the aircraft, flight simulator and classroom are all not only learning environments for the student, but also for the instructor.

I am grateful to those users of my training manuals who have provided me with specific recommendations towards improving the content, accuracy and effectiveness of my training manuals.

I wish to thank the airworthiness, pilot licensing and library staff at Transport Canada, in the Pacific, Prairie and Northern and Ontario Regional offices, who have always been courteous and helpful in handling my inquires.

My thanks to the aviation bookstores, aviation products wholesalers and pilot supply distributors who have carried and promoted our product line, and to the flying schools, clubs and instructors who have chosen to endorse my manuals and have adopted them in ground and flight training programs.

The aviation magazines and periodicals *COPA Flight*, *Airways*, *Pilot* and *Aviation Canada* have made favourable editorials for which I thank the editors and publishers.

Our current/past corporate memberships in the Canadian Owners and Pilot's Association (COPA) and in the Air Transport Association of Canada (ATAC) have been of enormous assistance in keeping current with developments in aviation in Canada. The ongoing hard work of COPA and ATAC is appreciated not only by myself but most certainly by all other persons connected with aviation in Canada: we benefit greatly from their energetic representation.

My friends Bill Walker, Capt. Pierre F. Gavillet, and Stuart and Laurel Walker have been energetic in their support and have provided useful comments and recommendations on the format and style used in my publications.

Permission to reproduce Canadian aeronautical charts or portions thereof has been granted by Geomatics Canada, Department of Natural Resources © 2002. Charts in this publication are not to be used for navigation purposes. Information on the procurement of current aeronautical charts and publications can be obtained by calling the Canada Map Office at 1-800-465-6277.

I thank Able Bookbinding, Budget Printing, Colour Time, Copies Plus, Kinko's Copies, Rainbow Press, Speedy Printing and Teldon who have all provided excellent service over the years in relation printing/binding/shrink wrapping our training manuals, book covers, annual catalogues and flyers.

This publication has been created using Adobe Pagemaker, Adobe Illustrator, and Adobe Photoshop software. My thanks to Apple Computers Inc. and to Adobe Systems Inc. for their reliable and user-friendly DTP products.

My family deserves mention for their support and assistance throughout: thanks Mom, Dad, Ruth, Steve, Dave, Martha, Tara, Laura, Noriko and Irena.

Above all and most importantly I wish to thank Patrizia Rossi who has been of invaluable assistance in the design, layout, production and completion of this publication.

MJC

Air Law and Procedures

1.01 Airline Transport Licence - Aeroplane

An applicant must be at least 21 years old, and have completed a Category 1 Medical Certificate (MC). The license holder may exercise the privileges of the Private Pilot License until the period for the Private Pilot medical lapses. The licence is maintained by a valid Category 1 MC.

At least 70% must be obtained in each of two written exams: SARON (Aeroplane Operations, Navigation General, Canadian Aviation Regulations - CARs, NOTAM, Human Factors) and SAMRA (Meteorology, Radio Aids to Navigation, Flight Planning).

The total flight time required is 1500 hours of which 900 hours shall have been acquired in aeroplanes. The total flight time shall also include a minimum of:

a) 250 hours flight time in aeroplanes as pilot-in-command (PIC), a maximum of 100 hours PIC under supervision (see s. 1.03 for PIC under supervision requirements, carry over requirements and limitations). The PIC time and/or PIC under supervision shall include 100 hours cross country flight time of which not less than 25 hours shall have been by night;
b) 100 hours night flight time in aeroplanes as PIC or as co-pilot of which a minimum of 30 hours shall have been in acquired in aeroplanes;
c) 200 hours cross-country flight as co-pilot in an aeroplane required to be operated with a co-pilot or, instead, 100 additional hours cross-country flight time as PIC which may have been part of the 250 hours specified in paragraph (a);
d) 75 hours instrument time of which up to a maximum of 25 hours may have been acquired in a flight simulator and a maximum of 35 hours may have been acquired in helicopters.
Instrument ground time cannot be applied to the 1500 hour flight time requirement.

Credit can be applied for up to a maximum of 50 hours toward the 1500 hour requirement where the applicant holds a pilot licence - Glider. Credit can be applied for up to a maximum of 50 hours toward the 1500 hour requirement where the applicant holds a Pilot Permit - Ultra Light Aeroplane.

An applicant shall have satisfactorily demonstrated in a multi-IFR aeroplane (excluding centre thrust) the ability to perform both normal and emergency flight maneuvers appropriate to the aeroplane in which the flight test is conducted, and to execute all maneuvers and procedures required for an Aeroplane Group 1 Instrument Rating.

Active and retired Canadian Forces personnel who hold a Canadian Forces Instrument Rating (unrestricted) in a multi-engined aeroplane (Group 1) shall be deemed to have met the skill requirement

[Reference: CARs 421.34]

1.02 Medical Requirements

No person shall exercise or attempt to exercise the privileges of a permit, licence or rating unless the person holds a valid medical. A Category 1 MC is the minimum standard for an ATPL or CPL. The requirements are:

a) ELECTROCARDIOGRAM (ECG) is required at first medical, within the two years preceding the examination between ages 30 and 40, and within the 12 months preceding the exam after age 40;
b) AUDIOGRAM is required at initial medical and at the first medical after age 55;
c) VISUAL REQUIREMENT shall be at least 20/30 in each eye separately, with or without the use of correcting lenses;
d) CHEST X-RAY is required at initial examination only in cases where there is a history or clinical evidence of respiratory problems;
e) VALIDITY of a Cat 1 medical is 12 months when under 40 years of age, and 6 months when 40 years of age or over. The medical validity period is calculated from the first day of the month following the date of medical examination.

No person shall exercise his/her licence privileges if any of the following exists:

a) the holder has any illness, injury or disability;
b) the holder is taking a drug;
c) the holder is receiving medical treatment;
d) the holder was involved in an aircraft accident that is in any way related to a), b) or c) above;
e) the holder has entered the 30th week of pregnancy or given birth in the preceding 6 weeks.

A medical examiner must re-examine the holder for those involved in a medical related accident, or for those having given birth.

[Reference: CARs 404.03, 404.04, 404.06, 404.10]

1.03 Pilot Licence Privileges

General Licencing Requirements

For aviation licencing purposes, CATEGORY refers to balloon, aeroplane, helicopter, ultra light aeroplane and glider. CLASS refers to single-engine land aeroplane, single-engine sea aeroplane, multi-engine land aeroplane, multi-engine sea aeroplane.

No person shall act as a flight crew member or exercise the privileges of a flight crew permit, licence or rating unless the proper permit, licence or rating has been obtained.

While exercising the associated privileges of a permit, licence or rating, the holder must have in his/her possession the appropriate permit, licence or rating and must be able to produce it along with an appropriate and valid MC.

No person shall act as a flight crew member or exercise the privileges of a flight crew licence in Canada in a aircraft registered in a contracting state other than Canada, unless the person holds, and can produce while so acting or while exercising such privileges, a valid Canadian flight crew licence or a flight crew licence or a foreign licence validation certificate issued under the laws of the contracting state.

Pilot Recency

5 Year Recency
No pilot shall exercise the privileges of his/her licence/permit/rating unless the holder has acted as PIC or co-pilot of an aircraft within the 5 years preceding the flight. To regain currency, within the 12 months preceding the flight, the holder must pass the appropriate TC written and complete a FLIGHT REVIEW (same as flight test) in the appropriate aircraft with a licenced flight instructor; the instructor certifies in the pilot's logbook that he/she meets required skills for the licence/permit/rating.

24 Month Recency
Within the 24 months preceding a flight, all licence/permit holders must complete a RECURRENT TRAINING PROGRAM, which is any of:

a) attendance at a TC safety seminar;
b) participation in a TC approved recurrent training program covering human factors, meteorology, flight planning and navigation and regulations;
c) completion TC annual self-paced study program found in the current *Aviation Safety Newsletter*;
d) completion of a training program for a PPC OR completion of a pilot permit/licence/rating including a night rating, VFR OTT rating, IFR rating, multi-engine rating, flight instructor rating, landplane or seaplane rating;
e) completion of permit/license/rating TC written.

6 Month Recency - Carrying Passengers
For any flight if passengers are carried, the pilot must have done at least 5 take-offs and landings in the same aircraft category and class within the previous 6 months, by day or night if the flight is by day, and by night if the flight is by night.

Pilot Logbooks

Anyone holding a pilot license must keep a logbook, to prove compliance with experience requirements and to prove dual/solo when applying for a license. Entries should be periodically certified by the CFI or owner. A logbook must have at least:

a) pilot's name;
b) date of flight;
c) type of aircraft and registration;
d) whether flight dual, PIC or Co-pilot;
e) flight conditions: day, night, VFR or IFR;
f) place of departure, arrival, intermediate stops;
g) flight time.

No person shall make an entry in a personal log unless the person is the holder of the log or has been authorized to make the entry by the holder of the log.

ATPL Privileges

A CPL license or ATP license is required to act as PIC or co-pilot of an aircraft operated in a commercial air service. The holder of an ATP license-aeroplane may exercise the privileges of a CPL pilot, and act as PIC in a commercial air service on any class and type of aeroplane for which his/her license is endorsed. The principal difference between the CPL and ATPL is that an ATPL holder can act as PIC in a commercial air service on an aeroplane with a two pilot minimum flight crew, while the CPL can only act as co-pilot on a two pilot minimum flight crew aircraft commercially.

Instrument Rating

To exercise the privileges of the ATP license, the pilot must maintain an IFR rating (see s. 1.06).

Flight Instructor Rating

Flight instructor rating privileges may be added to a CPL or ATPL. For aeroplanes, there are 4 classes of instructor rating: Class 1, 2, 3, and 4.

Crediting of Co-pilot Experience

An ATPL applicant may carry over a maximum of 100 hours of PIC flight time to meet the ATPL licencing requirements subject to these conditions:

a) the air operator must have instituted an approved AIRLINE TRANSPORT LICENCE TRAINING PROGRAM of supervision to allow co-pilots to credit flight time as PIC time;
b) the applicant must hold at least a CPL, multi engine rating and appropriate type rating, and Group 1 instrument rating. As well, the applicant must have accumulated a minimum of 150 hours PIC flight time in aeroplanes;
c) a maximum of 50% of the PIC under supervision flight time may be carried over to meet the ATPL requirements to a maximum of 100 hours.

No person shall record in a personal log the flight time acquired by a co-pilot while acting as PIC under supervision unless the flight time has been acquired an accordance with an airline transport licencing training program. The correct details should also be logged (i.e. minimum of one takeoff and one landing for each 10 hours of flight time, PIC duties performed etc.).

[Reference: CARs 401.03, 401.04, 401.05, 401.08, 401.10, 401.34, 401.61, 421.05, 421.10, 421.11]

1.04 Aeroplane Class Ratings

Seaplane Class Rating

Conversion training to add seaplane privileges are as follows:

a) 7 hours of seaplane training;
b) at least 5 hours dual instruction;
c) at least 5 take-offs and landings as sole occupant of aeroplane (as PIC for 2 crew aircraft);
d) training on taxiing, sailing, docking, takeoffs and landings;
e) experience on glassy water, rough water and crosswinds (recommended);
f) conversion training provided by holder of CPL or ATPL with at least 50 hours seaplane experience.

Landplane Class Rating

Conversion training for landplane privileges are:

a) 3 hours of landplane training;
b) at least 2 hours dual instruction;
c) training to include taxiing, landings, including crosswind and takeoffs;
d) at least 5 solo take-offs and landings;
e) conversion training provided by holder of CPL or ATPL with at least 50 hours PIC landplane experience.

Multi-Engine Class Rating

No specific number of hours of training are needed to obtain a multi engine rating. For training licenced pilots, the instructor must hold a valid CPL or ATPL endorsed for the type of aeroplane and at least 50 hours of multi-engine experience with at least 10 hours experience on type. The training shall be conducted in accordance with the *Flight Instructor Guide, Multi-Engine Class Rating - TP219E.*

A flight test must be successfully completed with a TC inspector or designated flight test examiner.

[Reference: CARs 421.38, 425.21]

1.05 Aircraft Type Ratings

Blanket Aircraft Type Ratings

A BLANKET AIRCRAFT TYPE RATING is rating endorsed on a pilot license that allows a pilot to operate any aircraft in a broad classification. A blanket type rating is obtained by completing the licensing requirements for a flying license. For example, a pilot who has completed his/her multi-engine rating and IFR rating on a Cessna 310 (or any other light twin) will be issued the following blanket type rating: "All single pilot, multi-engine, land aeroplanes excluding high performance". The blanket type rating only applies to single pilot minimum flight crew aircraft.

Individual Aircraft Type Ratings

An INDIVIDUAL AIRCRAFT TYPE RATING is issued for aircraft not included in a blanket type rating. It is indicated by an aircraft type designator (see Fig. 1-1 for type designators) endorsed on a licence. For aeroplanes, individual type ratings are required in the following cases:

a) each aeroplane with a minimum flight crew of at least two pilots;
b) each aeroplane with a minimum flight crew requirement of at least two pilots utilizing a cruise relief pilot;
c) each aeroplane type to be endorsed on a flight engineer licence;
d) each aeroplane type to be endorsed on a second officer rating;
e) each aeroplane type to be endorsed on a licence for which no class rating is issued;
f) each high performance aeroplane type to be endorsed on a pilot licence - aeroplane category.

Whether an aircraft requires one or two pilot minimum flight crew to operate it depends on the aircraft type certificate, flight manual, flight permit or aircraft type approval. Fig. 1-1 gives type designators and corresponding minimum flight crews.

Individual Type Rating, Two Crew Aeroplane

To obtain an individual type rating on an aeroplane having a minimum flight crew of at least two pilots, the following is required:

a) at least 250 hours total aeroplane flight time;
b) completion of course acceptable to Minister of ground school and flight training on type;
c) successful completion (70% or more) of ATPL writtens (if 750 hours total time) or IATRA written exam within previous 24 months;
d) successful completion of pilot proficiency check (PPC) on type within 12 months preceding the application for the rating.

Training for an individual aeroplane type rating can be given by the holder of a CPL or ATPL (aeroplane category) having at least 50 hours flight time on the aeroplane class, of which at least 10 hours are on the aeroplane type.

CREDITS FOR SIMILAR TYPES may be available where aircraft types share common operational characteristics. For example, a pilot having an individual type rating on a B-757 need only complete ground training on systems, limitations, engines and weight and balance differences to qualify for a type rating on a B-767. As well, an approved simulator can also be used for meeting the flight training requirement on some aircraft. Contact TC for specific requirements.

AERMACCHI		
MC6	AL60	1R/1
AERONCA MANUFACTURING		
AR11	All series 7 and 11 Chief/super Chief	1R/1
AR15	Sedan	1R/1
AR58	Champion Lancer	2R/1
AEROSPATIALE SOCAT-SNIAS		
S760	Paris MS760	2T/2
TB09	Tampico	1R/1*
TB10	Tobago	1R/1
TB20	Trinidad (models 20 & 21)	1R/1
TB70	TBM 700	1T/1*
AIR TRACTOR		
AT301	Model 301	1R/1
AT401	Model 401	1R/1
AIRBUS INDUSTRIE		
EA30	Airbus A300, except 600	2T/2
EA306	Airbus A300, series 600	2T/2
EA31	Airbus A310	2T/2
EA32	Airbus A320	2T/2
EA34	Airbus A340	4T/2
AUSTER		
AUST	Mark VI and VII	1R/1
AVIATION TRADERS		
AT98	Carvair ATL98	4R/2
AVION DE TRANSPORT REGIONAL		
AT42	ATR42	2T/2
AT72	ATR72	2T/2
AVIONS PIERRE ROBIN		
R200	All 2000 series models	1R/1
R300	All 3000 series models	1R/1
AYRES		
AYSC	Thrush SC2R-1340 (piston)	1P/1
AYSCT	Thrush SC2R-T15 (turbine)	1T/1
BARKLEY GROW		
T8P	Barkley Grow	2R/1
BEECH AIRCRAFT CORP.		
BE10	King Air 100	2T/1*
BE11	Kansan (all model 11)	2R/1
BE17	Stagger Wing (all model 17)	1R/1
BE18	Beech 18 (all model 18, C45, Expeditor)	2R/1
BE19	Sport 150	1R/1
BE20	King Air 200	2T/1*
BE23	Sundowner	1R/1
BE24	Sierra	1R/1
BE30	Beech 300	2T/2
BE35	Beech 350	2T/1*
BE36	Bonanza (includes 33/35)	1R/1
BE40	Beechjet	2T/2
BE50	Twin Bonanza (all model 50)	2R/1
BE55	Baron (all model 55, 56, 58)	2R/1
BE60	Duke (all model 60)	2R/1
BE76	Duchess (all model 76)	2R/1
BE80	Queen Air (all model 65, 65-80, 65B and 70)	2R/1
BE90	King Air 90	2T/1*
BE95	Travel Air (all model 95)	2R/1
BE99	Airliner (all model 99)	2T/1*
BE02	Commuter (Beechcraft 1900)	2T/2
BEST	Starship (model 2000)	2T/1*
BELLANCA AIRCRAFT CORP.		
BL14	Aircruiser	1R/1
BL26	Viking (all models)	1R/1
BL28	Scout	1R/1
BL31	Skyrocket	1R/1
CH10	Citabria (all models)	1R/1
BEST AVIATION		
W620	Weatherly 620B	1R/1
BOEING CO.		
B17	Fortress	4R/2
B707	Stratoliner (707-720, CC137, C and KC135)	4T/2
B727	Astrojet	3T/2
B737	737 100/200	2T/2
B7373	737/300	2T/2
B7374	737/400	2T/2
B7375	737/400	2T/2
B747	Super Jet (all 100, 200, 300)	4T/2
B7474	Super Jet series 400	4T/2
B75	Stearman	1R/1
B757	All models Boeing 757	2T/2
B767	All models Boeing 767	2T/2
B97	Stratofreighter (KC97)	4R/2
BRITISH AEROSPACE		
BA11	BAC111	2T/2
BA31	Jetstream 31	2T/2
BA32	Jetstream 3200 Series	2T/2
BA41	Jetstream 41	2T/2
BA46	146 (all)	4T/2
BRITISH AIRCRAFT CORP.		
BR31	Britannia 310	4T/2
BR70	Freighter (Wayfarer)	2R/2
VC8	Viscount (all VC2-700 and 800 series)	4T/2
VC9	Vanguard (all VC950 series)	4T/2
BRITTEN-NORMAN		
BN2	Islander (all model BN-2, BN-29)	2R/1
BN3	Trislander BN-2A	3R/1
CAMAIR CORP.		
CM48	Twin Navion (all models)	2R/1
CANADAIR LTD.		
CL21	Canadian CL215	2R/2
CL22	CL215 Turbine	2T/2
CL44	Yukon CL44	4T/2
CL60	Challenger (includes CL601)	2T/2
CL65	Regional Jet	2T/2
CASA		
CS12	Casa 212/200	2T/2
CESSNA AIRCRAFT CORP.		
CT50	Crane	2R/1
C120	Cessna 120 only	1R/1
C140	Cessna 140 only	1R/1
C150	Cessna 150 only	1R/1
C152	Cessna 152 only	1R/1
C170	Cessna 170 only	1R/1
C172	Cessna 172 only	1R/1
C175	Cessna 175 only	1R/1
C177	Cessna 177 & RG	1R/1
C180	Cessna 180 only	1R/1
C182	Cessna 182 only	1R/1
C185	Cessna 185 only	1R/1
C188	Cessna 188 only	1R/1
C190	Cessna 190 only	1R/1
C195	Cessna 195 only	1R/1
C205	Cessna 205 only	1R/1
C206	Cessna 206 only	1R/1
C207	Cessna 207 only	1R/1
C208	Caravan I	1T/1
C210	Cessna 210 all models	1R/1
C303	Cessna 303 only	2R/1
C305	Cessna 305 only	1R/1
C310	Cessna 310 only	2R/1
C320	Cessna 320 only (Sky Knight)	2R/1
C335	Cessna 335 only	2R/1
C336	Cessna 336 only	2R/1
C337	Cessna 337 only (Skymaster)	2R/1
C340	Cessna 340 only	2R/1
C401	Cessna 401 only	2R/1
C402	Cessna 402 only	2R/1
C404	Cessna 404 only	2R/1
C411	Cessna 411 only	2R/1
C414	Cessna 414 only	2R/1
C421	Cessna 421 only	2R/1
C425	Cessna 425 only (Corsair/ Conquest I)	2T/1*
C441	Cessna 441 only (Conquest II)	2T/1
C500	Citation I	2T/2
C501	Citation I/SP	2T/1*
C525	Citationjet 525 (some models single crew)	2T/2
C550	Citation II	2T/2
C551	Citation II/SP	2T/1*
C560	Citation V	2T/2
C650	Citation III	2T/2
CHANCE VOUGHT/GOODYEAR		
F4U	Corsair (all models)	1R/1*
CHRISTEN INDUSTRIES INC.		
CHCH	Eagle II (all models)	1P/1
C11	Pitts S-1 (all models)	1P/1
C12	Pitts S-2 (all models)	1P/1
CIA1	A-1 Huskey	1P/1
CIRRUS DESIGN CORP.		
VK30	Cirrus	1P/1
CONAIR LIMITED		
DHST	Turbo Firecat (turbine rebuild of Grumman & DHS2)	2T/1*
CONVAIR DIV. GENERAL DYNAMICS		
CV13	Valiant	1R/1
CV14	Canso (all model PBY, consolidated 28, 285, Catalina)	2R/2
CV44	Convair (all model 240/340/440; T-29, C-131)	2R/2
CV58	Convair (all models 540/580; CC109; Cosmopolitan)	2T/2
CV64	Convair (all models 600/640)	2T/2
CURTISS-WRIGHT CORP.		
CW20	Commando (all CW20 and C-46)	2R/2
DASSAULT-BREGUET		
DA10	Falcon 10	2T/2
DA20	Falcon 20	2T/2
DA21	Mystere Falcon 200	2T/2
DA50	Falcon 50	3T/2
DA90	Falcon 900	3T/2
DE HAVILLAND AIRCRAFT CANADA		
AT3	PZL powered Otter	1R/1
DH1	Chipmunk	1R/1
DH2	Beaver	1R/1
DH2T	Turbo Beaver	1T/1
DH3	Otter (CSR 123; U-1)	1R/1
DH3T	Turbo Otter	1T/1
DH4	Caribou CC-108	2R/2
DH5	Buffalo CC-115	2T/2
DH6	Twin Otter CC-138	2T/1
DH7	Dash 7	4T/2
DH8	Dash 8 (series 100/200/300)	2T/2
DH82	Tiger Moth	1R/1
DH83	DH83	1R/1
DH98	Mosquito	2R/1*
DHS2	Tracker CS2F1/2	2R/1*
HS25	Domonie (all model Hawker Siddeley HS-125 & Beech BH-125)	2T/2
DORNIER-WERKE		
DO27	Dornier 27	1R/1
DO28	Dornier 28 (all model A, B, H, Q and S)	2R/1
D228	Model 228 all series	2T/2
DO8D	Skyservant	2R/1
EMBRAER		
E110	Embraer Banderante (min. flt. crew 1 day VFR only)	2T/2
E120	Embraer Brasilia	2T/2
ERCO		
ER15	Ercoupe (415C, CD and D)	1R/1
FAIRCHILD AIRCRAFT LTD.		
FA11	Husky F-11	1R/1
FA27	Friendship (includes F-27, FH-227)	2T/2
FA62	Cornell	1R/1
FA71	Fairchild 71	1R/1
FA82	Fairchild 82	1R/1
SW2	All short-body models SA26, SA226T, SA227TT	2T/1*
SW3	All models SA226AT/TC	2T/1*
SW4	All modelsSA227AT/AC	2T/2
FAIRY AVIATION		
FF46	Firefly (model 4, 5 and 6)	1R/1
FLEET AIRCRAFT		
FL2	Fleet 2	1R/1
FL80	Canuck	1R/1
FOKKER-VFW		
FA27	Freindship (includes FH-227)	2T/2
FK10	Fokker 100	2T/2
FK28	Fellowship	2T/2
FK50	Fokker 50	2T/2
FOUND BROS.		
FB2	FBA2C	1R/1
GLOBE		
GC1B	Swift	1R/1
GRUMMAN AMERICAN AVIATION		
AA1	Trainer	1R/1
AA5	Traveller	1R/1
AA7	Couger	2R/1
GTBM	Avenger (all model TBM,TBF)	1R/1*
G159	Gulfstream I (G159)	2T/2
G164	Ag-cat	1R/1
G2	Gulfstream II (G1159)	2T/2
G21	Goose/Super Goose	2T/1
G3	Gulfstream III	2T/2
G30	S.C.A.N. Series	2R/1
G4	Gulfstream IV	2T/2
G44	Widgeon/Super Widgeon	2R/1
G73	Mallard	2R/1
G73T	Turbo Mallard	2T/2
HANDLEY PAGE		
HP13	Handley Page HP-137	2T/2
HP7	Herald (all model HPR7)	2T/2
HAWKER		
FURY	Sea Fury (all marks)	1R/1*
HURRI	Hurricane (all marks)	1R/1*
HAWKER-SIDDELEY		
HS04	Dove 104	2R/1
HS06	Comet (all models)	4T/2
HS06	Heron 114	4R/2
HS25	Domonie (all models Hawker Siddeley HS-125 and Beech BH-125 except series 1000)	2T/2
HS65	Argosy (AW650-100/200)	4T/2
HS74	Andover (HS748/780)	2T/2
HS8	Rapide 89	2R/1
HS251	BAE 125 series 1000	2T/2
HAWKER-SIDDELEY (AVRO)		
AV52	Anson (all marks)	2R/1
AV83	Lancaster	4R/2
AV85	York	4R/2
HELIO AIRCRAFT CORP.		
HE1	Courier (all model H-295, Mark II, U-10)	1R/1
HOAC AUSTRIA		
DV20	DV20 Katana (includes Dimona and DA20 Diamond)	1P/1
HK36	Super Dimona All models	1P/1
HOWARD AERO MANUFACTURING		
HW5	Howard 500	2R/1
HW8	Jobmaster (DGA815)	1R/1
ISRAEL AIRCRAFT INDUSTRIES		
AJ25	Astra	2T/2
JC21	Jet Commander 1121	2T/2
RV01	Arvava 101B	2T/1*
WW23	Westwind 1123	2T/2
WW24	Westwind 1124	2T/2
LAKE AIRCRAFT		
LA4	Buccaneer	1R/1
LA25	Renegade	1R/1
LEAR		
LR18	Learstar18	2R/2
LR23	Learjet 23	2T/2
LR24	Learjet 24	2T/2
LR25	Learjet 25	2T/2
LR28	Learjet 28	2T/2
LR29	Learjet 29	2T/2
LR31	Learjet 31	2T/2
LR35	Learjet 35	2T/2
LR36	Learjet 36	2T/2
LR54	Learjet 54	2T/2
LR55	Learjet 55	2T/2
LING-TEMCO-VOUGHT (LTV)		
LUSC	Luscombe 8	1R/1
LOCKHEED AIRCRAFT CORP		
L101	Tri Star L1011	3T/2
L12	Piston Electra (model 10, 12)	2R/1
L14	Hudson	2R/1*
L18	Lodestar (all model 18; C-57; C-58; C-60)	2R/2
L188	Electra (Orion, Aurora)	4T/2
L300	Lockheed 300	4T/2
L329	Jet Star	4T/2
L34	Ventura (all model 34; B-34; PV-1; PV-2)	2R/1*
L38	Lightning P-38	2R/1*
L382	Hercules 130	4T/2
L49	Constellation (all model 49; 649; 749; and C121)	4R/2
P2	Neptune	2R/2
T33	Shooting Star (all t-33; F-80)	1T/1*
MARTIN		
MJRM	Mars JRM	4R/2
MAULE AIRCRAFT CORP.		
ML4	Maule (includes M5)	1R/1
MCDONNEL-DOUGLAS		
B26	Invader (all models A-26, B-26)	2R/1*
DC3	Dakota (all model DC-3; C-47; CC-129; C-117; Skytrain)	2R/2
DC4	Skymaster (all model DC-4; C-54; C-5; North Star; Argonaut)	4R/2
DC6	Liftmaster (all model DC-6)	4R/2
DC7	Seven Seas (all model DC-7)	4R/2
DC8	DC-8	4T/2
DC9	DC-9	2T/2
DC10	DC-10	3T/2
MD11	MD-11	3T/2
MD80	MD80/81/82/83/87/88	2T/2
MESSERCHMITT-BOLKOW BLOHM		
HF20	Hansa HFB 320	2T/2
MITSUBISHI HEAVY INDUSTRIES		
MU2	Model MU2	2T/1*
MU3	Model MU300 (Diamond models only.)	2T/2
MOONEY AIRCRAFT CORP.		
MK20	Includes all models	1R/1
MORAVAN LTD.		
0242	Zlin model Z242L	1P/1
NEICO AVIATION INCORP.		
LC20	Lancair 235/320/360	1P/1
LC30	Lancair IV (all models)	1P/1*
NIHON AEROPLANE MANUF.		
YS11	All model YS11/YS11A	2T/2
NOORDUYN AVIATION LTD.		
N06	Norseman (all mark IV, V and VI; UC64)	1R/1
NORTH AMERICAN		
F86	Sabre	1T/1*
NAT6	Harvard (AT6, Texan)	1R/1
P51	Mustang	1R/1*
PARTENAVIA COSTRUZIONI AERONAUTICHE SPA		
PN68	Observer	2R/1
PIAGGIO RINALDO		
P136	Royal Gull (model P136)	2R/1
P180	Avanti	2T/1*
PILATUS		
PL6	Porter (all model PC-6 piston)	1R/1
PL6T	Turbo Porter (all model PT-6 Turbine)	1T/1
PIPER AIRCRAFT CORP.		
PA11	Cub	1R/1
PA12	Super cruiser	1R/1
PA14	Family Cruiser	1R/1
PA16	Clipper	1R/1
PA17	Vagabond (includes PA15)	1R/1
PA18	Super Cub	1R/1
PA20	Pacer	1R/1
PA22	Tri Pacer/Colt	1R/1
PA23	Apache	2R/1
PA24	Comanche	1R/1
PA25	Pawnee	1R/1
PA28	Cherokee	1R/1
PA3	Cub Trainer	1R/1
PA30	Twin Comanche	2R/1
PA31	Navajo	2R/1
PA32	Cherokee Six	1R/1
PA34	Seneca	2R/1
PA36	Brave	1R/1
PA38	Tomahawk	1R/1
PA39	Twin Comanche (CR)	2R/1
PA42	Cheyenne III/IV/400LS	2T/1*
PA44	Seminole	2R/1
PA46	Malibu	1R/1
PA60	Aerostar 600 Pressurized	2R/1
PARO	Cherokee Arow	1R/1
PAT4	PA-31T3 (T-1040)	2T/1
PAYE	Cheyenne I/II	2T/1
PAZT	Aztec	2R/1
PITTS		
PI2	Pitts Special (includes S1)	1R/1
POLSKIE ZAKLADY LOTNICZE (PZL)		
PO18	Dromander	1R/1
POW4	Wilga	1R/1
REPUBLIC		
RC3	Seabee	1R/1
ROCKWELL INTERNATIONAL		
ACB1	Ag Commander B-1 (Snipe Commander)	1R/1
ACS2	Ag Commander S-2 (Snow Commander, Aero Commander)	1R/1
AC10	Darter	1R/1
AC12	Aero Commander 112	1R/1
AC14	Aero Commander 114	1R/1
AC20	Aero Commander 200	1R/1
AC50	Commander 500 (all model 500 and Shrike Commander)	2R/1*
AC60	Grand Commander (all piston unpressurized 600 series)	2R/1*
AC68	Super Commander (all piston pressurized 600 series)	2R/1*
AC69	Turbo Commander (all turbine pressurized 600 series)	2T/1*
AC72	Alt-Cruiser	2R/1
AC84	Turbo Commander 840/ 980/1000	2T/1*
B25	Mitchell	2R/1*
JC21	Jet Commander 1121	2T/2
N145	Navion	1R/1
N265	Sabreliner (all series 265)	2T/2
T28	NA260 Trojan	1R/1
WW23	Westwind 1123	2T/2
WW24	Westwind 1124	2T/2
RUTAN AIRCRAFT FACTORY		
VZ10	Vari-EZE/Long EZE	1P/1
SAAB-SCANIA		
SF34	SAAB 340 (all)	2T/2
SAUNDERS AIRCRAFT CORP.		
ST27	Saunders ST27	2T/2
SCOTTISH AVIATION		
SC2	Twin Pioneer	2R/1
BA31	Jetstream	2T/2
SEQUOIA AIRCRAFT CORP.		
F8L	Falco	1P/1
SHORT BROS.AND HARLAND HD.		
SH33	Short SD3-30	2T/2
SH36	Short 360	2T/2
SH7	Skyvan SBH-SC7	2T/1
STINSON AIRCRAFT CO.		
ST10	All model 9 and 10	1R/1
ST75	Voyager (Station Wagon)	1R/1
ST77	Reliant (Vultee)	1R/1
STODDARD-HAMILTON AIRCRAFT		
GL20	Glasair II (all models)	1P/1
GL25	Glasair III (all models)	1P/1*
SUPERMARINE		
SPIT	Spitfire (all marks)	1R/1*
STRA	Stranraer	2R/2
TAYLORCRAFT		
TC19	Sportsman 19	1R/1
TED SMITH AIRCRAFT		
TS60	Aero Star (unpressurized)	2R/1
THURSTON, SCHWEIZER, TEAL CORP.		
TEAL	TSC-1A, TSC-1A1, TSC-1AW	1R/1
TRIDENT AIRCRAFT LTD.		
TR1	Trigull	1R/1

Fig. 1-1: Type designators. This is a summary of Canadian type approved aircraft indicating the type designators, engine classification and minimum flight crew. For each entry, following the aircraft manufacturer is provided the type designator, number of engines and type of engines, followed by the minumum flight crew. A minimum flight crew indicated as 1 means that certain types of this aircraft will be "high performance". For example, a Boeing 737 300 Series is classified as B7373 2T/2, meaning it has 2 turbine engines and minimum flight crew of 2 pilots. A Beech King Air 100 would have 2 reciprocating engines, a minimum flight crew of 1 pilot and would be likely classified as "high performance".*

Individual Type Rating, Two Crew, Cruise Relief Only

The following is required to obtain an individual aircraft type rating for cruise relief on two pilot minimum flight crew aircraft:

a) completion of program of ground and flight training on the aeroplane type;
b) for endorsement of CPL, at least 70% scored on either IATRA or ATPL writtens (SAMRA, SARON) within the 12 months preceding the application for the first endorsement of the rating;
c) completion of flight training on type and minimum of 250 hours flight time on aeroplanes;
d) successful completion of PPC on the aeroplane type, excluding take-off and landings, within the 12 months preceding the application for the rating.

Individual Type Rating, High Performance Aeroplane

A HIGH PERFORMANCE AEROPLANE has a minimum flight crew of one pilot and has a Vne of 250 Kts or greater, or a Vso (stall speed, landing configuration) of 80 K or greater. To obtain an individual type rating for a high performance aeroplane, the applicant must have at least 200 hours total flight time, ground school on type and flight training acceptable to the Minister. Upon completion of the training, the applicant must complete a successful "qualifying flight" under the supervision of a CPL or ATPL pilot having at least 50 hours on the aeroplane class and 10 hours on type.

Individual Type Rating, Second Officer

An applicant for an individual aircraft type rating for an aeroplane type associated with a Second Officer rating shall have passed, within the previous 12 months to the application, a course of training relating to that aircraft type that is administered by an institution recognized by the minister as being qualified to administer such training. An applicant shall have also passed a Second Officer PPC on the aeroplane type.

EXAMPLE: A pilot has a CPL license. His/her license has the following blanket type rating: "All single pilot, single and multi-engine, land aeroplanes excluding high performance". The pilot is offered a job in a commercial air service on a DC-3. Is the pilot qualified to accept the position?

ANSWER: No: the type is designated 2 pilot minimum flight crew. A CPL license allows the pilot to act as SIC on a 2-pilot aircraft, but an individual type rating is also required. If the pilot had at least 250 hours, passed the IATRA and the operator provides the necessary ground and air training, the pilot could complete a PPC and qualify for a DC-3 type rating. Once type endorsed for the DC-3, the pilot could act as SIC on the DC-3.

EXAMPLE: What is the minimum licensing standard to act as PIC on a DC-3 in a commercial air service carrying passengers?

ANSWER: ATP licence, individual type rating on DC-3, PPC.

[Reference: CARs 421.40, 425.21]

1.06 Instrument Rating

The holder of an instrument rating may exercise the privileges of his/her licence under IFR for the group of aircraft endorsed on his/her licence, and may exercise VFR OTT privileges.

An IFR rating may be obtained for the following aircraft groups:

a) Group 1 = all aeroplanes;
b) Group 2 = multi-engine, centre line thrust;
c) Group 3 = single engine aeroplane;
d) Group 4 = helicopters.

To following is required to complete an IFR rating:

a) successful completion (70%) of INRAT written;
b) 50 hours cross country PIC;
c) 40 hours instrument time of which 20 maximum may be simulator;
d) minimum 5 hours from holder of flight instructor rating;
e) 15 hours dual from IFR instructor;
f) one 100 NM simulated or actual IFR cross country at two airports to IFR minimums;
g) successful completion of IFR flight test.

Once issued, an instrument rating is valid for the period specified on the licence, not to exceed a period of 24 months.

The minimum qualifications to provide instruction for a group 1 instrument rating to licenced, multi-engine rated pilots is a CPL or ATPL with valid instrument rating and 500 hours PIC of which 100 hours are in the applicable group and 10 hours PIC on type.

[Reference: CARs 401.47, 401.48, 421.46; AIM 3.7.3, 3.7.4]

1.07 Regulation Making and Compliance

The Parliament of Canada has the constitutional authority of enacting laws and regulations affecting aeronautics. The AERONAUTICS ACT is federal legislation and is the overriding authority under which all other rules are set out. The Canadian Aviation Regulations (CARs) are made under the authority of the *Aeronautics Act*, and the Minister of Transport has authority to make standards and to publish them in the CARs.

The Minister shall not make a standard or an amendment to a standard unless the Minister has undertaken consultations with interested persons concerning the standard or the amendment in accordance with the procedures specified in *CARAC Management Charter & Procedures*. [CARAC refers to the Canadian Aviation Regulation Advisory Council - an industry/government advisory group involved in making recommendations for aviation regulations.] No standard or amendment to a standard shall come into effect less than 30 days after it is made, unless urgently required to ensure safety.

The owner or operator of an aircraft shall, on reasonable notice given by the Minister, make the aircraft available for inspection in accordance with the notice.

Canadian aviation document holders shall produce their document for inspection in accordance with the terms of a demand made by a peace officer, an immigration officer or the Minister.

Where a Canadian aviation document has been suspended or cancelled, the person to whom it was issued shall return it to the Minister immediately after the effective date of the suspension or cancellation.

Recording systems, including computer records and microfiche, that do not comprise entries on paper may be used to comply with the record-keeping requirements of the CARs, if measures are taken to ensure that the records contained in the recording systems are protected by electronic or other means, against inadvertent loss or destruction and against tampering and a copy can be printed on paper and given to the Minister on reasonable notice.

[Reference: CARs 103.01, 103.02, 103.03, 103.04]

1.08 Commercial Air Services

General

OPERATOR, in respect of an aircraft, means the person that has possession of the aircraft, as owner, lessee, or otherwise.

A SMALL AEROPLANE means an aeroplane having a maximum permissible take-off weight of 5,700 kgs (12,566 lbs) or less. A LARGE AEROPLANE means an aeroplane with an MCTOW of more than 5,700 kgs (12,566 lbs).

An AIR TRANSPORT SERVICE means a commercial air service that is operated for the purpose of transporting persons, personal belongings, baggage, goods or cargo in an aircraft between two points.

No person shall operate an air transport service unless the person holds an complies with the provisions of an AIR OPERATOR CERTIFICATE (AOC) that authorizes the person to operate that service.

A COMMERCIAL AIR SERVICE means any use of aircraft for hire or reward.

The following commercial operations have their own unique regulatory requirements in the CARs depending on the nature of the operation:

a) Aerial Work;
b) Air Taxi;
c) Commuter;
d) Airline.

FLIGHT DUTY TIME means the period that starts when a flight crew member reports for a flight, or reports as a flight crew member on standby, and finishes at "engines off" at the end of the final fight, and includes the time required to complete any duties assigned by the air operator or delegated by the Minister prior to the reporting time.

FLIGHT CREW MEMBER ON STANDBY means a flight crew member who has been designated by an air operator to remain at a specified location in order to be available to report for flight duty on notice of one hour or less.

FLIGHT CREW MEMBER ON CALL means a flight crew member who has been designated by an air operator to be available to report for flight duty on notice of one hour or less.

FLIGHT CREW MEMBER ON RESERVE means a flight crew member who has been designated by an air operator to be available to report for flight duty on notice of more than one hour.

MINIMUM REST PERIOD means a period during which a flight crew member is free from all duties, is not interrupted by the air operator, and is provided with an opportunity to obtain not less than 8 consecutive hours of sleep in suitable accommodation, time to travel to and from that accommodation and time for personal hygiene and meals.

SUITABLE ACCOMMODATION means a single-occupancy bedroom that is subject to a minimal level of noise, is well ventilated and has facilities to control the levels of temperature and light, or where such a bedroom is not available, an accommodation that is suitable for the site and season, is subject to a minimal level of noise and provides adequate comfort and protection from the elements.

UNFORESEEN OPERATIONAL CIRCUMSTANCE means an event, such as unforecast adverse weather, or an equipment malfunction or ATC delay, that is beyond the control of an air operator.

Monitoring System

Every air operator shall establish a system that monitors the flight time, flight duty time and rest periods of each of its flight crew members and shall include in its Company Operations Manual (COM) the details of that system. Where a person becomes aware that an assignment by an air operator to act as a flight crew member on a flight would result in the maximum flight duty time being exceeded, the person shall so notify the air operator.

Flight Time Limitations

No air operator shall assign a flight crew member for flight time, and no flight crew member shall accept such an assignment, if the flight crew member's total flight time in all flights conducted by the flight crew member will, as a result, exceed

a) 1,200 hours in any 365 consecutive days;
b) 300 hours in any 90 consecutive days;
c) 120 hours in any 30 consecutive days, or in the case of a flight crew member on call, 100 hours in any 30 consecutive days;
d) in an air taxi operation, 60 hours in any 7 consecutive days;
e) in an airline or commuter operation, 40 hours in any 7 consecutive days;
f) where the flight crew member conducts single-pilot IFR flights, 8 hours in any 24 consecutive hours.

An air operator may assign a flight crew member for flight time, and a flight crew member may accept such an assignment, where the flight crew member's flight time will, as a result, exceed the flight time referred to above if authorized in the AOC.

Flight Duty Time Limitations and Rest Periods

No air operator shall assign a flight crew member for flight duty time, and no flight crew member shall accept such an assignment, if the flight crew member's flight duty time will, as a result, exceed 14 consecutive hours in any 24 consecutive hours.

Extension of Flight Duty Times with Rest Periods

Where flight duty time includes a rest period, flight duty time may be extended beyond the maximum flight duty time by one-half the length of the rest period to a maximum of 3 hours if the air operator provides the flight crew member with advance notice of the extension of flight duty time, it is of at least 4 consecutive hours in suitable accommodation and the flight crew member's rest is not interrupted by the air operator during the rest period. This minimum rest period following flight duty time and prior to the next flight duty time shall be increased by an amount at least equal to the extension to the flight duty time.

Specific operators may be given permission to increase the above times - see the AOC.

Extension of Flight Duty Times: Unforeseen Operational Circumstances

Flight duty time may be extended beyond the maximum flight duty time referred to, to a maximum of 3 hours, where the flight is extended as a result of unforeseen operational circumstances, and the PIC after consultation with other flight crew members considers it safe to do so.

Delayed Reporting Time

Where a flight crew member is notified of a delay in reporting time before leaving a rest facility and the delay is in excess of 3 hours, the flight crew member's flight duty time is considered to have started 3 hours after the original reporting time.

Requirements for Time Free From Duty

An air taxi operator shall provide each flight crew member with one period of time free from duty of at least 24 consecutive hours 13 times within each 90 consecutive days and 3 times within each 30 consecutive days. A commuter or airline operator shall provide one period of at least 36 consecutive hours within each 7 consecutive days or one period of at least 3 consecutive calendar days within each 17 consecutive days. Where the flight crew member is on call, one period free from duty of at least 36 consecutive hours is required within each 7 consecutive days or one period of at least 3 consecutive calendar days within each 17 consecutive days is required. Other times free of duty time may be applied in specific cases depending on the AOC.

Flight Crew Positioning

Where a flight crew member is required by an air operator to travel for the purpose of positioning after the completion of flight duty time, the air operator shall provide the flight crew member with an additional rest period at least equal to one-half the time spent travelling that is in excess of the flight crew member's maximum flight duty time.

Flight Crew Members on Reserve

The air operator must provide reserve pilots the opportunity to sleep 8 consecutive hours in any 24 hours by

a)   giving 24 hours notice of time of rest period commencement, which cannot shift more than 3 hours earlier or later than the preceding rest period, nor more than a total of 8 hours in any 7 consecutive days;
b)   the pilot must have a least 10 hours notice of the assignment and shall not be assigned any duty for these 10 hours;
c)   no interruptions to rest period or assignments from 22:00 - 06:00 local.

If the operator cannot comply with the above rest period requirements, and the flight crew member is notified to report for flight duty or the reporting time occurs between 22:00 and 06:00 local time, then the maximum flight duty time shall be 10 consecutive hours and the subsequent minimum rest period shall be increased by at least 1/2 the length of the preceding flight duty time.

Long Range Flights

A flight or series of flights that terminates more than 4 one-hour time zones from the point of departure, other than flights conducted entirely within the NDA, shall be limited to 3 sectors and shall be followed by a rest period that is at least equal to the length of the preceding flight duty time. Where the flight is a transoceanic flight, the maximum number of sectors that may be completed after the transoceanic sector is one, excluding one unscheduled technical stop.

Controlled Rest on the Flight Deck

An air operator may institute a program of controlled rest on the flight deck if the program is authorized in the AOC and specifically outlined in the COM. Basically, one pilot is permitted to "catnap" for a maximum of 45 minutes while the other pilot remains alert. Such rest periods can only take place during the cruise portion of flight and shall be completed at least 30 minutes before planned TOD, workload permitting.

[Reference: Aeronautics Act s. 3, CARs 700.02; CARs 700.14 - 700.23, 720.23]

1.09 Air Taxi Operations

NOTE: The following outline of the CARs regulatory requirements for Air Taxi operations is given in extensive detail so as to introduce terminology and regulatory requirements that will repeat for other levels of operators (Commuter, Airline, etc.).

Air Taxi Operator: Certification Requirements

An AIR TAXI OPERATION applies to a Canadian air operator in an air transport service or in aerial work involving sight-seeing operations, of single engined aircraft or multi-engine aircraft other than turbo-jet having a MCTOW of 19,000 lbs (8,618 kgs) or less and a seating configuration, excluding pilot seats of 9 or less. Air taxi operators must have an AOC.

Once an air operator has obtained certification, it indicates that the air operator has demonstrated the ability to maintain an adequate organizational structure, to maintain an operational control system, to meet training program requirements, to comply with maintenance requirements, to meet the required standards and to conduct the operation safely. An air taxi AOC contains the following:

a) legal name of operator;
b) number of AOC;
c) effective date of AOC;
d) date of issue of AOC;
e) general conditions: compliance with operations manual, maintenance of organizational structure, training in accordance with its approved training program, only qualified crew members may operate aircraft, proper aircraft equipment for area of operation, proper maintenance, operational support services and equipment, required managerial personnel, Minister to be notified within 10 days of change in legal or trade name or change of base of operations or managerial personnel, requirement to conduct safe operation;
f) specific conditions authorizing areas of operation and base, types of service, types of aircraft and other restrictions;
g) specific operations specifications for aircraft performance, equipment and emergency equipment, instrument approach procedures, enroute aerodrome authorizations, special weather minima authorizations, authorizations concerning flight crew member complement, pilot training and PPCs, maintenance control systems, leasing arrangements and any other conditions deemed necessary for aviation safety.

The required managerial personnel are a chief pilot, operations manager, and maintenance engineer.

Air Taxi Operator: Operations Manager

Operations Manager Qualifications
The operations manager shall have or have held the appropriate licence and ratings which a PIC is required to hold for one of the aeroplanes operated, or at least two years related supervisory experience with an air operator having similar flight operations. The operations manager shall demonstrate knowledge to the Minister with respect to the content of the COM, the AOC and operations specifications, and knowledge of the regulations and standards to carry out the required duties and responsibilities and to ensure safety.

Operations Manager Responsibilities
The Operations Manager is responsible for:

a) control of operations and operational standards of all aeroplanes operated;
b) identification of factors which impact on operational control such as maintenance, crew scheduling, load control, equipment scheduling;
c) supervision, organization, function and manning of flight operations, cabin safety, crew scheduling and rostering, training programs and flight safety;
d) the contents of the COM;
e) the supervision of and the production and amendment of the COM;
f) liaison with any external agencies which may affect air operator operations;
g) liaison with TC on all matters concerning flight

operations, including any variations to the AOC;
h) ensuring that operations comply with current regulations, standards and air operator policy;
i) ensuring that crew scheduling complies with flight and duty time regulations;
j) ensuring that all crew members are kept informed of any regulatory changes;
k) receiving, reviewing and taking action on aeroplane safety information;
l) qualifications of flight crew members;
m) maintenance of a current operations library.

Air Taxi Operator: Chief Pilot Requirements

Chief Pilot Qualifications
a) For a VFR only operator, the chief pilot must hold a valid ATPL or a valid CPL appropriate to the aeroplane (s) used. For a Day and Night VFR operator, or for an IFR operator, the chief pilot must hold a valid ATPL or CPL and a valid IFR rating appropriate to the aeroplane (s) used;
b) type rating for one of the types operated (if a type rating applies);
c) at least 500 hours flight time of which 250 hours shall be as PIC within preceding 3 years on same category and class of aircraft operated;
d) qualified as PIC on one of the types operated;
e) show the Minister knowledge of COM, training manuals, SOPs (if applicable), company check pilot manual (if applicable), and of regulations and standards of the position.

Chief Pilot Responsibilities
The chief pilot has the following responsibilities:

a) developing SOPs;
b) setting up and implementing flight crew training programs;
c) issuing directives, notices, accident/incident reports to flight crews;
d) acting on crew reports;
e) supervising flight crews;
f) other duties delegated by operations manager.

Air Taxi Operator: Flight Operations

Operating Instructions
An air operator shall ensure that all operations personnel are properly instructed about their duties and about the relationship of their duties to the operation as a whole, as specified in the COM.

Scheduled Air Service Requirements
Air operators having a scheduled air service transporting persons shall operate the service between airports or between an airport and a military aerodrome. Specific authorization in the AOC is required to operate a scheduled air service for transporting persons between an airport and an aerodrome other than a military aerodrome or between two aerodromes.

Operational Control System
All air operators must have an OPERATIONAL CONTROL SYSTEM under the control of the operations manager. Operational control for a flight is delegated to the PIC, while the operations manager retains day to day responsibility. Each aeroplane shall be equipped with communications equipment to allow the PIC to communicate with a ground radio station for the purposes of FLIGHT FOLLOWING - to monitor a flight's

progress and notify the air operator and SAR if the flight is overdue or missing. Flight following procedures are in the COM. The ground station can be operated by the air operator, government or private agency. A person, qualified and knowledgeable in the air operator's flight alerting procedures and delegated by the operations manager, shall be on duty or available from the commencement to termination of IFR or night VFR flights, including intermediate stops.

Flight Authorization
No person shall commence a flight unless the flight has been authorized in accordance with the procedures specified in the COM.

Operational Flight Plan
All flights require an OPERATIONAL FLIGHT PLAN. The PIC shall ensure that a copy of the operational flight plan is left at a point of departure, in accordance with the procedures in the COM. An operational flight plan gives detailed information concerning the flight including standard information that would go on any flight plan and other information such as: ground speed during cruise, fuel weights for taxi, destination, alternate, contingency, holding reserve, routing to alternate aerodrome, signature of PIC or alternate means of certifying acceptance, distance to destination etc. The operational flight plan format shall allow the crew to record the fuel state and flight progress relative to the plan. It may be computer generated or produced manually working from charts and tables provided by the flight crew, with blank spaces for the required entries.

Fuel Requirements
Air taxi operators require fuel reserves in accordance with the normal regulations - see s. 1.18 in this manual. Additionally, for IFR flights with an air operator, fuel is required to allow aircraft to descend from any cruise point to the lower of the single engine service ceiling or 10,000 feet, to cruise to a suitable alternate aerodrome, to conduct an approach and missed approach and to hold for 30 minutes at an altitude of 1,500 feet above the elevation of the suitable aerodrome.

Admission to Pilot's Compartment
The PIC shall give a TC inspector free access to the cockpit, and make available for the inspector the seat he/she requires.

Transport of Passengers in Single Engine Aircraft
No air operator shall operate a single engined aircraft with passengers in IFR or night VFR flight. Special authorization is available for some types of single engine turbine powered airplanes having specified equipment, for single engine IFR or night VFR with passengers in limited geographical areas with specific pilot training.

Aircraft Operating over Water
No aeroplane air operator shall, except when taking off or landing, operate a land aircraft over water, beyond a point where the land aircraft could reach shore in the event of an engine failure.

Number of Passengers in Single Engine Aircraft
No aeroplane air operator shall operate a single-engined aircraft with more than 9 passengers.

VFR Flight Obstacle Clearance Requirements
Except when conducting a take-off or landing, no person shall operate an aircraft in VFR flight at night at less than 1,000 feet above the highest obstacle located within a horizontal distance of three miles, or during the day, at less than 300 feet AGL or at a horizontal distance of less than 300 feet from any obstacle.

VFR Flight Minimum Visibility - Uncontrolled Airspace
Where an aeroplane is operated in day VFR flight within uncontrolled airspace at less than 1,000 feet AGL, a person may operate at less than the required 2 miles flight visibility with specific authorization (aircraft must have A/I, H/I, GPS, 500 hours pilot experience with carrier, specified pilot training etc. - not available for BC coast).

VFR Weather Conditions
No person shall commence a VFR flight unless current weather reports and forecasts, if obtainable, indicate that the weather conditions along the route to be flown and at the destination aerodrome will be such that the flight can be conducted in compliance with VFR.

Enroute Limitations
No person shall operate a multi engined aircraft with passengers in IFR flight or in night VFR flight if the weight of the aircraft is greater than the weight that will allow the aircraft to maintain, with any engine inoperative, the MOCA of the route flown.

VFR OTT Flight
No person shall operate an aircraft in VFR OTT flight unless the person is authorized to do so in the AOC. For multi engined aeroplanes, where the pilot holds a valid IFR rating, the flight shall be operated under conditions allowing descent under VMC or continuation of the flight under IFR or VMC if its critical engine fails. For multi engined aeroplanes where the pilot does not hold a valid IFR rating, or that cannot continue flight on failure of critical engine, and for single engined aeroplanes, the flight shall be operated allowing descent under VMC or continuation under VMC (if critical engine fails for multi engine, or for engine failure for single engine).

Routes in Uncontrolled Airspace
No person shall, in uncontrolled airspace, conduct an IFR flight or a night VFR flight on a route other than an air route unless the operator establishes the specific route's safety and navigational accuracy and obtains permission to use it.

Weight and Balance Control
No person shall operate an aircraft unless, during every phase of the flight, the load restrictions, weight and CG of the aircraft conform to the limitations specified in the Aircraft Flight Manual (AFM). An air operator shall specify in its COM its weight and balance system and instructions to employees regarding the preparation and accuracy of weight and balance forms.

Passenger and Cabin Safety Procedures
An air operator shall use these procedures so that passengers move to and from the aircraft safely:

a) aeroplanes should be parked in a location that avoids passenger exposure to hazards;
b) passengers are to be alerted to hazards;
c) guidance, or if necessary an escort, is required to ensure passengers are directed along a safe route to or from the aeroplane;
d) smoking restrictions should be enforced;
e) "walkman" type headsets that reduce traffic aware-

ness or audible warning signal information should not be worn.

Fuelling of Aircraft with Passengers on Board

No air operator shall permit an aircraft with passengers on board to be fuelled unless:

a) the pilot supervises the fuelling and remains near the aeroplane main exit to immediately communicate with and assist the evacuation of passengers in an emergency;

b) all exits are clear of obstruction and available for passenger evacuation;

c) the aeroplane engines are not running (unless the aircraft utilizes a propeller brake that is set);

d) electrical power supplies are not being connected or disconnected, and any equipment likely to produce sparks or arcs is off;

e) smoking is not permitted in the aeroplane or in the vicinity of the aeroplane;

f) fuelling is suspended when there are lightning discharges within 8 km of the aeroplane;

g) combustion heaters in the aeroplane or in the vicinity of the aeroplane are not operated;

h) high energy equipment such as HF radios and weather radar are off, unless used in accordance with the AFM;

i) photographic equipment is not used within 10 feet of the fuelling equipment or the fill or vent points of the aeroplane fuel systems.

Portable Electronic Devices

No aircraft operator shall permit the use of a portable electronic device (cell phones, modems, fax machines, computers, beepers, pagers, radios etc) on board an aircraft, where the device may impair the functioning of the aircraft's systems or equipment. No person shall use a portable electronic device on board an aircraft except with the permission of the operator.

Briefing of Passengers

The PIC shall ensure that passengers are given a STANDARD SAFETY BRIEFING. Where the standard safety briefing is insufficient for a passenger because of that passenger's physical, sensory or comprehension limitations or because that passenger is responsible for another person on board the aircraft, the PIC shall ensure that the passenger is given an INDIVIDUAL SAFETY BRIEFING that is appropriate to the passenger's needs. The PIC shall ensure that each passenger who is seated next to an emergency exit is made aware of how to operate that exit. A SAFETY FEATURES CARD shall be provided to every passenger. The PIC shall ensure that in the event of an emergency, circumstances permitting, all passengers are given an EMERGENCY LANDING BRIEFING.

Standard Safety Briefing

A standard safety briefing is an oral briefing given by a flight crew member or audio/visual means covering the following as applies to type or operation:

a) Prior to take off: when, where and how carry-on baggage is required to be stowed; fastening, unfastening, adjusting and general use of safety belts or safety harnesses; when seat backs must be secured in the upright position and tables stowed; location of emergency exists and for passengers seated next to an exit, how that exit operates; the location, purpose of, and advisability of reading the safety features card; the regulatory requirement to obey crew instructions regarding seat belts and no smoking instructions or Fasten Seat Belts and No

Smoking signs and the location of these signs; the location of any emergency equipment the passenger may have a need for in a emergency, such as ELT, fire extinguisher, survival equipment, including the key if in a locked compartment, first aid kit and life raft; the use of passenger operated portable electronic devices; the location and operation of the fixed passenger oxygen system, including the location and presentation of the masks, the action to be performed by the passenger in order to obtain the mask, activate the flow of oxygen and correctly don and secure the mask, including a demonstration mask location, donning, operation, and instruction on the priority for persons assisting others; location and use of life preservers, including how to remove from stowage or packaging and a demonstration of their location, method of donning and inflation, and when to inflate life preservers.

b) After take-off, if not included in the pre-take-off briefing, that smoking is prohibited and the advisability of using safety-belts or safety harnesses during flight.

c) In-flight because of turbulence, when the use of seat belts is required and the requirement to stow carry-on baggage.

d) Prior to passenger disembarkment, the safest direction and most hazard-free route for exiting the aeroplane, and any associated dangers for the aeroplane type such as pitot tube locations, propellers, or engine intakes.

e) Where no additional passengers have embarked the flight for subsequent take-offs on the same day, the pre-take-off and after-take-off briefing may be omitted provided a crew member has verified that all carry-on baggage is properly stowed, safety belts or harnesses are properly fastened, and seat backs and chair tables are properly secured.

Individual Safety Briefing

The individual safety briefing shall include:

a) Any information from the standard safety briefing and the safety features card that the passenger would not be able to receive during the normal conduct of that safety briefing.

b) Additional information applicable to the needs of that person including the most appropriate brace position for that passenger in consideration of his or her condition, injury, stature, and/or seat orientation and pitch; the location to place any service animal that accompanies the passenger; for a mobility restricted passenger who needs assistance in moving expeditiously to an exit during an emergency a determination of the most suitable time and mode of egress; specific appropriate information for the visually or hearing impaired.

c) For a passenger who is responsible for another person on board, information for the needs of the other person as applicable in the case of an infant, seat belt instructions, method of holding infant for take-off and landing, instructions for the use of a child restraint system, oxygen mask donning instructions, recommended brace position, location and use of life preservers; in the case of non-infant's oxygen mask, donning instructions; instructions for the use of a child restraint system and evacuation responsibilities.

d) For an unaccompanied minor, instructions to pay close attention to the normal safety briefing and to follow all instructions.

e) A passenger that has been provided with an

individual safety briefing need not be re-briefed following a change in crew if the crew member that provided the individual safety briefing has advised a member of the new crew of the contents of that briefing and any information respecting the passenger's special needs.

f) A passenger may decline an individual safety briefing.

Passenger preparation for an emergency landing shall include the following instructions, where time and circumstances permit:

a) safety belts or safety harnesses;
b) seat backs and tables;
c) carry-on baggage;
d) safety features cards;
e) brace position (when to assume, time to remain);
f) life preservers (as applicable).

Safety Features Card
A safety features card shall contain the following information (depending on type and equipment):

a) General safety information including smoking is prohibited on board the aeroplane; each type of safety belt or safety harness installed for passenger use, including when to use, and how to fasten, tighten and release; when and where carry-on baggage must be stowed and any other related requirements and restrictions on the particular aeroplane; correct positioning of seat backs and tables for take-off and landing;
b) Emergency procedures and equipment including fixed passenger oxygen system showing mask location and presentation, the actions to be performed by the seated passenger to obtain the mask, activate the flow of oxygen and correctly don and secure the mask and priority for persons assisting others with oxygen; location of first aid kits; location of passenger accessible fire extinguishers; location of ELTs; location of survival equipment and if the stowage compartment is locked, the means of access or location of the key; passenger brace position for impact, as appropriate for each type of seat and restraint system installed for passenger use, including the brace position for an adult holding an infant; the location, operation and method of using each exit on the aeroplane; the safest direction and most hazard-free escape route for passenger movement away from the aeroplane following evacuation; the attitude of the aeroplane while floating; the location of life preservers and correct procedures for removal from stowage or packaging, donning and use of the life preservers for adult, child and infant users including when to inflate; location and use of life rafts (as applicable); location, removal and use of flotation devices.
c) The safety features card shall bear the name of the air operator and the aeroplane type and shall contain only safety information.
d) The safety information provided by the card shall be accurate for the aeroplane type and configuration in which it is carried and in respect of the equipment carried and be depicted in a clear and distinct manner, with clear separation between each instructional procedure. All actions required to complete a multi-action procedure are to be presented in correct sequence and the sequence of actions are to be clearly identified.

Air Taxi Operator - Aircraft Equipment Requirements

General Requirements
No person shall operate a multi-engined aircraft with passengers on board in IMC unless the aircraft is equipped with:

a) a vacuum instrument power failure warning device;
b) an alternate static source for the altimeter, ASI and VSI;
c) two generators, each driven by a separate engine or by a rotor drive train;
d) two independent sources of energy, at least one of which is an engine driven pump or generator, and each of which is able to drive all gyroscopic instruments and is installed so that the failure of one instrument or one source of energy will affect neither the energy supply to the remaining instruments nor the other source of energy;
e) no person shall operate an aircraft with passengers at night unless the aircraft is equipped with at least one landing light.

Airborne Thunderstorm Detection and Weather Radar
No person shall operate an aircraft with passengers on board in IMC when current weather reports or forecasts indicate that thunderstorms may reasonably be expected along the route to be flown, unless the aircraft is equipped with thunderstorm detection equipment or weather radar equipment.

Additional Equipment for Single-Pilot Operations
No person shall operate an aircraft on a single pilot operation in IMC unless the aircraft is equipped with an auto-pilot that is capable of operating the aircraft controls to maintain flight and manoeuvre the aircraft about the lateral and longitudinal axes, a headset with a boom microphone and transmit button on the control column and a chart holder that is placed in a easily readable position with a means of illumination for the chart holder.

Protective Breathing Equipment
No air operator shall operate a pressurized aircraft unless protective breathing equipment with a 15 minute oxygen supply at a pressure altitude of 8,000 feet is readily available at each flight crew member position. This equipment may be used to meet the crew member oxygen equipment (see s. 1.35).

First Aid Oxygen
No air operator shall operate an aircraft with passengers on board above FL 250 unless the aircraft is equipped with oxygen dispensing units and an undiluted supply of first aid oxygen sufficient to provide at least one passenger with oxygen for at least one hour or the entire duration of the flight at a cabin pressure altitude above 8,000 feet, after an emergency descent following cabin depressurization, whichever period is longer.

Shoulder Harnesses
No person shall operate an aircraft unless the pilot seat and any seat beside the pilot seat are equipped with a safety belt that includes a shoulder harness.

Air Taxi Operator - Emergency Equipment

Survival Equipment

For flights over land, the COM shall show how the operator will comply with the requirement to carry survival equipment required under the CARs (see s. 1.33). As well, a list of survival equipment shall be carried on board with information on how to use it, with a survival manual appropriate for the season and climate, and crew members shall have training on survival concepts and use of survival equipment.

Where life rafts are required to be carried (see s. 1.33), they shall be equipped with an attached survival kit containing at least the following: pyrotechnic signalling device, radar reflector, life raft repair kit, bailing bucket and sponge, signalling mirror, whistle, raft knife, inflation pump, dye marker, waterproof flashlight, two day supply of water based on overload capacity of raft and one pint or water per day per person or means of distilling or desalting salt water in an equivalent amount, fishing kit, sea survival book and first aid kit containing antiseptic swabs, burn dressing compresses, bandages and anti-motion sickness pills.

First Aid Kits

A first aid kit is required to be stowed inside the aircraft in locations that are appropriately and conspicuously marked and are readily accessible to the occupants of the aircraft, as close as practicable to an emergency exit, and be dust and moisture proof. A first aid kit must contain at least antiseptic swabs, disposable applicators, bandages and gauze, triangular bandage, safety pins, scissors, burn dressing, current first aid manual, sterile compress dressings and gauze, hand cleansing towelettes, eye pad, first aid record, splints and padding, tweezers, adhesive surgical tape, one pair of latex gloves.

Air Taxi Operator - Personnel Requirements

Minimum Crew and Single Pilot IFR operations

No air operator shall operate an aircraft with passengers in IFR flight with fewer than two pilots. TC may authorize IFR flight with passengers without a SIC under the following conditions:

a) pilot shall have at least 1,000 hours flight time including, if the type flown is multi engine, 100 hours multi time. The pilot shall also have 50 hours of simulated or actual IMC flight time, and at least 50 hours flight time on type;

b) the PPC shall be in the aeroplane type flown and the following shall be added to the PPC renewal: knowledge of the auto-pilot and limitations, performance of normal and emergency procedures without assistance, passenger briefing with respect to emergency evacuation and demonstration of the use of the auto-pilot during appropriate phases of flight;

c) flight in pressurized aeroplanes shall be conducted at or below FL 250;

d) a pilot's single PPC, if still valid, is transferrable between air operators which have an AOC authority to conduct such operations and utilize the same type and model of aeroplane.

Designation of PIC and SIC

An air operator shall designate for each flight a PIC and, if the crew includes two pilots, a PIC and SIC.

Flight Crew Member Qualifications

No air operator shall permit a person to act and no person shall act as a flight crew member in an aircraft unless:

a) pilot holds the required licences and ratings;

b) within the previous 90 days, pilot has completed at least three take-offs and three landings in an aircraft of that type (where a type rating is required) or in an aircraft of that category and class (TC can approve use of advanced flight simulators in place of aircraft);

c) pilot has successfully completed a PPC or a competency check, that remains valid. For PIC of multi engine aircraft or PIC of single engine aircraft operated night VFR or IFR with passengers, PPC required in type (remember: TC approval required for operator for single engine night VFR or IFR carrying passengers). For an SIC of a multi-engined aircraft, PPC or a competency check for type required. For PIC of a single engined aeroplane that is not operated night VFR or IFR with passengers and where the person is not a chief pilot, competency check;

d) the pilot has completed the air operator's ground and flight training program.

With TC permission, an air operator may group similar types as a single type for PPC purposes (ie Piper Cheyenne I, II and III may be grouped).

No person shall act as a PIC of an aircraft with a person other than a flight crew member on board in night VFR unless the person acting as the PIC holds an IFR rating for that class of aircraft.

No air operator shall permit a person to act and no person shall act as the PIC of an aircraft with passengers on board unless the person has acquired, prior to designation as PIC, on the type and basic model of aircraft and in the PIC position, 5 hours for a single engined aeroplane and 15 hours for a multi engined aeroplane (the flight time required for this provision may be reduced by one hour for each take-off and landing completed, to a maximum of 50 percent).

The PPC and training requirements in paragraphs a) to d) above need not be completed in the case of ferry flights, training flights or positioning flights.

PPC Requirements

The validity period of a PPC, a competency check and the annual carrier ground and flight training expires on the first day of the 13th month following the month completed. Where a PPC, competency check or annual training is renewed within the last 90 days of its validity period, its validity period is extended by 12 months. The Minister may extend the validity period of a PPC, competency check or annual training by up to 60 days where the Minister is of the opinion that aviation safety is not likely to be affected. Where the validity of a PPC, competency check or annual training has been expired for 24 months of more, the person must requalify by meeting all initial training requirements.

A PPC shall allow the pilot to demonstrate knowledge and skill respecting the aeroplane, its systems and components, proper control of airspeed, direction, altitude, attitude and configuration of the aeroplane, in normal, abnormal and emergency procedures and limitations set out in the Aeroplane Operating Manual (AOM, where applicable), the AFM, the air operator's

COM, the air operator's SOPs, checklists, and any other information relating to the operation of the type. As well, the pilot shall demonstrate the required skill and knowledge respecting departure, enroute and arrival instrument procedures if it is an IFR PPC.

A PIC check shall be completed in the seat normally occupied by the PIC and a SIC check shall be completed in the seat normally occupied by the SIC.

A TC inspector or an approved company check pilot shall determine whether a person has demonstrated the knowledge and skill and the pilot's adherence to approved procedures and the pilot's airmanship in selecting a course of action.

During the PPC check the person conducting the check may request any maneuver or procedure from the PPC schedule (CARs 723.88, see PPC items below) as required to determine the candidate's skill.

A PPC must include a demonstration of IFR proficiency if the candidate posseses a valid IFR rating, and if the candidate conducts commercial IFR operations on the aeroplane in which the PPC is conducted. Where a pilot successfully completes the full PPC, the pilot also successfully completes the requirements for the renewal of an IFR rating.

Where an AOC authorizes single engine operations in IFR flight, the PPC shall include all items which are relevant to a single engined aeroplane.
Approval can be given by TC for flight simulators to be used in a PPC or in training instead of aeroplanes, otherwise, training and checking procedures shall be completed in the aeroplane.

Competency Check
For pilots flying single engine aeroplanes operated in day VFR, the chief pilot, or a pilot delegated by the chief pilot, shall be responsible for the training and will certify the competency of each pilot on the most complex single engined aeroplane to be flown.

For pilots flying as SIC on multi engined aeroplanes operating under IFR or VFR,

a) where the aeroplane is type certified for two pilot operation, the SIC shall complete a PPC;
b) where the operation of the aeroplane requires a type rating, and the SIC does not possess the required rating, he/she shall complete an initial PPC to obtain that type rating. The chief pilot, or a pilot delegated by the chief pilot, shall then be responsible for annual recurrent training and will certify the competency of the pilot on each multi engined aeroplane type to be flown. If SIC already possesses the required type rating, chief pilot (or delegate) is responsible for initial and recurrent training and competency certification for each type flown;
c) for all other multi engined aeroplanes, the chief pilot, or a pilot delegated by the chief pilot, shall be responsible for the training and will certify the competency of the pilot on each multi-engined aeroplane type to be flown.

A pilot shall be certified as competent in the PPC performance for single engined aeroplanes, multi engined aeroplanes, wheels, floats or skis, as appropriate for the operation to be conducted.

PPC Items
The PPC consists of a pre-flight phase and a flight phase covering the following specific items:

a) flight planning and equipment examination;
b) aeroplane inspection;
c) taxiing;
d) engine checks;
e) take-off (normal, IFR, crosswind, simulated engine failure on take off, rejected take off);
f) IFR procedures (unless operator is day VFR only or pilot is assigned only to day VFR flights);
g) in-flight maneuvers (steep turns, approaches to stalls, landings and approaches to landings, including cross wind landing);
h) normal procedures (including knowledge of systems ie anti icing, de icing, auto pilot, stall warning devices, airborne radar, and other aids);
i) abnormal, emergency procedures (system failures, at least two simulated engine failures).

Air Taxi Operators - Training

General
Every air operator shall establish an maintain a ground and flight training program that is designed to ensure that each person who receives training acquires the competence to perform the assigned duties and that has been approved by the Minister. An air operator's ground and flight training program shall include:

a) company indoctrination training;
b) upgrading training;
c) initial and annual training including training on aircraft type, aircraft servicing and ground handling, emergency procedures, operational control personnel and aircraft surface contamination;
d) any other training to ensure a safe operation.

An air operator is required to include a detailed syllabus of its ground and flight training program in its COM and to ensure that qualified personnel provide the required training. As well, air operators must establish and maintain a safety awareness program on aircraft surface contamination.

Every air operator shall, for each person who is required to receive training, establish and maintain TRAINING RECORDS that include the person's name, licence number, type and ratings, medical category, dates while the person was employed and completed training including PPC, competency check or examination. The records shall include whether a pass or fail was obtained on exams. Training records shall be retained for at least three years. An air operator shall retain a copy of the most recent written examination completed by each pilot for each type of aircraft for which the pilot has a qualification.

Training Program Standards
Ground training programs shall provide a means of evaluating the trainee after completion of the syllabus by completion of examinations with a review and correction of any errors.

Company Indoctrination Training
This training is required upon employment for all persons assigned to an operational control function including base managers, pilots and persons responsible for flight watch or flight following. The program shall ensure that persons involved in control of flight

operations are aware of their responsibilities, know company reporting relationships and are competent to fulfil their assigned duties relating to flight operations. Company indoctrination shall include:

a) CARs;
b) AOC and operations specifications;
c) company organization, reporting relationships and communication procedures including duties and responsibilities of flight crew members and relationship of duties to other crew members;
d) flight planning and operating procedures;
e) fuelling procedures, fuelling with passengers on board and fuel contamination precautions;
f) critical surface contamination and safety awareness program;
g) passenger safety briefings and safe movement of passengers to and from the aeroplane;
h) COM use and status, maintenance release procedures and accident or incident reporting;
i) use of MELs (if applicable);
j) windshear, aeroplane icing, and other meteorological training appropriate to operation;
k) navigation procedures;
l) on board passenger medical emergency;
m) handling of disabled passengers;
n) operational control system;
o) weight and balance system procedures.

Technical Ground Training, Initial and Recurrent
This training shall ensure that each flight crew member is knowledgeable with respect to aeroplane systems and all normal, abnormal and emergency procedures, and includes the following:

a) aeroplane systems operation and limitations as contained in the AFM, AOM and SOPs;
b) operation of all installed equipment in all operator's aeroplanes of same type;
c) equipment differences installed in aeroplanes of the same type in air operator's fleet;
d) applicable SOPs for pilot flying and pilot not flying duties for normal, abnormal and emergency procedures for the aeroplane;
e) weight and balance procedures.

Technical ground training shall be conducted annually.

Synthetic Flight Training Device Training Programs
Depending on whether the air operator has a Full Flight Simulator (FFS) or Flight Training Device (FTD), TC may approve a "Level A" or "Level B" training program, depending on the simulator. Much of the required flight training for initial upgrade and recurrent training can be conducted on an FFS or FTD. However, some specific flight training in the aeroplane will always be required.

Aeroplane Flight Training Program
Any simulated failures of aeroplane systems shall only take place under operating conditions which do not jeopardize safety of flight.

SOP procedures training is required for normal, abnormal and emergency operations of the aeroplane systems and components including:

a) use of checklists (interior and exterior pre flight checks);
b) ground maneuvering;
c) cabin crew coordination and command;
d) normal take-off, circuit and landing;
e) simulated aeroplane and cargo fire, ground, air;
f) simulated engine fire and failure;
g) briefings on effects of airframe and engine icing and anti-ice operation;
h) take-off, landing and flight with critical engine simulated inoperative, including driftdown and engine inoperative performance capabilities;
i) for 3 and 4 engine aeroplanes, in flight procedures including approach and landing with two engines inoperative (PIC only);
j) simulated loss of pressurization and emergency descent;
k) no electronic glide slope approach and landing;
l) simulated systems failures including hydraulic, electrical, flight control, navigation, communications, or other system failures;
m) pilot incapacitation - recognition and response;
n) briefing on recovery from turbulence and windshear on take-off and approach;
o) approach to the stall and recovery procedure simulating ground contact-imminent and ground contact-not-a-factor (clean, take-off and landing configuration);
p) buffet onset boundary, steep turns (45° bank) and other flight characteristics (initial and upgrade only);
q) aeroplane performance for climb, cruise, holding, descent and landing;
r) normal and performance limited take-offs;
s) crosswind take-off and landing, and briefing on contaminated runway take-off and landing;
t) take-off and landing data calculations;
u) simulated rejected take-off procedures (at or below 60 kts) and rejected landings;
v) crew and passenger evacuation procedures;
w) other specialized equipment if applicable;
x) flight planning and IFR procedures where the air operator is authorized for IMC or night VFR flight including departure, enroute, holding and arrival procedures, and all types of IFR approaches and missed approaches in simulated minimum visibility conditions, circling approaches using available automation levels.

Emergency Procedures Training for Pilots
This training is required on an annual basis and shall include instruction in the location and operation of all emergency equipment. Where practical training is required, it shall be completed on initial training and every three years thereafter.

The following training should be provided:

a) aeroplane fire in the air and on the ground;
b) use of fire extinguishers, practical training;
c) operation and use of emergency exits including practical training;
d) passenger preparation for an emergency landing or ditching (as applicable), practical training;
e) evacuation procedures, practical training;
f) donning and inflation of life preserves (when equipped), practical training;
g) removal from stowage, deployment, inflation and boarding of life rafts/slide rafts (where equipped) including practical training;
h) pilot incapacitation, practical training;
i) hijacking, bomb threat and security concerns;
j) passenger on board medical emergency;
k) special emergency procedures for MEDEVAC operations and patient emergency evacuation.

Regaining Qualifications Training
For pilots who have not maintained their three take offs and landings within previous 90 days recency, for a period between 90 and 180 days, a briefing is required on changes that have occurred to the aeroplane or its operation since the last flight, and three take-offs and landings are required (may be carried out as part of a PPC if it has come due).

Regaining Qualifications after PPC Expiry
Where a PPC has expired for less than 6 months, to regain type qualification, emergency procedures training must be completed, plus any recurrent training, including a PPC, which may have come due during the absence from flying duties.

Where a PPC has expired from between 6 and 24 months, to regain type qualification, emergency procedures training must be completed, plus any recurrent training, including a PPC, which may have come due during the absence from flying duties. As well, a technical ground training course consisting of an aeroplane system review is required.

Where the PPC has expired for a period greater than 24 months, a complete initial aeroplane type training course shall be carried out.

Right Seat Conversion Training
For a left seat qualified pilot to operate an aeroplane from the right seat, the pilot must be qualified and current on the aeroplane type for left seat duties, ground training on right seat duties must be completed, and the pilot must receive on an annual basis sufficient flight or FFS training to enable a company check pilot/chief pilot to certify the pilot's competency to carry out duties from the right seat.

Upgrade Training and Checking
Upgrade training and checking for pilots qualified as a SIC on that aeroplane type shall include:

a) successful completion of PIC training in all areas of aeroplane handling and operating as outline in the air operator's approved initial course;
b) command and decision making;
c) specialized operations qualification training (ie lower take-off limits etc);
d) completion of initial PPC on that type of aeroplane conducted by TC inspector an approved company check pilot.

Upgrade training and checking for pilots whose PPC as SIC on that aeroplane has expired within the previous 24 months shall consist of the above upgrade training, and as well, a briefing on changes that have occurred to the aeroplane or its operation since the last flight, and three take-offs and landings (which may be carried out as part of the PPC).

Pilots who have not held a valid PPC on that aeroplane type as SIC for a period greater than 24 months shall be given a complete initial aeroplane type training course as well as the upgrade training above.

Flight Follower Training
An approved initial and annual recurrent training program is required for company personnel responsible for flight following of company aeroplanes. The training program shall consist of duties and responsibilities, communication procedures, applicable regulations and standards, flight preparation procedures as applicable to assigned duties, procedures in the event of an emergency or overdue aircraft, accident and incident reporting procedures, requirements of COM as applicable to the duties and responsibilities.

Aeroplane Surface Contamination Training
An approved surface contamination initial and recurrent training program is required for all operations personnel to ensure they are aware of the hazards and procedures for ice, frost and snow critical contamination on aircraft. The training program shall include responsibility of PIC and other operations personnel, regulations related to operations in icing conditions, weather conducive to ice, frost and snow contamination, inspection before flight and removal of contamination, in-flight icing recognition and hazards related to critical surface contamination of ice, frost and snow.

Minimum Equipment List Training (MEL)
When an MEL has been approved for use on an aeroplane type (see s. 1.28) the air operator shall provide the following MEL training to crew members, maintenance personnel and to any persons exercising operation control as applicable:

a) training for maintenance personnel on those sections of the MCM which deal with the MEL, placarding of inoperative equipment, maintenance release of an aeroplane, dispatching, and any other related procedures;
b) training for pilots and operational control personnel shall include instruction on purpose and use of an MEL, air operator MEL procedures, elementary maintenance procedures and responsibility of the PIC;
c) recurrent training shall be conducted annually to ensure air operator personnel are aware of any changes to the MEL or MEL procedures.

Transportation of Dangerous Goods
Training on the specific procedures on labelling, handling and loading of dangerous goods is required where the operator transports dangerous goods.

High Altitude Training
High altitude training is required for all flight crew members operating aeroplanes above 13,000 feet ASL before the first assignment on a pressurized aeroplane and every three years thereafter. The training shall cover physiological phenomena in a low pressure environment such as respiration, hypoxia, duration of consciousness at altitude without supplemental oxygen, and gas expansion and gas bubble formation. As well, training on other factors associated with rapid loss or pressurization including most likely causes, noise, cabin temperature gauge, cabin fogging, effects on objects located near the point of fuselage failure, and actions of crew member immediately following the event and the likely resultant attitude.

Air Taxi Operators - Manuals

Company Operations Manual (COM)
Every air operator shall establish and maintain a COM. The COM and amendments shall be submitted to the Minister. The COM shall be amended by the air operator when required. A COM, which may be issued in separate parts corresponding to specific aspects of an operation, shall include instructions and information necessary to enable the personnel concerned to perform their duties safely. The COM must be such that all

parts of the manual are consistent and compatible in form and content, and the manual can be readily amended.

An air operator shall provide a copy of the appropriate parts of its COM, including any amendments, to each of its crew members and to its ground operations and maintenance personnel. A copy of the COM may be placed in each aircraft operated, instead of providing a copy to each crew member, if the air operator has established in its COM procedures for amending that manual. Every person who has been provided with a copy of the appropriate parts of a COM shall keep it up to date with the amendments and shall ensure that the appropriate parts are accessible when the person is performing assigned duties.

For air operators utilizing multi-engined aeroplanes or single engined aeroplanes operating IFR or night VFR, the COM shall consist of:

a) preamble relating to use and authority of COM, table of contents; amending procedures, amendment record sheet, distribution list and list of effective pages;
b) copy of AOC and operations specifications;
c) chart of management organization;
d) duties, responsibilities and management command succession and operations personnel;
e) description of operational control system including flight authorization and flight preparation procedures, preparation of operational flight plan and other flight documents, procedures to ensure the flight crew are advised, prior to dispatch, of any aeroplane defects that have been deferred (by MEL or other means), flight watch, flight following and communication requirements, dissemination procedures for operational information and acknowledgment, fuel and oil requirements, weight and balance system, accident/incident or overdue aircraft reporting procedures, use of checklists, maintenance discrepancy reporting and requirements on flight completion, retention period of operational flight plans, sample operational flight plan, weight and balance form and retention period;
f) FDR and CVR procedures, if applicable, policy regarding GPWS and TCAS (if applicable);
g) operating weather minima including alternate and applicable requirements for IFR, VFR, VFR at night, VFR OTT;
h) instrument and equipment requirements;
i) IFR approach procedures (including company approaches), alternate aerodrome procedures;
j) procedures for establishing company routes in uncontrolled airspace;
k) procedures pertaining to enroute operation of navigation and communication equipment (including collision avoidance procedures);
l) operations in hazardous conditions such as icing, thunderstorms, whiteout, windshear;
m) aeroplane performance limitations, securing of cargo, passenger briefing procedures;
n) use of AFM, AOM, SOPs and MELs;
o) aeroplane ice, frost and snow critical surface contamination procedures;
p) procedures for carriage of dangerous goods;
q) fueling procedures including fuel contamination precautions, bonding requirements, fuelling with engine running (not permitted with passengers on board), fuelling with passengers on board;
r) list of emergency survival equipment carried on the

aeroplane and how to use equipment, emergency procedures for ELT, passenger preparation for emergency landing/ditching, emergency evacuation, ground emergency coordination procedures and unlawful interference, inspection details and frequency of inspection of emergency equipment carried on board;
s) minimum flight crew members required and flight crew member qualifications, flight duty time limitations and rest requirements, training programs including copy of company training and qualification record forms;
t) use of oxygen;
u) operations support services and equipment, passenger and cabin safety procedures;
v) policy on occupation of observer seat (if applicable);
w) requirement to make runway analysis charts;
x) procedures for reduced VFR limits in uncontrolled airspace (if applicable);
y) copies of all forms utilized including sufficient instruction on form completion;
z) other information related to safety.

Standard Operations Procedures (SOPs)
Every air operator shall, for each of its aircraft that is required to be operated by two or more pilots, establish and maintain SOPs that enable the crew members to operate the aircraft within the limitations specified in the AFM. If required, SOPs is to be on board the aircraft. SOPs shall contain the following for each type of two pilot aeroplane operated (may form part of COM):

a) general: table of contents, list of effective pages, amending procedure, preamble, use of checklists, communications, crew coordination, standard briefings and calls;
b) normal procedures: weight and balance, ramp - gate procedures, battery/APU engine starts, taxi, take-off and climb, cruise, descent, approaches (IFR, visual, VFR, circling), landing, missed approach, balked landing procedures, stall recovery, refuelling with passengers on board, use of navigation and alerting aids, checklists;
c) abnormal and emergency procedures: emergency landing/ditching (with time to prepare and without time to prepare), pilot incapacitation two communication rule (2 pilot crew), bomb threat and hijacking, engine fire/failure/shutdown, propeller overspeed (as applicable), fire - internal/external, smoke removal, rapid decompression (as applicable), flapless approach and landing (as applicable), rejected take-off, other abnormal and emergency procedures that are specific to the type of aeroplane;
d) diagrams: normal take-off, engine out take-off, precision approach with all engines operating and with engine out, non precision approach with all engines operating and with engine out, approach with engine out, go around with and without all engines operating, VFR circuits, partial flaps/slats approach, flapless approach.

[Reference: CARs 703, 723]

1.10 Aerial Work Operations

AERIAL WORK means a commercial air service other than an air transport service or a flight training service. Aerial work covers operations such as aerial product dispersal and banner towing.

AERIAL WORK ZONE means an area, delineated in an aerial work zone plan, in which aerial work is being conducted and that is over a built-up area of a city or town or over or adjacent to an area where persons may assemble.

No air operator shall operate an aircraft in aerial work operations unless the operator has obtained and complies with an AOC. Aerial work operators must also have an approved COM.

A flight authorization based on the procedures in the COM is required before any flight is commenced.

All flights must have an operational flight plan in accordance with the procedures outlined in the COM.

No air operator shall allow a person, other than a flight crew member or a person whose presence on board is essential during the flight, to be carried on board an aircraft unless the air operator is authorized to do so in its AOC.

An air operator operating an aeroplane under day VFR within uncontrolled airspace less than 1,000 feet AGL may operate at less than a flight visibility of two miles if authorized in the AOC.

Unless otherwise authorized in its AOC, no air operator shall operate an aircraft at night, in VFR OTT or IFR flight while towing, while dispersing products, or where the aircraft is single engined.

Unless otherwise authorized in its AOC, no air operator shall operate an aircraft at night with persons other than flight crew members on board unless the PIC has an instrument rating.

Unless otherwise authorized in its AOC, no air operator shall, except when conducting a take-off or landing, operate a land aircraft over water, beyond a point where the land aircraft could reach shore in the event of an engine failure.

In an aerial work zone, if specifically authorized by the Minister, or if so authorized in its AOC, an air operator may conduct a take-off, approach or landing in an aircraft within a built-up area of a city or town at a place other than an airport or a military aerodrome.

The PIC shall ensure that persons, other than flight crew members, who are on board the aircraft are given a complete safety briefing.

No person shall operate an aircraft at night unless the aircraft is equipped with at least one landing light.

No person shall operate a multi-engined aircraft in IMC unless the aircraft is equipped with two alternators/ generators, each of which is driven by a separate engine. As well, two independent sources of energy are required, at least one of which is not a battery, and each of which is able to drive all flight instruments requiring a source of energy and is installed so that the failure of one instrument or one source of energy will affect neither the energy supply to the remaining instruments nor the other source of energy.

No air operator shall operate an aircraft on a single-pilot operation in IFR flight unless the aircraft is equipped with:

a) an autopilot that is capable of operating the aircraft controls to maintain flight and maneuver the aircraft about the lateral and longitudinal axes;
b) a headset with a boom microphone or equivalent and a transmit button on the control column;
c) a chart holder that is equipped with a light and that is placed in an easily readable position.

No air operator shall operate an aircraft unless the pilot seat and any seat beside the pilot seat are equipped with a safety belt that includes a shoulder harness.

An air operator shall designate for each flight a PIC and, where the crew includes two pilots, a PIC and an SIC.

No air operator shall permit a person to act and no person shall act as a flight crew member in an aircraft unless the person

a) holds the required licences and ratings;
b) where the aircraft is operated in IFR flight and persons other than flight crew members are on board, has successfully completed a PPC, the validity period of which has not expired, for that type of aircraft;
c) has fulfilled the requirements of the air operator's ground and flight training program.

A PPC shall be conducted by the Minister.

The normal validity period of a PPC expires on the first day of the 25th month following the month in which the PPC was completed. Where a PPC is renewed within the last 90 days of its validity period, its validity period is extended by 24 months. Where a competency check or annual training is renewed within the last 90 days of its validity period, its validity period is extended by 12 months.

The Minister is authorized to extend the validity period of a PPC, a competency check or annual training by up to 60 days where the Minister is of the opinion that aviation safety is not likely to be affected.

Where the validity period of a PPC, a competency check or annual training has been expired for 24 months or more, the person must requalify meeting all initial training requirements.

An air operator shall provide a copy of the appropriate parts of its COM, including any amendments to those parts, to each of its crew members and to its ground operations and maintenance personnel. An air operator may place a copy of the appropriate parts of its COM in each aircraft that it operates, instead of providing a copy to each crew member, if the air operator has established in its COM procedures for amending it. Every person who has been provided with a copy of the appropriate parts of a COM shall keep it up to date with the amendments provided and shall ensure that the appropriate parts are accessible when the person is performing assigned duties.

Every air operator shall, for each of its aircraft that is required to be operated by two or more pilots, establish and maintain Standard Operating Procedures (SOPs) that enable the crew members to operate the aircraft within the limitations specified in the AFM and that meet CARs Commercial Air Service Standards. An air operator that has established SOPs for an aircraft shall ensure that a copy of the SOPs is carried on board the aircraft.

[Reference: CARs 702]

1.11 Commuter Operations

Commuter Operator: General

A COMMUTER OPERATION applies to a Canadian air operator in an air transport service or in aerial work involving sightseeing operations to the following aircraft:

a) a multi-engined aircraft that has a MCTOW of 8,618 kgs (19,000 lbs) or less and a seating configuration, excluding pilot seats, of 10 to 19 inclusive;

b) a turbo-jet powered aeroplane that has a maximum zero fuel weight of 22,680 kgs (50,000 lbs) or less and for which a Canadian type certificate has been issued authorizing the transport of not more than 19 passengers.

Commuter Operator: Operating Instructions

An air operator shall ensure that all operations personnel are properly instructed about their duties and about the relationship of their duties to the operation as a whole. The operations personnel of an air operator shall follow the procedures specified in the air operator's COM in the performance of their duties.

Commuter Operator: General Operational Information

Every air operator shall establish a system for the timely dissemination of general operational information that includes a means for each crew member to acknowledge receipt of such information.

Commuter Operator: Flight Authorization

No person shall commence a flight unless the flight has been authorized in accordance with the procedures specified in the air operator's COM.

Commuter Operator: Operational Flight Plan

No air operator shall permit a person to commence a flight unless an operational flight plan has been prepared in accordance with the procedures in the COM. The PIC shall ensure that one copy of the operational flight plan is left at a point of departure, in accordance with the procedures specified in the COM, and that another copy is carried on board the aircraft until the aircraft reaches the final destination of the flight. An air operator shall retain a copy of the operational flight plan, including any amendments to that plan, for the period specified in the COM.

Commuter Operator: Checklist

Every air operator shall establish an aircraft checklist for each aircraft type that it operates and shall make the appropriate parts of the checklist readily available to the crew members. Each crew member shall follow the checklist in performance of assigned duties.

Commuter Operator: Fuel Requirements

No air operator shall authorize a flight and no person shall commence a flight unless the aircraft carries sufficient fuel to meet the normal CARs fuel requirements for VFR or IFR flight (see s. 1.18). As well, for IFR flights, additional fuel is required to allow descent at any point along the route to the lower of the single-engine service ceiling or 10,000 feet, to cruise at that altitude to a suitable aerodrome, to conduct an approach and missed approach and to hold for 30 minutes at an altitude of 1,500 feet above the elevation of the aerodrome.

Commuter Operator: Simulation of Emergency Situations

No person shall, where passengers are on board an aircraft, simulate emergency situations that could affect the flight characteristics of the aircraft.

Commuter Operator: VFR Flight Obstacle Clearance Requirements

Except when conducting a take-off or landing, no person shall operate an aircraft in VFR flight

a) at night, at less than 1,000 feet above the highest obstacle located within a horizontal distance of three miles from the route to be flown; or

b) during the day, at less than 500 feet AGL or at a horizontal distance of less than 500 feet from any obstacle.

Commuter Operator: VFR Flight Weather Conditions

No person shall commence a VFR flight unless current weather reports and forecasts, if obtainable, indicate that the weather conditions along the route to be flown and at the destination aerodrome will be such that the flight can be conducted in compliance with VFR.

Commuter Operator: Take-Off Minima

No person shall conduct a take-off in an aircraft in IMC where weather conditions are at or above the take-off minima, but below the landing minima for the runway to be used unless the take-off is authorized in the AOC.

A person may conduct a take-off in an aircraft in IMC where weather conditions are at or above the take-off minima, but below the landing minima, for the runway to be used, if the weather conditions are at or above the landing minima for another suitable runway at that aerodrome, taking into account the aircraft performance limitations.

A person may conduct a take-off in an aircraft in IMC where weather conditions are below the take-off minima specified in the IAP if the person is authorized in the AOC.

Commuter Operator: No Alternate Aerodrome - IFR Flight

A person may conduct an IFR flight where an alternate aerodrome has not been designated in the IFR flight plan or in the IFR flight itinerary, if the person is authorized to do so in the AOC.

Commuter Operator: Routes in Uncontrolled Airspace

No person shall, in uncontrolled airspace, conduct an IFR flight or a night VFR flight on a route other than an air route unless the air operator has established the route to TC standards.

Commuter Operator: Instrument Approach Procedures

No person shall conduct a CAT II or CAT III instrument approach unless the air operator is authorized to do so in its AOC and the approach is conducted in accordance with the *Manual of All Weather Operations - Categories II and III*.

No person shall terminate an instrument approach with a landing unless, immediately prior to landing, the PIC ascertains, by means of radiocommunication or visual inspection, the condition of the intended landing surface and the wind direction and speed.

Commuter Operator: Weight and Balance Control

No person shall operate an aircraft unless, during every phase of the flight, the load restrictions, weight and CG of the aircraft conform to the limitations specified in the AFM. An air operator shall specify in its COM its weight and balance system and instructions to employees regarding the preparation and accuracy of weight and balance forms.

Commuter Operator: Apron and Cabin Safety Procedures

An air operator shall establish procedures to ensure that

a) passengers move on the apron and embark and disembark safely, in accordance with procedures specified in the COM;
b) passengers are seated and secured as required;
c) each seat back is in the upright position and all chair tables are stowed during movement on the surface, take-off and landing and at such other times as the PIC considers necessary for the safety of the persons on board the aircraft;
d) seats located at emergency exits are not occupied by passengers whose presence in those seats could adversely affect the safety of passengers or crew members during an emergency evacuation;
e) the flight crew can exercise supervisory control over passengers by visual and aural means.

An air operator may, for the transportation of any passenger who has been certified by a physician as unable to sit upright, allow the back of the seat occupied by such a passenger to remain in the reclining position during movement on the surface, take-off and landing if the passenger is seated in a location that will not restrict the evacuation of other passengers from the aircraft, the passenger is not seated in a row that is next to or immediately in front of an emergency exit and the seat immediately behind the passenger's seat is vacant.

No air operator shall assign a person to perform duties on board an aircraft unless that person has completed his/her required training program.

No air operator shall permit an aircraft with passengers on board to be fuelled unless it is carried out in accordance with the procedures in the COM.

No air operator shall permit the use of a portable electronic device on board an aircraft unless the air operator has established procedures to be applied for such devices in the COM.

Commuter Operator: Briefing of Passengers

The PIC shall ensure that passengers are given the required safety briefings.

Commuter Operator: Take-Off Weight Limitations

No person shall conduct a take-off in an aircraft if the weight of the aircraft

a) exceeds the maximum take-off weight specified in the AFM for the pressure altitude and the ambient temperature at the aerodrome where the take-off is to be made;
b) after allowing for planned fuel consumption during the flight to the destination aerodrome or alternate aerodrome, exceeds the landing weight specified in the AFM for the pressure altitude and ambient temperature at the destination or alternate aerodrome.

Unless otherwise authorized in the AOC, in the determination of the maximum take-off weight above, for a small aeroplane, the required accelerate-stop distance shall not exceed the ASDA and the all-engines operating take-off distance shall not exceed the TODA.

Unless otherwise authorized in the AOC, in the determination of the maximum take-off weight above, for a large aeroplane, the required accelerate stop distance shall not exceed the ASDA, the required take-off run shall not exceed the TORA, and the required take-off distance shall not exceed the TODA.

For the purposes of determining the maximum take-off weight, the following factors shall be taken into account:

a) the pressure altitude at the aerodrome;
b) the ambient temperature;
c) the runway slope in the direction of take-off;
d) no more than 50 % of the reported headwind component or not less than 150 % of the reported tailwind component.

Commuter Operator: Net Take-Off Flight Path

Unless otherwise authorized in the AOC, no person shall conduct a take-off in a large aeroplane if the weight of the aeroplane is greater than the weight specified in the AFM as allowing a net take-off flight path that clears all obstacles by at least 35 feet vertically or at least 200 feet horizontally within the aerodrome boundaries, and by at least 300 feet horizontally outside those boundaries.

In determination of the maximum weight, minimum distances and flight path referred to above, corrections shall be made for

a) the runway to be used;
b) the runway slope in the direction of take-off;
c) the pressure altitude at the aerodrome;
d) the ambient temperature;
e) the wind component at the time of take-off, where not more than 50 % of the reported headwind component or not less than 150% of the reported tailwind component is considered.

As well, calculations shall be based on the pilot not banking the aeroplane before reaching an altitude of 50 feet, if authorized in the AOC a bank of 15° at or below 400 feet, and using no more than 25° of bank thereafter, aircraft speed and configuration permitting.

Commuter Operator: Enroute Limitations with One Engine Inoperative

No person shall operate a multi-engined aircraft with passengers on board if the weight of the aircraft is greater than the weight that will allow the aircraft to maintain, with any engine inoperative, the following altitudes:

a) when operating in IMC or in IFR flight on airways or air routes, the MOCA of the route to be flown;
b) when operating in IMC or in night VFR flight on routes established by an air operator, the MOCA of the route to be flown;
c) when operating in VFR flight, at least 500 feet above the surface.

Commuter Operator: Dispatch Limitations - Landing at Destination and Alternate Aerodromes

No person shall dispatch or conduct a take-off in a turbo-jet powered aeroplane or in a large aeroplane unless the weight of the aeroplane on landing at the destination and alternate aerodromes will allow a full-stop landing within 60 % of the LDA (turbo-jet) or 70% of the LDA (propeller). In making this determination, consideration shall be given to

a) the pressure altitude at the destination aerodrome and at the alternate aerodrome;
b) not more than 50% of the reported headwind component or not less than 150% of the reported tailwind component;
c) the aeroplane must be landed on a suitable runway, considering the wind speed and direction, the ground handling characteristics of the aeroplane, and other conditions such as landing aids and terrain.

Commuter Operator: Dispatch Limitations - Wet Runway, Turbo-Jet Aeroplanes

When weather reports or forecasts indicate that the runway may be wet at the estimated time of arrival, no person shall dispatch or conduct a take-off in a turbo-jet powered aeroplane unless the LDA at the destination airport is at least 115% of the landing distance required full stop landing limit (as determined by the dispatch landing limitations above).

This requirement does not apply if the AFM includes specific information about landing distances on wet runways.

Commuter Operator: Aircraft Equipment Requirements

General
No person shall operate an aircraft in IMC unless the aircraft is equipped with

a) at least two generators, each of which is driven by a separate engine, and at least half of which have a sufficient rating to supply the electrical loads of all instruments and equipment necessary for the safe emergency operation of the aircraft;
b) two independent sources of energy and a means of selecting either source, at least one source of energy being an engine-driven pump or generator, and each source of energy being able to drive all gyroscopic instruments and being installed so that the failure of one instrument or one source of energy will affect neither the energy supply to the remaining instruments nor the other source of energy;
c) no person shall operate an aircraft at night unless the aircraft is equipped with at least one landing light.

Operation in Icing Conditions
No person shall conduct a take-off or continue a flight when icing conditions are reported to exist or are forecast to be encountered along the route to be flown unless the aircraft is equipped to be operated in those conditions and the aircraft type certificate authorizes flight in those conditions. No person shall operate an aeroplane in icing conditions at night unless the aeroplane is equipped with a means to illuminate or otherwise detect the formation of ice.

Airborne Thunderstorm Detection and Weather Radar Equipment
No person shall operate an aircraft with passengers on board in IMC when current weather reports or forecasts indicate that thunderstorms may reasonably be expected along the route to be flown, unless the aircraft is equipped with thunderstorm detection equipment or weather radar equipment.

Additional Equipment for Single Pilot Operations
No person shall operate an aircraft on a single-pilot operation in IMC unless the aircraft is equipped with

a) an auto-pilot that is capable of operating the aircraft controls to maintain flight and maneuver the aircraft about the lateral and longitudinal axes;
b) a headset with a boom microphone or equivalent and a transmit button on the control column;
c) a chart holder that is equipped with a light and that is placed in an easily readable position.

Protective Breathing Equipment
No air operator shall operate a pressurized aircraft unless protective breathing equipment with a 15 minute supply of breathing gas at a pressure altitude of

8,000 feet is readily available at each flight crew member position. This equipment can be used to meet the standard crew member oxygen requirements (see s. 1.35).

First Aid Oxygen
No air operator shall operate an aircraft with passengers on board above FL 250 unless the aircraft is equipped with oxygen dispensing units and an undiluted supply of first aid oxygen sufficient to provide at least one passenger with oxygen for at least one hour or the entire duration of the flight at a cabin pressure altitude above 8,000 feet, after an emergency descent following cabin depressurization, whichever period is longer.

Shoulder Harnesses
No person shall operate an aircraft unless the pilot seat and any seat beside the pilot seat are equipped with a safety belt that includes a shoulder harness.

Commuter Operator: Emergency Equipment
No air operator shall operate an aircraft with passengers on board unless at least one HAND-HELD FIRE EXTINGUISHER is readily accessible for immediate use and is located in the passenger compartment. All required emergency equipment (see s. 1.33) must be inspected regularly in accordance with the COM.

Commuter Operator: Personnel Requirements

Minimum Crew
No air operator shall operate an aircraft with fewer than two pilots, where the aircraft is an aeroplane carrying 10 or more passengers or is carrying passengers in IFR flight.

Designation of PIC and SIC
An air operator shall designate for each flight a PIC and, where the crew includes two pilots, a PIC and an SIC.

Flight Crew Member Qualifications
No air operator shall permit a person to act and no person shall act as a flight crew member in an aircraft unless the person:

a) holds the required licences and ratings;
b) within the previous 90 days, has completed at least three take-offs and landings in an aircraft of that type (where a type rating is required) or in an aircraft of that category and class (TC can approve use of advanced flight simulators in place of aircraft);
c) has successfully completed a PPC that remains valid for that type of aircraft;
d) has fulfilled the requirements of the air operator's ground training program and, except where undergoing line indoctrination training, the air operator's flight training program.

If authorized in its AOC, an air operator may group similar aircraft as a single type for PPC purposes.

No person shall act as the PIC of an aircraft with passengers on board in IFR flight unless the person has acquired at least 1,200 hours of flight time as a pilot.

No person shall act as the PIC of an aircraft in VFR flight unless the person has acquired at least 500 hours of flight time as a pilot.

No person shall act as the PIC of an aircraft with a person other than a flight crew member on board in night VFR flight unless the person acting as the PIC holds an IFR rating for that class of aircraft.

The PPC training requirements in paragraphs a) to d) above need not be completed in the case of ferry flights, training flights or positioning flights.

PPC Validity Periods
The validity period of a PPC and of the annual training required expires on the first day of the 13th month following the month in which the PPC or training was completed. Where a PPC or annual training is renewed with the last 90 days of its validity period, its validity period is extended by 12 months. The Minister may extend the validity period of a PPC, competency check or annual training by up to 60 days where the Minister is of the opinion that aviation safety is not likely to be affected. Where the validity of a PPC or annual training has been expired for 24 months or more, the person must requalify by meeting all initial training requirements.

Commuter Operator: Manuals
Air operators must produce an approved COM, and provide a copy of the appropriate parts of its COM, including amendments, tos each of its crew members and to its ground operations and maintenance personnel. Every person receiving parts the COM shall keep it up-to-date with amendments and that the appropriate parts are accessible while the person is performing his or her duties.

An air operator may establish and maintain an AOM for the use and guidance of crew members in the operation of its aircraft. An AOM shall contain the aircraft operating procedures, and where the AFM is not carried on board the aircraft, the aircraft performance data and limitations specified in the AFM, which shall be clearly identified as AFM requirements. An air operator that has established an AOM shall ensure that a copy is carried on board each aircraft to which it relates.

Every air operator shall, for each of its aircraft that is required to be operated by two or more pilots, establish and maintain SOPs that enable the crew members to operate the aircraft within the limitations specified in the AFM. An air operator that has established SOPs for an aircraft shall ensure that a copy is carried on board the aircraft. Where an air operator has established an AOM, the SOPs for the aircraft shall form part of that manual.

[Reference: CARs 704]

1.12 Airline Operations

Airline Operator: Application
An AIRLINE OPERATION applies to a Canadian air operator in an air transport service or in aerial work involving sightseeing operations, of an aeroplane, other than a commuter aeroplane, that has a MCTOW of more than 8,618 kgs (19,000 lbs) or for which a Canadian type certificate has been issued authorizing the transport of 20 or more passengers.

Airline Operator: Exceptions
Normal passenger seating and safety briefing requirements do not apply where there are 9 or fewer persons

on board and each person is either an air operator employee or a person whose presence is necessary for flight safety, safe handling of animals or dangerous goods, security of valuable cargo or preservation or handling of cargo. Such persons can be carried on board the aircraft where normally the type certificate does not authorize the transport of passengers. The air operator must provide for the safety of such persons including seating, safety belts, a means of communication with flight crew members, a safety briefing before take-off and established procedures for the transport of such persons.

Airline Operator: Opertational Control System
No air operator shall operatre an aircraft unless the air operator has an operational control systems that meets CARs Commercial Air Service Standards and is under the control of its operations manager.

Airline Operator: Flight Authorization
No person shall commence a flight unless the flight has been authorized in accordance with the procedures specified in the air operator's COM.

Airline Operator: Operational Flight Plan
No air operator shall permit a person to commence a flight unless an operational flight plan has been prepared in accordance with the procedures specified in the COM. The PIC shall ensure that one copy of the operational flight plan is left at a point of departure, in accordance with the procedures specified in the COM, and that another copy is carried on board the aircraft until the aircraft reaches the final destination of the flight. An air operator shall retain a copy of the operational flight plan, including any amendments to that plan, for the period specified in the COM.

Airline Operator: Aircraft Maintenance
No air operator shall permit a person to conduct a take-off in an aircraft that has not been maintained in accordance with the air operator's maintenance control system.

Airline Operator: Checklist
Every air operator shall establish the required checklist for each type that it operates and shall make the appropriate parts of the checklist readily available to the crew members. Every crew member shall follow the checklist in the performance of the crew member's assigned duties.

Airline Operator: Fuel Requirements
Unless otherwise authorized in its AOC, no air operator shall authorize a flight and no person shall commence a flight unless the aircraft

a) when operating VFR, carries sufficient fuel to fly to the destination aerodrome and thereafter to fly for 45 minutes at normal cruising speed;

b) when operating IFR on designated routes or over designated areas such as flights exclusively within or terminating within the NDA or certain international routes (ie USA), carries an enroute fuel reserve of 5% of the fuel required to the destination aerodrome;

c) when operating in IFR flight, except when complying with ETOPS criteria, carries sufficient fuel to allow the aircraft to descend at any point along the route to the lower of the one- engine inoperative service ceiling or 10,000 feet ASL, to cruise at that altitude to a suitable aerodrome, to conduct an approach and a missed approach, and to hold for

30 minutes at an altitude of 1,500 feet above the elevation of the aerodrome.

Airline Operator: Extended Range Twin-Engined Operations (ETOPS)
No air operator shall operator a twin-engined aeroplane on a route containing a point that is farther from an adequate aerodrome than the distance that can be flown in 60 minutes at the one-engine inoperative cruise speed, unless the flight is conducted wholly within CDA. Such routes can be flown for operators using turbine-powered aeroplanes having ETOPS approval in their AOC and complying with *Safety Criteria for Approval of Extended Range Twin Engine Operations Manual.*

Airline Operator: Admission to Flight Deck
Where a TC air carrier inspector presents an official identity card to the PIC, the PIC shall give the inspector free and uninterrupted access to the flight deck, making available for the inspector the observer seat determined by the inspector to be the most suitable. Only flight crew members, TC inspectors or other persons authorized in the COM shall be admitted to the flight deck of an aircraft unless there is a seat available for that person in the passenger compartment.

Airline Operator: Seats for Cabin Safety Inspectors
An air operator shall provide a cabin safety inspector who is performing an in-flight cabin inspection with a confirmed passenger seat in the passenger compartment.

Airline Operator: Flight Crew Members at Controls
Flight crew members who are on flight deck duty shall remain at their duty stations with their safety belts fastened and, where the aircraft is below 10,000 feet ASL, with their safety belts, including their shoulder harnesses, fastened. Flight crew members may leave their duty stations for the performance of duties in connection with the operation of the aircraft, for physiological needs or for taking an authorized rest period.

Airline Operator: Simulation of Emergency Situations
No person shall, where passengers are on board an aircraft, simulate emergency situations that could affect the flight characteristics of the aircraft.

Airline Operator: Crew Member Briefing
The PIC shall ensure that, prior to each flight or series of flight segments, the crew members of the aircraft are given the required pre-flight briefing. The pre-flight briefing covers anticipated weather and flying conditions, flight time, altitudes, review of selected communication procedures, review of selected emergency procedures, review of selected safety procedures, additional information relating to unserviceable equipment or abnormalities that may affect passengers. The PIC shall also give a pre-flight briefing to the in-charge flight attendant and cabin crew.

Airline Operator: VFR Flight Obstacle Clearance Requirements
Except when conducting a take-off or landing, no person shall operate an aeroplane in VFR flight

a) during the day, at less than 1,000 feet AGL or at a horizontal distance of less than 1,000 feet from any obstacle or

b) at night, at less than 1,000 feet above the highest obstacle located within a horizontal distance of 5 miles from the route to be flown or, in designated mountainous regions, at less than 2,000 feet above the highest obstacle located within a horizontal distance of 5 miles from the route flown.

Airline Operator: VFR Flight Weather Conditions

No person shall commence a VFR flight unless current weather reports and forecasts, if obtainable, indicate that the weather conditions along the route to be flown and at the destination aerodrome will be such that the flight can be conducted in compliance with VFR.

Airline Operator: Take-Off Minima

No person shall conduct a take-off in an aircraft in IMC where weather conditions are at or above the take-of minima, but below the landing minima, for the runway to be used unless an alternate aerodrome is specified in the operational flight plan and that aerodrome is located

a)   in the case of a twin-engined aircraft, within the distance that can be flown in 60 minutes at the one-engine inoperative cruise speed, or
b)   in the case of a 3 or 4 engined aircraft or where an air operator is authorized in its AOC for ETOPS with the type of aircraft operated, within the distance that can be flown in 120 minutes at the one engine inoperative cruise speed.

A person may conduct a take-off in an aircraft in IMC where weather conditions are at or above the take-off minima, but below the landing minima, for the runway to be used, if the weather conditions are at or above the landing minima for another suitable runway at that aerodrome, taking into account the required aircraft performance operating limitations.

A person may conduct at take-off in an aircraft in IMC where weather conditions are below the take-off minima specified in the IAP if authorized in the AOC. For example, for take-off with RVR 1200, in addition to one-engine inoperative cruise speeds above, take-off runway must have high intensity runway lights and centre line lights/visible centreline markings, PIC must be satisfied RVR 1200 exists before commencing take-off, PIC and SIC attitude indicators must have increments above and below horizontal to 15°, essential instrument failure warning systems are functional (AI's, DG's and HSI's), specific training has been provided to flight crew members, chief pilot has certified in pilot's training file that PIC is competent to conduct an RVR 1,200 feet take-off, and PIC has completed at least 100 hours PIC on type. For RVR 600 take-offs, in addition to RVR 1200 take-off requirements, at least two transmissometers must be available at runway, each reading not less than RVR 600 and approach end and mid-point end of runway, PIC must be satisfied at least RVR 600 exists, and both PIC and SIC must be certified to conduct RVR 600 take-offs by company check pilot or TC inspector in training files.

Airline Operator: No Alternate Aerodrome - IFR Flight

A person may conduct an IFR flight where an alternate aerodrome has not been designated in the IFR flight plan or in the IFR flight itinerary if authorized in the AOC.

Airline Operator: Routes in Uncontrolled Airspace

No person shall, in uncontrolled airspace, conduct an IFR flight or a night VFR flight on a route other than an air route unless the air operator has established that route with TC.

Airline Operator: Instrument Approach Procedures

No person shall conduct a CAT II or CAT III instrument approach unless the air operator is so authorized in its AOC and the approach is conducted in accordance the *Manual of All Weather Operations (Cats II and III)*. No person shall terminate an instrument approach with a landing unless, immediately prior to landing, the PIC ascertains, by means of radiocommunication or visual inspection, the condition of the intended landing surface and the wind direction and speed.

Airline Operator: Weight and Balance Control

No person shall operate an aircraft unless, during every phase of the flight, the load restrictions, weight and CG of the aircraft conform to the limitations specified in the AFM. An operator shall establish a detailed weight and balance system to show how the operator will be responsible for accurate weight and balance calculations. The weight and balance calculations may be incorporated in the operational flight plan or by a separate form. An air operator shall specify in its COM its weight and balance system an instructions to employees regarding the preparation and accuracy of weight and balance forms.

Airline Operator: Passenger and Cabin Safety Procedures

An air operator shall establish procedures to ensure that passengers move to and from the aircraft and embark and disembark safely as specified in the COM. All passengers must be seated and secured as required by the CARs.

Each seatback must be in the upright position and all chair tables and carry-on baggage must be stowed during movement on the surface, take-off and landing and at such other times as the PIC considers necessary for the safety of the persons on board the aircraft.

Seats located at emergency exits and seats that are not located on the main deck of an aircraft must not be occupied by passengers whose presence in those seats could adversely affect the safety of passengers or crew members during an emergency evacuation.

An air operator may, for the transportation of any passenger who has been certified by a physician as unable to sit upright, allow the back of the seat occupied by such a passenger to remain in the reclining position during movement on the surface, take-off and landing if

a)   the passenger is seated in a location that will not restrict the evacuation of other passenger from the aircraft;
b)   the passenger is not seated in a row that is next to or immediately in front of an emergency exit;
c)   the seat immediately behind the passenger's seat is vacant.

No air operator shall permit an aircraft with passengers on board to be fuelled unless the fuelling is carried out in accordance with the required procedures in the COM.

No air operator shall permit the use of a portable electronic device on board an aircraft unless the air operator has established procedures that has been specified in the COM.

Airline Operator: Carry-On Baggage

Every air operator shall establish an approved carry-on baggage control program. Carry-on baggage must be with the control program requirements and stowed in an approved compartment or overhead rack, under a passenger seat or otherwise restrained in accordance with airworthiness standards. All carry-on baggage shall be stowed so that it does not obstruct access to safety equipment, exits or aircraft aisles.

No air operator shall allow the passengers entry doors of an aircraft to be closed for departure until a crew member has verified that all carry on baggage is stowed in the proper approved location.

All carry-on baggage shall be safely stowed prior to movement of the aircraft on the surface and during take-off, periods of in-flight turbulence and landing.

No carry-on baggage that may cause injury to passengers in the event of turbulence or an emergency shall be stowed in an overhead rack unless that rack is equipped with approved restraining devices or doors.

Airline Operator: Briefing of Passengers

An air operator shall ensure that passengers are given a safety briefing. For detailed passenger briefing requirements, refer to s. 1.09.

Airline Operator: Take-off Weight Limitations

No person shall conduct a take-off in an aircraft if the weight of the aircraft

a) exceeds the maximum take-off weight specified in the AFM for the pressure altitude and the ambient temperature at the aerodrome where the take-off is to be made, or
b) after allowing for planned fuel consumption during the flight to the destination aerodrome or alternate aerodrome, exceeds the landing weight specified in the AFM for the pressure altitude and the ambient temperature at the destination or alternate aerodrome.

In the determination of the maximum take-off weight above,

a) the required accelerate-stop distance shall not exceed the ASDA;
b) the required take-off run shall not exceed the TORA;
c) the required take-off distance shall not exceed the TODA.

The following factors shall be taken into account in determining the maximum take-off weight:

a) the pressure altitude at the aerodrome;
b) the ambient temperature;
c) the runway slope in the direction of take-off;
d) not more than 50% of the reported headwind component or not less than 150% of the reported tailwind component.

Airline Operator: Net Take-Off Flight Path

No person shall conduct a take-off in an aeroplane if the weight of the aeroplane is greater than the weight specified in the AFM as allowing a net take-off flight path that clears all obstacles by at least 35 feet vertically or at least 200 feet horizontally within the aerodrome boundaries, and by at least 300 feet horizontally outside those boundaries.

In making the above determination, corrections shall be made for

a) the runway to be used;
b) the runway slope in the direction of take-off;
c) the pressure altitude at the aerodrome;
d) the ambient temperature;
e) the wind component at the time of take-off, where not more than 50 % or the reported headwind component or not less than 150% of the reported tailwind component is considered.

Calculations shall also be based on the pilot

a) not banking before reaching altitude of 50 feet;
b) using 15° or less of bank at or below 400 feet and using no more than 25° of bank thereafter, aircraft speed and configuration permitting (a bank angle of greater than 15° may be used if authorized in AOC).

Airline Operator: Enroute Limitations with One Engine Inoperative

No person shall conduct a take-off in an aeroplane if the weight of the aeroplane is greater than the weight that will allow the aeroplane to attain, with any engine inoperative, a net flight path that

a) has a positive slope at 1,000 feet above all terrain and obstructions within 5 NM on either side of the intended track, at all points along the route or planned diversion therefrom or
b) will permit flight from the cruising altitude to an aerodrome that meets landing weight/LDA requirements, and clears vertically, by at least 2,000 feet, all terrain and obstructions within 5 NM on either side of the intended track.

For these purposes, the following factors shall be taken into account after an engine failure:

a) the effects of wind and temperature on the net flight path;
b) the effects of fuel jettisoning, where the jettisoning is conducted in accordance with procedures set out in the COM and sufficient fuel remains to complete a landing with the required fuel reserves.

Airline Operator: Enroute Limitations with Two Engines Inoperative

a) No person shall operate an aeroplane having three or more engines unless all points along the intended track are located at a distance that can be flown in 90 minutes or less, with all engines operating at cruise power, from an aerodrome that meets the landing weight/LDA requirements; or
b) the weight of the aeroplane is not greater than the weight that, according to the two-engines inoperative enroute net flight path data shown in the AFM, will allow the aeroplane to clear vertically, by at least 2,000 feet, all terrain and obstructions within 5 NM on either side of the intended track, and thereafter to continue flight to an aerodrome that meets the weight/LDA requirements.

In applying b) above, the following factors shall be taken into account after the failure of two engines:

a) the effects of wind and temperature on the net flight path;

b) the effects of fuel jettisoning, where the jettisoning is conducted in accordance with procedures set out in the COM and sufficient fuel remains to arrive at the destination aerodrome at 1,500 feet AGL with a fuel reserve sufficient to fly for 15 minutes thereafter at cruise power.

Airline Operator: Dispatch Limitations - Landing at Destination and Alternate Aerodromes

No person shall dispatch or conduct a take-off in an aeroplane unless the weight of the aeroplane on landing at the destination and alternate aerodromes will allow a full-stop landing within 60 % of the LDA (turbojet) or 70% of the LDA (propeller).

In determining whether an aeroplane can be dispatched or a take-off can be conducted based on the LDA limits above, the following shall be taken into account:

a) the pressure altitude at the destination aerodrome and at the alternate aerodrome;
b) no more than 50% of the reported headwind component or not less than 150% of the reported tailwind component;
c) the aeroplane must be landed on a suitable runway, considering the wind speed and direction, the ground handling characteristics of the aeroplane, and other conditions such as landing aids and terrain (if destination aerodrome does not meet these conditions but alternate does, the aeroplane may be dispatched and take-off conducted).

Airline Operator: Dispatch Limitations, Wet Runway - Turbo-Jet Aeroplanes

When weather reports or forecasts indicate that the runway may be wet at the estimated time or arrival, no air operator shall dispatch or conduct a take-off in a turbo-jet powered aeroplane unless the LDA at the destination aerodrome is at least 115% of the LDA required (the landing distance required is determined based on dispatch limitations above).

This requirement does not apply if the aircraft flight manual includes specific information about landing distances on wet runways (subject to dispatch limitations above).

Airline Operator: Aircraft Equipment Requirements

General

No person shall operate an aircraft unless the aircraft is equipped with:

a) two independent static pressure systems;
b) a windshield wiper or rain removal system for each pilot station;
c) heating or de-icing equipment for each carburetor or an alternate air source for each pressure carburetor or fuel injection system;
d) a placard on each door that provides passenger access to a passenger emergency exit, stating that the door must be secured or locked open during take-off and landing;
e) a means for the crew, in an emergency, to unlock each door that leads to a compartment that is normally accessible to passengers and that can be locked by passengers.

Landing Lights

No person shall operate an aircraft at night unless the aircraft is equipped with at least two landing lights.

Operation of Aircraft in Icing Conditions

No person shall conduct a take-off or continue a flight in an aircraft when icing conditions are reported to exist or are forecast to be encountered along the route to be flown unless the aircraft is equipped to be operated in those conditions and the aircraft type certificate authorizes flight in those conditions.

No person shall operate an aeroplane in icing conditions at night unless the aeroplane is equipped with a means to illuminate or otherwise detect the formation of ice.

Weather Radar Equipment

No person shall operate an aircraft with passengers on board in IMC when current weather reports or forecasts indicate that thunderstorms may reasonably be expected along the route to be flown, unless the aircraft is equipped with weather radar equipment.

Protective Breathing Equipment

No air operator shall operate a pressurized aircraft unless, at each required station, protective breathing equipment with a 15 minute supply of breathing gas at a pressure altitude of 8,000 feet is provided. The protective breathing equipment shall be conveniently located and readily available at required locations. In addition to a fixed or portable breathing gas supply for use by each flight crew member on the flight deck, a portable breathing gas supply for use by crew members in combating fires is required as follows:

a) one unit in each Class A, B and E cargo compartment that is accessible to crew members in the cabin during flight,
b) one unit for each hand held fire extinguisher in each isolated galley;
c) one unit on the flight deck;
d) one unit located within one metre of each hand held fire extinguisher required in the passenger compartment (unless otherwise authorized by the Minister);
e) the number of units of protective breathing equipment used to satisfy these requirements shall not be less than the number of flight attendants required for the flight.

First Aid Oxygen

No air operator shall operate a pressurized aircraft with passengers on board unless the aircraft is equipped with oxygen dispensing units and an undiluted supply of first aid oxygen sufficient to provide 2% of the occupants, and in any case at least one person, with oxygen for one hour or the entire duration of the flight at a cabin pressure altitude above 8,000 feet, after an emergency descent following cabin depressurization, whichever period is longer.

Interphone System

No person shall operate an aircraft unless the aircraft is equipped with an interphone system that can be operated independently of the public address system except for handsets, headsets, microphones, selector switches and signalling devices.

Public Address System

No person shall operate an aircraft with passengers on board unless the aircraft is equipped with a public address system that can be operated independently of the interphone system except for handsets, headsets, microphones, selector switches and signalling devices.

Crew Member Shoulder Harnesses
No person shall operate an aircraft unless all pilot seats and seats for flight attendants are equipped with a safety belt that includes dual upper torso straps with a single point release.

Lavatory Fire Protection
No person shall operate an aircraft unless:

a) each lavatory in the aircraft is equipped with a smoke detector system or equivalent that provides a warning light in the cockpit or a warning light or audible warning in the passenger compartment that can be readily detected by a flight attendant, taking into consideration the positioning of flight attendants throughout the passenger compartment during the flight;
b) each lavatory in the aircraft is equipped with a built in fire extinguisher for each waste disposal receptacle that is installed in the lavatory, and each extinguisher is designed to discharge automatically into the disposal receptacle on the occurrence of a fire in that receptacle;
c) a readily visible sign that clearly displays a symbol indicating that smoking is prohibited or the words NO SMOKING/DEFENSE DE FUMER is installed above the door handle on both sides of the door to each lavatory in the aircraft;
d) a readily visible sign that clearly displays a symbol indicating that cigarette disposal is prohibited or the words NO CIGARETTE DISPOSAL and DEFENSE DE JETER DES CIGARETTES is installed adjacent to the opening of each waste disposal receptacle that is located in a lavatory in the aircraft;
e) a self contained, removable ashtray is installed on or near the outside of the door to each lavatory in the aircraft or in some other location or locations where it is readily visible to the users of each lavatory from outside the lavatory.

Floor Proximity Emergency Escape Path Markings
No person shall operate, with passengers on board, an aeroplane for which an initial type certificate was issued after January 1, 1958 unless the aeroplane is provided with approved floor proximity emergency escape path markings.

Flashlight Stowage
No person shall operate an aircraft unless it is equipped with flashlight stowage provisions that are accessible from each required flight attendant seat.

Airline Operator: Emergency Equipment

Megaphones
No person shall operate, with passengers on board, an aeroplane for which a type certificate has been issued authorizing transport of 60 or more passengers, unless the following number of portable battery powered megaphones are carried on board the aeroplane and are conveniently located and readily available for use by the flight attendants:

a) for each passenger deck, at least one megaphone;
b) 61 - 99 passenger seats, one megaphone;
c) 100 or more passenger seats, two megaphones.

First Aid Kits
No person shall operate an aircraft unless the following number of approved first aid kits are carried on board the aircraft:

a) 0 - 50 passengers seats: 1 kit;
b) 51 - 150 passengers seats: 2 kits;
c) 151 - 250 passengers seats: 3 kits;
d) 251 or more passengers seats: 4 kits.

First aid kits shall be distributed throughout the aircraft cabin, readily available to crew members and passengers, and clearly identified and marked with the date of the last inspection. Where the aircraft is equipped with only one first aid kit, it is to be located as close as practicable to an emergency exit. A stowage compartment that contains a first aid kit shall be clearly marked as to its contents.

Emergency Medical Kit
No person shall operate an aircraft that has a seating configuration, excluding crew seats, of more than 100 unless an approved emergency medical kit is carried on board the aircraft. The emergency medical kit contains more advanced medical equipment (ie stethoscope) and drugs (epinephrine, nitroglycerin tablets) for use by physician/paramedic for serious medical emergencies such as heart attack.

Crash Axe
No person shall operate an aircraft unless a crash axe is carried on board the aircraft.

Hand Held Fire Extinguishers
No person shall operate an aircraft unless hand held fire extinguishers for use in the flight deck, passenger compartment and cargo compartment are carried on board the aircraft. The type and quantity of extinguishing agent shall be suitable for extinguishing fires that are likely to occur in the flight deck, passenger compartment or cargo compartment where the extinguisher is intended to be used and, in the case of the extinguishing agent for extinguishers intended to be used in the passenger compartment, shall be designed to minimize the hazard of toxic gas concentrations.

At least one hand held fire extinguisher shall be conveniently located and readily available for immediate use in each Class E cargo compartment that is accessible to crew members during flight, and at least one hand held fire extinguisher shall be located in each isolated galley.

At least one hand held fire extinguisher shall be conveniently located on the flight deck and readily available for immediate use by the flight crew members.

The following number of hand held fire extinguishers shall be conveniently located, readily available for immediate use and uniformly distributed throughout the passenger compartment on each deck:

a) 60 or fewer passenger seats: two extinguishers;
b) 61 - 200 passenger seats, three extinguishers;
c) 201 or more passenger seats, one extra extinguisher for each additional unit of 100 passenger seats.

At least two hand held fire extinguishers shall contain Halon 1211 (bromochlorodiflouromethane) or its equivalent.

A stowage compartment or stowage container that contains a hand held fire extinguisher shall be clearly marked as to its contents.

Portable Oxygen
No person shall operate a pressurized aircraft above FL 250 unless there is readily available to each flight attendant on board portable oxygen equipment with a 15 minute supply of oxygen or sufficient portable oxygen units with masks, or spare outlets and masks, to ensure an immediate supply of oxygen to each flight attendant are distributed throughout the cabin.

Airline Operator: Personnel Requirements

Designation of Pilot In Command and Second in Command
An air operator shall designate for each flight a PIC and a SIC.

Flight Attendant Requirements
Unless otherwise authorized in its AOC, no air operator shall operate an aircraft with passengers on board unless the crew includes at least the following number of flight attendants:

a)   1 to 40 passengers on board, one attendant;
b)   41 - 80 passengers on board, two attendants;
c)   81 or more passengers on board, one attendant for each unit of 40 passengers or portion thereof.

Regardless of the above, no air operator shall operate an aircraft with passengers on board with fewer flight attendants than the number required to satisfy the following requirements:

a)   the air operator shall, for each type and model of aircraft operated, assign to each flight attendant the duties to be performed in an emergency, including an emergency evacuation, and shall show that the performance of those duties adequately meets any emergency that may be reasonable anticipated, including the possible incapacitation of another flight attendant;
b)   the air operator shall ensure that the duties assigned to flight attendants are described in its COM.

Pilot Qualifications
No air operator shall permit a person to act and no person shall act as the PIC, SIC or cruise relief pilot of an aircraft unless the person

a)   holds the required licences, ratings and endorsements;
b)   within the previous 90 days has completed at least three take-offs and three landings as the pilot at the controls and one sector assigned to duty as a flight crew member in an aircraft of that type, has completed 5 sectors assigned to duty as a flight crew member in an aircraft of that type, or has fulfilled the required training requirements;
c)   has successfully completed a PPC, the validity period of which has not expired, for that type of aircraft;
d)   has successfully completed or is undergoing a line check or line indoctrination training, the validity period of which has not expired, for that type of aircraft;
e)   has fulfilled the requirements of the air operator's training program.

NOTE: "SECTOR" means a flight composed of a take-off, departure, arrival and landing including at least a 50 NM enroute segment.

An air operator may permit a person to act and a person may act as the PIC or SIC of an aircraft where the person does not meet the above requirements if the aircraft is operated on a training, ferry or positioning flight or the air operator is so authorized in its AOC. Upon completion of a successful PPC, except when an SIC is upgrading to PIC on same type, pilots are also required to meet the following consolidation period requirements:

a)   50 hours in 60 days;
b)   75 hours in 90 days; or
c)   100 hours in 120 days.

Flight Engineer and Second Officer Qualifications
No air operator shall permit a person to act and no person shall act as a flight engineer or a second officer on board an aircraft unless

a)   the person holds the required licence and endorsements;
b)   the air operator has determined, by means of a check in-flight or in an approved flight simulator, that the person meets the required standards for that type of aircraft, or the person has, within the previous 6 months, completed at least 50 hours of flight time as a flight engineer on an aircraft of that type;
c)   the person has successfully completed or is undergoing line indoctrination training for that type;
d)   the person has fulfilled the requirements of the air operator's training program.

A person who is qualified to act as a PIC or SIC may act as a second officer on board an aircraft during the cruise portion of a flight if the person has received initial and annual recurrent training in normal and emergency procedures pertaining to the cruise portion of the flight, and the air operator has determined, by means of a check that the person meets the required standards.,

Crew Pairing
Crew pairing restrictions establish minimum experience requirements for flight crew when any of the following situations apply to either the PIC or SIC:

a)   initial appointment to SIC;
b)   first upgrade from SIC to PIC;
c)   transition from reciprocating powered aeroplane to turbo-prop or turbo-jet;
d)   transition from turbo-prop to turbo-jet;
e)   transition to an aeroplane whose control systems use a technology or present information differing significantly from pilot's experience;
f)   upon completion of training on a second aeroplane type when the pilot will be flying both types in service.

Crew pairing restrictions remain in effect for the consolidation period above.

When after the completion of the line indoctrination, crew pairing restrictions apply to one of the flight crew members, the other flight crew member shall have completed the consolidation period.

When, after completion of their individual line indoctrination, crew pairing restrictions apply to the PIC and to the SIC, a training pilot shall occupy the jump seat.

Route and Aerodrome Qualifications

No air operator shall permit a person to act and no person shall act as the PIC of an aircraft on a flight along a route or into an aerodrome unless within the previous 12 months, the person has acted as a flight crew member or has been on the flight deck as an observer on a flight along that route and into that aerodrome or the person has received the required aerodrome and area qualification training and demonstrated adequate knowledge.

Validity Period

The validity period of a line check or training program expires on the first day of the 13th month following the month in which the check was completed.
The validity period of a PPC expires:

a) on the first day of the 7th month following the month in which the check was completed;
b) on the first day of the 13th month following the month in which the check was completed, where the pilot successfully completes the approved 6 month recurrency training as a substitute for the PPC and that is identified in the COM; or
c) at the end of the validation period, where the air operator has TC approval and the pilot completes a proficiency evaluation within the evaluation period authorized for the air operator.

Where a PPC, line check or training is renewed within the last 90 days of its validity period, its validity period is extended by 6 or 12 months, as appropriate.

The Minister may extend the validity period of a PPC, line check or any training by up to 60 days where the Minister is of the opinion that aviation safety is not likely to be affected.

Where the validity period of a PPC, a line check, or annual or semi-annual training has been expired for 24 months or more, the person shall requalify by meeting all initial training requirements.

Training Program

Every air operator shall establish and maintain a training program that is designed to ensure that each person who receives training acquires the competence to perform the person's assigned duties and is approved by the Minister.

An airline operator's training program shall include, for flight crew members,

a) company indoctrination training;
b) line indoctrination training;
c) upgrading training, where applicable;
d) initial and annual training, including aircraft type training, aircraft servicing and ground handling training, emergency procedures training, and aircraft surface contamination training.

Every air operator shall include a detailed syllabus of its training program in its COM.

Training and Qualification Records

Every air operator shall, for each person who is required to receive training under CARs, establish and maintain a record of

a) the person's name and personnel licence number, type and ratings;
b) the person's medical category and expiry;
c) the dates on which the persons, while in the air operator's employ, successfully completed any training, PPC or examination or qualification required;
d) information relating to any failure of the person, while in the air operator's employ, to successfully complete any training, PPC or examination or qualification required;
e) the type of aircraft or flight training equipment used for any training, PPC, line check or qualification required.

An air operator shall retain training records as above for at least three years. An air operator shall retain a copy of the most recent written examination completed by each person for each type of aircraft for which the person has a qualification.

Airline Operator: Manuals

Company Operations Manual (COM)

An air operator shall provide a copy of the appropriate parts of its COM, including any amendments to those parts, to each of its crew members and to its ground and maintenance personnel.

An air operator may place a copy of the appropriate parts of its COM in each aircraft that it operates, instead of providing a copy to each crew member, if all amendments to the manual are included in the system for the dissemination of general operational information required.

Every person who has been provided with a copy of the appropriate parts of a COM shall keep it up-to-date with the amendments provided and shall ensure that the appropriate parts are accessible when the person is performing assigned duties.

Aircraft Operating Manual (AOM)

An air operator may establish and maintain an AOM for the use and guidance of crew members in the operation of its aircraft.

An AOM shall contain the aircraft operating procedures and where the AFM is not carried on board the aircraft, the aircraft performance data and limitations specified in the AFM, which shall be clearly identified as AFM requirements.

An air operator that has established an AOM shall submit a copy of the manual, and any amendments to it, to the Minister for approval. An air operator that has established an AOM shall ensure that a copy of it is carried on board each aircraft to which it relates.

Standard Operating Procedures (SOPs)

Every air operator shall, for each of its aircraft, establish and maintain SOPs that enable the crew members to operate the aircraft within the limitations specified within the AFM. An air operator shall submit a copy of its aircraft SOPs for TC approval. An air operator shall ensure that a copy of the SOPs for an aircraft is carried on board the aircraft. Where an air operator has established an AOM, the SOPS for the aircraft shall form part of that manual.

[Reference: CARs 705]

1.13 Private Operator Passenger Transportation

Where a private operator uses turbine powered aircraft or large aeroplanes (over 12,566 lbs), for the transport of passengers, a private operator certificate or operator certificate is required. This is for situations such as a company owned aircraft transporting company passengers/executives on company business only (not general public). Under the CARs, for private operators, a similar regimen of rules and procedures that parallels commercial air operator requirements is applied.

[Reference: CARs 604]

1.14 VFR Flight

VFR Minimums

Review the following:

Airspace	VFR Weather Minimums
Control Zones, Controlled airspace	Visibility 3 SM. Distance from cloud: 1 mile horizontally, 500 feet vertically. Distance from ground or water: minimum 500 feet.
Uncontrolled Airspace (below 1000 feet)	Flight visibility 2 SM day, 3 SM night. Distance from Cloud: clear of cloud.
Uncontrolled Airspace (above 1000 feet)	Visibility 1 SM day, 3 SM night. Distance from cloud: 2,000 feet horizontally, 500 feet vertically.

Special VFR Weather Minimums

SVFR is authorized only in control zones. An SVFR clearance permits aircraft to operate in IFR weather conditions without compliance with the instrument flight rules. SVFR flight does not relieve pilots of responsibility for avoiding other aircraft, avoiding weather conditions beyond their own flight capabilities, or avoiding weather conditions beyond the capability of the aircraft.

Requirements for SVFR (aeroplanes)

a) 2-way radio;
b) flight visibility 1 SM;
c) aircraft to remain clear of cloud;
d) aircraft must remain at least 500 feet AGL (unless taking off or landing).

VFR Over-The-Top (VFR OTT) Rules

During the cruise portion of a day VFR flight, an aircraft may be flown VFR OTT provided that:

a) aircraft must be at least 1,000 feet vertically from cloud;
b) if the aircraft is between two cloud layers, the vertical distance between the layers must be at least 5,000 feet;
c) flight visibility is at least 5 NM;
d) the destination weather is forecast to have a sky condition of scattered cloud or clear and ground visibility of 5 SM or more with no forecast of precipitation, fog, thunderstorms or blowing snow, and these conditions are forecast to exist for the period from one hour before to two hours after the ETA based on the TAF (if no TAF, and weather forecast is based on GFA, then for period from one hour before to three hours after ETA).

Additional aircraft equipment is required for VFR OTT operations (see below). A CPL or ATPL licence holder may fly VFR OTT (VFR OTT operations are not permitted for RPP holders, and PPL pilots need a VFR OTT rating).

Day VFR Flying Equipment

DAYLIGHT is the period from 1/2 hour before sunrise to 1/2 hour after sunset, or when the centre of the Sun's disc is less than 6° below the horizon. The minimum equipment for day VFR are as follows:

a) airspeed indicator;
b) altimeter;
c) magnetic compass;
d) a reliable and functioning timepiece;
e) engine tachometer;
f) oil pressure indicator;
g) oil temperature gauge;
h) manifold pressure gauge if equipped with variable pitch propeller system;
i) fuel quantity gauges;
j) two-way radio if operating in Class B, Class C or Class D airspace or within an MF area or ADIZ;
k) appropriate radio navigation equipment if the aircraft is operated within Class B airspace.

For VFR OTT operations, the aircraft must also have a pitot de-ice control, gyroscopic direction indicator, attitude indicator, turn and slip indicator, two-way radio and adequate radio navigation equipment (ie VOR, ADF).

VFR Night Flying Equipment and Aerodrome Lighting

NIGHT is from 1/2 hour after sunset to 1/2 hour before sunrise or when the centre of the Sun's disc is more than 6 degrees below the horizon. In addition a ASI, altimeter, and magnetic compass, for night VFR operations the following are also required:

a) turn and bank indicator, or turn coordinator;
b) gyroscopic heading indicator;
c) a means of illuminating flight instruments;
d) a means of determining the aircraft heading that is independent of magnetic north, when operating in the NDA;
e) navigation lights (projection requirements: 110° for left (red) and right (green) position lights, 140° for rear position light (white), 30° above and below horizontal plane for rotating beacon (flashing red or white/red strobes);
f) landing light;
g) two-way radio when in controlled airspace;
h) flashlight at each crew member's station;
i) spare fuses.

To use an aerodrome at night, it must be lighted, and have an illuminated wind indicator. Many aerodromes have a flashing white beacon to help find it. For an aerodrome used by aeroplanes, the beacon flashes 20 - 30 times per minute at a constant rate. For a helicopter-only aerodrome, the beacon flashes the Morse code for H (4 dots), 3 - 4 times per minute. For aerodromes other than major airports, the minimum aerodrome lighting required is two parallel rows of white lights (no threshold and end lights required). Red lights mark unserviceable areas.

VFR 2-Way Communications Failure

VFR Comm Failure within Class B/C/D Airspace
Where there is a two-way radio communication failure between the controlling ATC unit and a VFR aircraft while operating in Class B, C or D airspace, the PIC shall leave the airspace via the shortest route (land at the airport if already within a control zone). If transponder equipped, squawk 7600 and inform ATC as soon as possible of the actions taken.

The pilot should then land at a suitable aerodrome following the NORDO arrival procedures:

a) approach the traffic circuit of aerodrome from upwind side, cross mid-runway at circuit height and join circuit on downwind leg. Keep vigilant for traffic and if at a controlled airport, look for light signals;
b) before turning final, look for traffic;
c) landing clearance should be given on final approach, if not, pull up and make another circuit;
d) taxi to ramp (look for light signal at controlled airport).

VFR Comm Failure Outside Class B/C/D Airspace
If a VFR 2-way radio failure is experienced, comply with the following:

a) continue to fly in VFR weather conditions and land at the nearest suitable aerodrome;
b) if transponder equipped, select code 7600;
c) complete a NORDO arrival as above.

If no suitable nearby aerodrome is available, the pilot may continue into and enter Class B, C or D airspace, squawk 7600, use the arrival procedures above and advise ATC afterwards.

[Reference: CARs 602.88, 602.114 to 602.117, 602.138, 604.14, 604.15, 301.07, 605.16; AIM AGA 7.1, 7.2, 7.3; AIM RAC 4.4.8, 4.5.8, 6.3.2.2]

1.15 Altimeter Setting Procedures

Altimeter Setting Region

Within the Altimeter Setting Region, the aircraft altimeter is to be set to the current altimeter setting of the nearest station. When approaching an aerodrome for landing, set the aircraft altimeter to that of the aerodrome if it is obtainable.

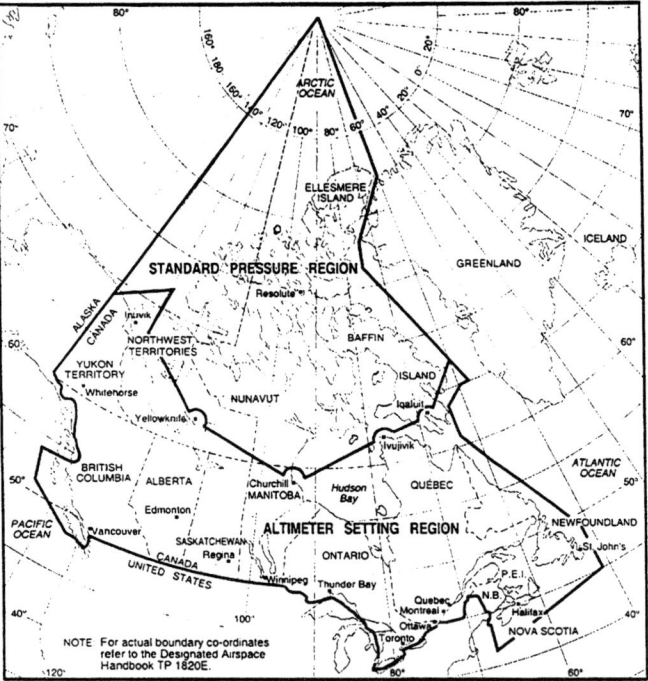

Fig. 1-2: Altimeter Setting Region, Standard Pressure Region.

When flying from the Altimeter Setting Region into the Standard Pressure Region, set the altimeter to 29.92" immediately following entry into the Standard Pressure Region. NOTE: if pressure is 31.00" Hg or more, all enroute aircraft below 18,000 feet should set 31.00 on the altimeter subscale. For IFR aircraft unable to set current altimeter above 31.00, for flight planning alternate/ destination requirements, add 100 feet to ILS DH and 1/4 SM for each .10" Hg over 31.00. For approach, use published DH/MDA with 31.00 on altimeter subscale (aircraft will be higher than indicated!).

Standard Pressure Region

Within the Standard Pressure Region, set the altimeter to 29.92", except in the following cases:

a) Departure: set to airport altimeter setting (airport elevation if unavailable). Immediately prior to reaching cruise flight level, set 29.92".
b) Arrival: immediately prior to commencing a descent for landing at an aerodrome located in the Standard Pressure Region, set the altimeter to the current altimeter setting of the aerodrome of intended landing, if it is available. When conducting a holding procedure prior to landing at an aerodrome within the Standard Pressure Region, do not set the aircraft altimeter to the current altimeter setting of the aerodrome of intended landing until immediately prior to descending below the lowest flight level from which the holding procedure is conducted. CURRENT ALTIMETER SETTING is from previous 90 minutes observation. When flying from the Standard Pressure Region into the Altimeter Setting Region, set the altimeter to the current altimeter setting immediately prior to entry into the altimeter setting region.

[Reference: CARs 602.35 to 602.37; AIM RAC 2.11]

1.16 Altimeter and Static Line Testing

IFR aircraft or VFR in Class B airspace, must have had an altimeter inspection and static pressure system check within the preceding 24 months.

[Reference: CARs 625, Appendix C]

1.17 Transponder Operation

General

Operate the transponder in accordance with ATC, except in a emergency, communications failure or hijack. Use STANDBY for taxi, ON as late as possible before take-off, and OFF as soon as possible after landing. Operate IDENT only as directed by ATC. Codes: 7500 - Hijack, 7600 - Comm Failure, 7700 - Emergency (selection of these codes activates an alarm at ATC, so use caution when selecting so as not to accidentally activate).

When the transponder or the automatic pressure altitude reporting equipment (Mode C) fails during flight where its use mandatory, an aircraft may be operated to the next airport of intended landing and, thereafter, to complete an itinerary or to a repair base, if authorized by ATC.

Transponder Airspace

Aircraft shall be equipped with a functioning transponder incorporating an automatic pressure reporting device when operating in the following airspace:

a) all Class A airspace;
b) all Class B airspace;
c) all Class C airspace;
d) all Class D TCA's at the following aerodromes: Winnipeg International, Ottawa/Macdonald-Cartier International, Quebec/Jean Lesage International, St-Hubert and Halifax International;
e) all Class E airspace of defined dimensions at the following aerodromes: Saskatoon/John G. Diefenbaker, Regina, Thunder Bay, Moncton, Gander, and St. John's;
f) all Class E airspace between 10,000 feet and 12,500 feet ASL within radar coverage - see Fig. 1-3.

Aircraft operating within controlled, high level airspace must be equipped with a serviceable transponder incorporating an automatic pressure altitude reporting device. The transponder should be operated as directed by ATC, if no direction is given, adjust to reply Mode A Code 2000 plus Mode C.

During IFR flight in controlled low level airspace, operate the transponder as directed by ATC, and if no direction is given, squawk Mode A 1000 plus Mode C (if available). If the transponder equipment fails, the aircraft may be operated to the next intended airport of landing, and thereafter, to complete a planned itinerary if so authorized by ATC.

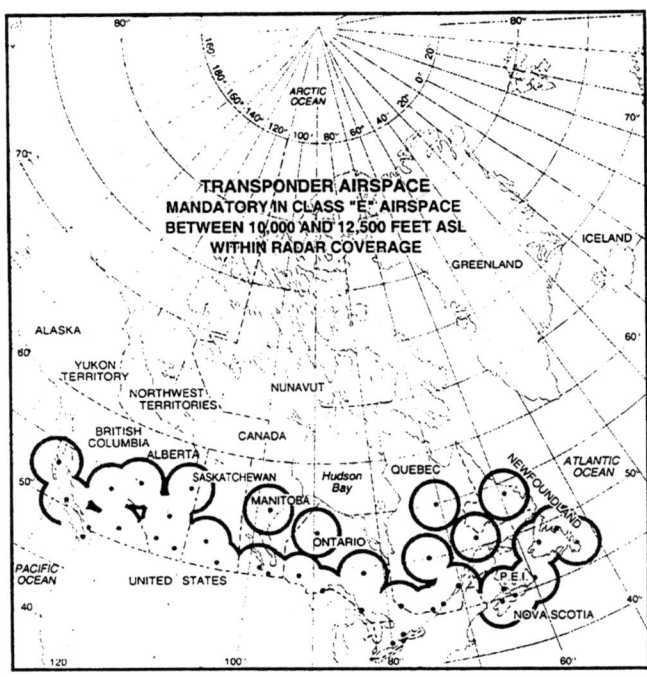

Fig. 1-3: Transponder Airspace.

If an IFR flight plan is cancelled, the transponder should be adjusted to reply on the appropriate VFR code. VFR codes: as directed by ATC, where no directions, reply Mode A 1200 when at or below 12,500 ft, 1400 when above 12,500 ft ASL. When climbing above 12,500 ft ASL, VFR aircraft should use Code 1200 until leaving 12,500 ft ASL, then select 1400. When descending from above 12,500 ft ASL, pilots should select Code 1200 upon reaching 12,500 ft ASL.

[Reference: CARs 605.35; AIM RAC 1.9]

1.18 Flight Preparation, Flight Plans, Flight Itineraries

General

No PIC shall operate an aircraft in VFR flight unless a VFR flight plan or a VFR flight itinerary has been filed, except where the flight is conducted within 25 NM of the departure aerodrome.

No PIC shall operate an aircraft in IFR flight unless an IFR flight plan has been filed. A PIC may file an IFR flight itinerary instead of an IFR flight plan where the flight is conducted in part or in whole outside controlled airspace or facilities are inadequate to permit the communication of flight plan information to an ATC unit, FSS or CARS station.

Filing

A flight plan shall be filed with an ATC unit, a FSS station or a CARS station. A flight itinerary shall be filed with a responsible person, an ATC unit, an FSS station or a CARS station. A flight plan or flight itinerary shall be filed by sending, delivering or otherwise communicating the flight plan or flight itinerary or the information contained therein and receiving acknowledgement that the flight plan or flight itinerary or the information contained therein has been received.

A RESPONSIBLE PERSON means an individual who has agreed with the person who has filed a flight itinerary to ensure that an ATC unit, FSS/CARS station or rescue coordination centre will be notified of the flight itinerary information if the aircraft is overdue.

For departures from locations having no means of communicating the departure time, the flight plan or flight itinerary will be activated at the proposed time of departure.

To assist ATC, pilots are encouraged to file IFR flight plans or flight itineraries as early as practicable, preferable at least 30 minutes prior to the proposed departure time.

IFR flight itineraries are limited to one departure from and one entry into controlled airspace - multiple exits and entries into controlled airspace will not be accepted by ATS.

Changes to IFR Flight Plan or Flight Itinerary

The PIC of an aircraft for which an IFR flight plan or an IFR flight itinerary has been filed shall notify as soon as practicable an ATC unit or the responsible person, as the case may be, and where the flight is being conducted in controlled airspace, receive an ATC clearance before making the intended change where the pilot intends to make a change in any of the following:

a)   cruising altitude or cruising flight level;
b)   route of flight;
c)   destination aerodrome;
d)   in the case of a flight plan, the TAS at the cruising altitude or cruising flight level, where the change intended is 5% or more of the TAS specified in the IFR flight plan or the Mach number, where the change intended is .01 or more of the Mach number that has been included in the ATC clearance.

The PIC of an aircraft for which a VFR flight plan or a VFR flight itinerary has been filed shall notify as soon as practicable an ATC unit, an FSS/CARS station or the responsible person where the pilot intends to make a change in

a)   the flight route;
b)   the flight duration;
c)   the destination aerodrome.

Closing

To close a flight plan, an ARRIVAL REPORT must be submitted to an ATS unit no later than the search and rescue time specified in the flight plan, and if not specified, within one hour of the last reported ETA. If a flight plan is not closed, search and rescue will be initiated. An arrival report must include the registration marks, type of flight plan/itinerary, departure aerodrome, date and time of arrival, and arrival aerodrome.

To close a flight itinerary, an arrival report must be filed as above, either with ATC/FSS/CARS or the responsible person, not later than the SAR time specified or where not specified, within 24 hours after the last reported ETA.

Overdue Aircraft Notification

An aircraft is considered OVERDUE on a flight plan immediately after any search and rescue notification time specified on the flight plan, and if no time has been specified, within one hour after the last reported ETA. An aircraft is considered overdue on a flight itinerary immediately after any search and rescue notification time specified in the flight itinerary, and if no time has been specified, within 24 hours after the last reported ETA. Any person who assumes responsibilities with respect to an aircraft and who has reason to believe that the aircraft is overdue, or any other person who has been directed by that person to do so, shall immediately, by the quickest means available, notify an ATC unit, FSS/CARS or a rescue coordination centre and provide, to the best of the person's knowledge, all the available information concerning the overdue aircraft that may be requested.

Composite Flight Plan or Flight Itinerary - VFR and IFR

A COMPOSITE FLIGHT PLAN/ITINERARY may be filed that describes parts of the route as operating under VFR and parts of the route as operating under IFR. All rules governing VFR or IFR apply to that portion of the route of flight.

Pilots who file IFR for the first part of a flight and VFR for the next part will be cleared by ATC to the point within controlled airspace at which the IFR part of the flight ends. Pilots who file VFR for the first part of a flight and IFR for the next part are expected to contact the appropriate ATC unit for clearance prior to approaching the point where the IFR portion of the flight commences. If direct contact with an ATC unit is not possible, the pilot may request the ATC clearance through an FSS. It is important that the pilot remains in weather conditions that will continue to permit VFR flight until such time as ATC issues and the pilot acknowledges an appropriate ATC clearance for IFR flight within controlled airspace.

Refer to Fig. 1-4 for an example composite flight itinerary.

Transborder Flight Plan Requirements

A VFR or IFR flight plan must be filed prior to conducting any flight between Canada and a foreign state. If the flight is to any country other than the USA, an ICAO flight plan must be filed.

In the case of transborder flights to the USA where the point of departure is in close proximity to the boundary, flight plans should be filed at least one hour in advance in order to facilitate adequate co-ordination and data transfer.

Intermediate Stops

Intermediate stops may not be included in a single IFR flight plan. Except for transborder flights, a single VFR flight plan or an IFR or VFR flight itinerary including one or more intermediate stops en route may be filed provided:

a) for VFR flight plans, the stop will be of short duration (for purposes such as boarding passengers, refuelling, etc.);
b) for IFR flight itineraries, the stop will be in uncontrolled airspace;
c) each intermediate stop is indicated by repeating the name of the stopping point and its duration in the route section of the flight plan/itinerary.

EXAMPLE: for a 30 minute stop at Montreal, Dorval, enter in Route section "CYUL (0+30) CYUL".

When intermediate stops are planned, the Estimated Elapsed Time (EET) must be calculated as the total time to the final destination, including the duration of the intermediate stop (s). It should be noted that SAR action would only be initiated at the specified SAR time or, in the event that a SAR time is not indicated, 60 minutes for a flight plan and 24 hours for a flight itinerary after the ETA at the final destination. Pilots wishing SAR action based on every leg of a flight should file one flight plan or flight itinerary for each stop.

Consecutive IFR Flight Plans

Consecutive IFR flight plans may be filed at the initial point of departure provided that:

a) initial point of departure and en route stops must be in Canada except that one flight plan will be accepted for a departure point within United States controlled airspace;
b) the sequence of stops will fall within one 24 hour period;
c) the flight planning unit must be provided with at least the following items of information for each stage of the flight: point of departure, altitude, route, destination, proposed departure time, EET, alternate, fuel on board, TAS, number of persons on board and where arrival report will be filed.

Cross Country IFR Training Flights

A cross country IFR training flight is one in which there are no intermediate stops and one or more instrument approaches are made en route. For example, an aircraft departs Airport A, completes a practice approach at Airport B and either lands at destination Airport C or returns to land at Airport A. For such flights:

a) a single flight plan is filed;
b) those en route locations at which instrument approaches and overshoots are requested shall be listed in the "Other Information" portion of the flight plan form, together the estimated period of time to carry out each approach;
c) the EET of the flight plan form is NOT to include the estimated time to carry out approaches at the en route locations;
d) ATC will normally clear the aircraft to final destination;
e) if it is not practicable to clear the aircraft to final destination or to assign an operationally suitable altitude with the initial clearance, a time or specific location for the aircraft to expect further clearance to the destination or to a higher altitude will be issued with the initial clearance;

f) when an en route approach clearance is requested, a missed approach clearance will be issued to the aircraft prior to the commencement of the approach;
g) if traffic does not permit an approach, holding instructions will be issued to the aircraft if requested by the pilot.

Flight Plan/Flight Itinerary Format

Review Fig. 1-4 which shows a sample IFR Flight Plan. The same form is used for a Canadian VFR Flight Plan/Itinerary, IFR Flight Plan/Itinerary, VFR/IFR Composite Flight Plan/Itinerary and ICAO Flight Plan. The following is a summary of the details for completing a VFR or IFR flight plan/itinerary.

Aircraft Identification
The aircraft identification (call sign) is inserted in Box 7. EXAMPLE: CF-ABC

Flight Rules and Type of Flight
In Box 8, first enter the flight rules. V = VFR, I = IFR, Y = IFR first then VFR, Z = VFR first then IFR. If Y or Z has been entered, in the route section, specify the point (s) at which a change in flight rules is planned. Where there is more than one change in the type of flight rules, enter Z if the first rules will be VFR and Y if IFR. EXAMPLE: for a composite IFR/VFR/IFR flight, enter Y.

Next, enter the type of flight in the first character: C = Controlled VFR, D = Defence VFR, E = Defence Flight Itinerary, F = Flight Itinerary. Second Character - enter G = General Aviation, S = scheduled air service, N = non-scheduled air transport operation, M = Military, X for other. EXAMPLE: FG = flight itinerary, general aviation.

Number and Type of Aircraft, Wake Turbulence Category
The first part of Box 9 is for the number of aircraft (for VFR flights leave blank). Then enter the type designator. For wake turbulence category, use no entry for non-TCAS aircraft with a maximum take- off weight of less than 300,000 lbs. T = TCAS, MCTOW less than 300,000 lbs. H = non TCAS, MCTOW 300,000 lbs or more. B = TCAS, MCTOW 300,000 lbs or more.

Equipment
In Box 10, enter the communications, navigation and transponder equipment in the order COM - NAV - SSR. For COM/NAV equipment, N = no navigation or communications equipment or the equipment is unserviceable, S = Standard: VHF communications radios, ADF, VOR, and ILS. Other codes: C = LORAN C, D = DME, F = ADF, G = GPS, H = HF RTF, I = INS, J = Data Link, K = MLS, L = ILS, M = OMEGA, O = VOR, R = RNP type certification, T = TACAN, U = UHF, V = VHF, W = RVSM certification, X = MNPS certification, Y = CMNPS certification, Z = other. For SSR equipment, N = Nil, A = Mode A, C = Mode C, X = Mode S without A/C ident and without PA transmission, P = Mode S with PA but not AC ident transmission, I = Mode S with AC ident but not PA transmission, S = Mode S with both AC ident and PA transmission.

EXAMPLE: A Cessna 310 with VHF radios, VOR, ADF, GPS and Mode C transponder would be abbreviated to C310 under "type" and V/FOG/C under "equipment" in Box 10.

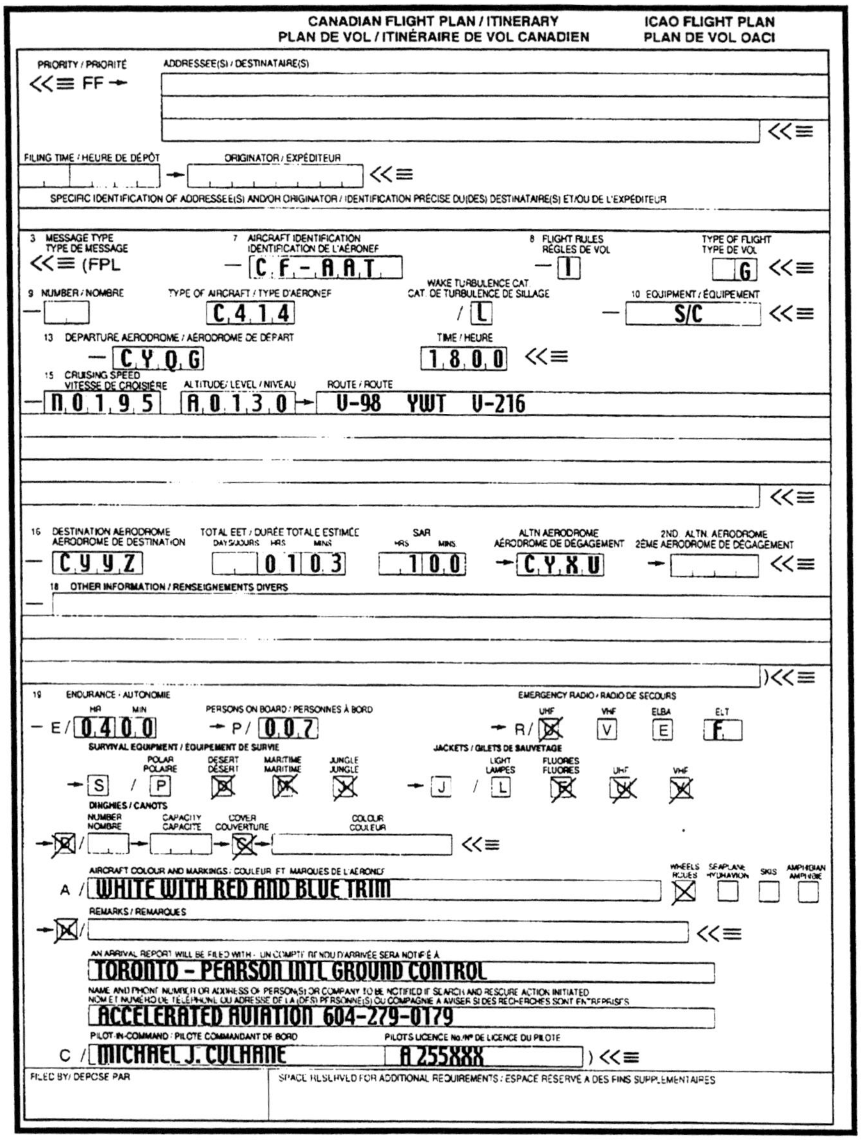

Fig. 1-4 (a): Sample IFR Flight Plan.

Departure Aerodrome
In Box 13, enter the departure aerodrome using the three or four character location from the CFS. EXAMPLE: London, ON = CYXU, Valcourt, QC = SQ3.

Departure Time
Enter anticipated time of departure from departure aerodrome in hours and minutes UTC. Pilots may file a flight plan or flight itinerary up to 24 hours in advance of the departure time.

Cruising Speed, Altitude, Level, Route
In Box 15, "cruising speed", insert flight planned TAS in knots. N = knots, M = Mach number. EXAMPLE: N0090 = 90 Knots TAS, M080 = Mach .8.

For "cruising level", insert the planned cruising level for the first or the whole portion of the route to be flown. A = altitude in hundreds of feet ASL. F = Flight Level. On designated airways and air routes, IFR flights may be operated at the published MEA/MOCA except that in winter, when temperatures are much lower than ISA,

aircraft should be operated at least 1,000 feet above the published MEA/MOCA. EXAMPLE: A060 = 6,000 feet ASL, F330 = Flight Level 330.

For "route", include changes of speed, altitude changes, and airway numbers or points along route. Flight is assumed to be direct unless DCT is used for points between other points named. Pilots are encouraged to file preferred IFR routes where available - refer to the CFS. ATS routing should be used based on IFR charts (LO, HE). Where any changes are to be made in speed or level, or change of ATS route or flight rules are changed, enter the details. EXAMPLE: YMX/ N0200A170 IFR.

Destination Aerodrome, EET, SAR, Alternate
In Box 16, enter the 3 or 4 letter code for the destination aerodrome as found in the CFS. Enter the Estimated Elapsed Time (EET), which may also include the number of days. The pilot can insert own SAR time, to a maximum of 24 hours beyond EET. If no location identifier has been assigned, insert ZZZZ and specify alternate aerodrome in Item 18 followed by ALTN/. No alternate is required on a VFR fight plan/itinerary.

Other Information
Enter "0" in Box 18 if no other information applies. Enter ADCUS and number US/ Canadian citizens for a transborder flight and pilot's name.

Supplemental Information
Enter fuel endurance in hours and minutes in Box 19. Enter total number of persons on board. For "emergency radio", enter "X" over the "U" if not equipped with UHF emergency frequency (243.0 MHz), "X" through the V if not equipped with VHF emergency frequency (121.5 MHz). If ELT is not available, cross out the "E" under ELBA. For ELT, insert ELT type.

Type A or AD ELT = automatic ejectable or automatic deployable. Type F or AF = fixed or automatic fixed ELT. Type AP = automatic portable. Type W or S = water activated or survival ELT. Type P = Personnel (non fixed).

Under Survival Equipment, Jackets and Dinghies, cross out those types of survival equipment not carried. Under arrival report, include with who or what facility the flight plan will be closed. Enter pilot name, licence number and person to be notified if SAR initiated.

Computer Flight Plan
In addiiton to the TC flight plan format, operators may utilize computer flight plans for flight operations purposes. A sample is provided at Fig. 1-4 (b). Definitions for the various codes used is provided at Fig. 1-4 (c).

```
PLAN 1510                    CYAM TO CYOW CES2 HSC/F   IFR  08/24/00
NONSTOP COMPUTED  1209Z      FOR ETD 1700Z  PROGS  2400ADF  CFZZZ   LBBS

              FUEL TIME  DIST  ARRIVE     TAKEOFF     LAND      AV PLD    OPNLWT
POA CYOW     001475      01/09 0386 1809Z 013703      012228    000457    008446
ALT CYND     000369      00/13 0013 1822Z
HLD          000000      00/00
RES          002956      03/16
TOT          004800      04/38

CYAM . . SSM . . YYB J513 SMARE YOW314 YOW . . CYOW

WIND P035  MXSH  1/SMARE
FL 330

WPT          MTR  TTR   T   TAS  G/S  DR  ZD  DREM  ZT   CTR   ZF  FREM  AFR  ETTA

SSM          125.5 118  ..  ..   ..   ..  009 0377  ./..  ./..  ..  ..   .

TOC          093.1 089  ..  ..   ..   ..  069 0308  0/20  0/49  004 0043

YYB          093.1 089  -48 372  403  R05 134 0174  0/20  0/29  004 0039

SMARE        102.9 092  -48 373  410  R05 053 0121  0/07  0/22  001 0038

TOD          131.3 118  -48 374  423  R01 035 0086  0/05  0/17  001 0037

YOW          131.3 118  ..  ..   ..   ..  074 0012  ./..  ./..  ..  ..   .

CYOW         140.5 126  ..  ..   ..   ..  012 0000  0/17  0/00  000 0033

CYAM    N46291W084306   SSM  N46247W084189   YYB    N46218W0792622
SMARE   N46196W078098   YOW  N45265W075538   CYOW   N45194W0754022

FIRS       KZMP/0000  CZYZ/0004  CZUL/0103

(FPL-I
-C550/L
-CYAM1700
-N0372F330 DCT SSM DCT YYB J513 SMARE YOW314 YOW DCT
-CYOW0109  CYND
-EET/KZMP0000 CZYZ0004 CZUL0103
 SEL/
-E/0438 P/ R/ S/ J/ D/ C
A/ )

IN . . . . . .DOWN . . . . .ZFW . . . . . .
OUT. . . . .UP. . . . . . .R/FUEL. . . . .
FLT. . . . .AIR . . . . . . .T/O WT . . . . .
```

Fig. 1-4 (b): Sample Computer Flight Plan.

```
CODES: Sample Computer Flight Plan

POA        -    Point of Arrival
ALT        -    Alternate
HLD        -    Holding
RES        -    Reserve
TOT        -    Total
AV PLD     -    Average Payload
OPNLWT     -    Operational Weight

NOTE: Weight and balance calculation computed sepa-
rately take precedence over these weight calculations.

CYAM . . . SSM   -   CYAM Direct to SSM
YOW 314 YOW      -   314° Radial to YOW
WIND P035        -   Wind Push of 35 kts
FL330            -   Flight Level 330
WPT              -   Waypoint
MTR              -   Magnetic Track
T                -   Temperature
TAS              -   True Airspeed
G/S              -   Ground Speed
DR               -   Drift
ZD               -   Zone (leg) Distance
DREM             -   Distance Remaining
ZT               -   Zone (leg) Time
CTR              -   Time Remaining
ZF               -   Zone (leg) Fuel
FREM             -   Fuel Remaining
AFR              -   Actual Fuel Remaining
ETA              -   Estimated Time of Arrival
CYAM             -   CYAM Latitude and longitude
FIRS             -   FIR Boundary Times
FPL-I            -   Instrument Flight Plan
TOC              -   Top of Climb
TOD              -   Top of Descent
```

Fig. 1-4 (c): Codes used in sample computer flight plan.

Carry-on Baggage, Equipment and Cargo

No person shall operate an aircraft with carry-on baggage, equipment or cargo on board unless the carry-on baggage, equipment and cargo are

a) stowed in a bin, compartment, rack or other location that is certified in accordance with the aircraft type certificate in respect of the stowage of carry-on baggage, equipment or cargo; or

b) restrained so as to prevent them from shifting during movement of the aircraft on the surface and during take-off, landing and in-flight turbulence.

No person shall operate an aircraft with carry-on baggage, equipment or cargo on board unless

a) the safety equipment, the normal and emergency exits that are accessible to passengers and the aisles between the flight deck and a passenger compartment are not wholly or partially blocked by carry-on baggage, equipment or cargo;

b) all of the equipment and cargo that are stowed in a passenger compartment are packaged or covered to avoid possible injury to persons on board;

c) where the aircraft is type-certificated to carry 10 or

more passengers and passengers are carried on board, no passenger's view of any "seat belt" sign, "no smoking" sign or exit sign is obscured by carry-on baggage, equipment or cargo except if an auxiliary sign is visible to the passenger or another means of notification of the passenger is available, all passenger service cars and trolleys are securely restrained during movement of the aircraft on the surface, take-off and landing, and during in-flight turbulence where the PIC or in charge flight attendant has directed that the cabin be secured, and all of the video monitors that are suspended from the ceiling of the aircraft and extend into an aisle are stowed and securely restrained during take-off and landing;

d) all of the cargo that is stowed in a compartment to which crew members have access is stowed in such a manner as to allow a crew member to effectively reach all parts of the compartment with a hand-held fire extinguisher.

Crew Member Instructions

The PIC of an aircraft shall ensure that each crew member, before acting as a crew member on board the aircraft has been instructed with respect to the duties that the crew member is to perform and the location and use of all of the normal and emergency exits and of all of the emergency equipment that is carried on board the aircraft.

Fuel Requirements

No PIC of an aeroplane shall commence a flight or, during flight, change the destination aerodrome set out in the flight plan or flight itinerary, unless the aeroplane carries the required fuel.

An aeroplane operated in VFR flight shall carry an amount fuel that is sufficient to allow the aeroplane, when operated during the day, to fly to the destination aerodrome and then to fly for a period of 30 minutes at normal cruising speed, or when operated at night, to fly to the destination aerodrome and then to fly for a period of 45 minutes at normal cruising speed.

An aeroplane operated in IFR flight shall carry an amount of fuel that is sufficient to allow the aeroplane,

a) where an alternate aerodrome is specified in the flight plan or flight itinerary, to fly to and execute an approach and a missed approach at the destination aerodrome, to fly to and land at the alternate aerodrome and then to fly for a period of 45 minutes;

b) where an alternate aerodrome is not specified in the flight plan or flight itinerary, to fly to and execute an approach and a missed approach at the destination aerodrome and then to fly for a period of 45 minutes;

c) in the case of a turbo-jet-powered aeroplane, where an alternate aerodrome is specified in the flight plan or flight itinerary, to fly to and execute an approach and a missed approach at the destination aerodrome, to fly to and land at the alternate aerodrome and then to fly for a period of 30 minutes;

d) in the case of a turbo-jet-powered aeroplane where no alternate aerodrome is specified in the flight plan or flight itinerary, to fly to and execute an approach and a missed approach at the destination aerodrome and then to fly for a period of 30 minutes.

Every aircraft shall carry an amount of fuel that is sufficient to provide for

a) taxiing and foreseeable delays prior to take-off;
b) meteorological conditions;
c) foreseeable air traffic routings and traffic delays;
d) landing at a suitable aerodrome in the event of loss of cabin pressurization or, in the case of a multi-engined aircraft, failure of any engine, at the most critical point during the flight; and
e) any other foreseeable conditions that could delay the landing of the aircraft.

Passenger Briefings
The PIC of an aircraft shall ensure that all of the passengers on board the aircraft are briefed before take-off with respect to the following, where applicable:

a) the location and means of operation of emergency and normal exits;
b) the location and means of operation of safety belts, shoulder harnesses and restraint devices;
c) the positioning of seats and the securing of seat backs and chair tables;
d) the stowage of carry-baggage;
e) where the aircraft is unpressurized and it is possible that the flight will required the use of oxygen by the passengers, the location and means of operation of oxygen equipment;
f) any prohibition against smoking.

The PIC of an aircraft shall ensure that all of the passengers on board the aircraft are briefed

a) in the case of an over-water flight where the carriage of life preservers, individual flotation devices or personal flotation devices is required, before commencement of the over-water portion of flight, with respect to the location and use of those items;
b) in the case of a pressurized aircraft that is to be operated at an altitude above FL 250, before the aircraft reaches FL 250, with respect to the location and means of operation of oxygen equipment.

The PIC of an aircraft shall, before take-off, ensure that all of the passengers on board the aircraft are provided with information respecting the location and use of

a) first aid kits and survival equipment;
b) where the aircraft is a small aeroplane (under 12,500 lbs), any ELT that is required to be carried;
c) any life raft that is required to be carried.

[Reference: CARs 602.70 to 602.79, 602.86, 602.87, 602.88, 602.89, 602.145; AIM RAC 3.0]

1.19 IFR Flight

General Requirements - IFR
No PIC shall operate an aircraft in IMC in any class of airspace, except in accordance with IFR. No PIC of an aircraft shall conduct an IFR flight within controlled airspace unless the aircraft is operated in accordance with an ATC clearance.

Alternate Aerodrome Requirements
Except as otherwise authorized by the Minister in an air operator certificate or in a private operator certificate, no PIC shall operate an aircraft in IFR flight unless the IFR flight plan or IFR flight itinerary that has been filed includes an alternate aerodrome having a landing area suitable for use by that aircraft.

Alternate Aerodrome Weather Minima
No PIC of an aircraft shall include an alternate aerodrome in an IFR flight plan or IFR flight itinerary unless available weather information indicates that the ceiling and visibility at the alternate aerodrome will, at the expected time of arrival, be at or above the alternate aerodrome weather minima specified in the CAP.

Minimum Altitudes to Ensure Obstacle Clearance
The PIC of an IFR aircraft shall, except when taking off or landing, or when being radar-vectored by an ATC unit, ensure that the aircraft is operated at or above the MOCA, when the aircraft is on an airway or air route, and at or above the minimum altitude established by the Minister to ensure obstacle clearance and specified on an IFR chart, when the aircraft is within airspace in respect of which such a minimum altitude has been established. When an IFR aircraft is not being operated on an airway or air route or within airspace in respect of which a minimum altitude has been established, the PIC shall ensure that the aircraft is operated at or above 1,000 feet above the highest obstacle located within a horizontal distance of 5 NM from the estimated position of the aircraft in flight.

Enroute IFR Position Reports
The PIC of an IFR aircraft shall transmit position reports over compulsory reporting points, unless advised by the appropriate ATC unit that the aircraft is radar-identified.

Take-Off Minima
No PIC of an aircraft shall conduct a take-off if the take-off visibility is below the minimum take-off visibility specified in the air operator certificate, the private operator certificate or the CAP. The take-off visibility is defined for this purpose as

a) the RVR of the runway, if the RVR is reported to be at or above the minimum take-off visibility specified in the operator certificate/CAP;
b) the ground visibility of the aerodrome for the runway, if the RVR is reported to be less than the minimum take-off visibility specified in the operator certificate/CAP, or is reported to be fluctuating above and below the minimum take-off visibility specified in the operator certificate/CAP, or is not reported;
c) the visibility for the runway as observed by the PIC if the RVR is not reported and the ground visibility of the aerodrome is not reported.

Instrument Approaches
Unless otherwise authorized by the appropriate ATC unit, the PIC of an IFR aircraft shall, when conducting an approach to an aerodrome or a runway, ensure that the approach is made in accordance with the instrument approach procedure. No PIC of an IFR aircraft shall commence an instrument approach procedure unless the aircraft altimeter is set to an altimeter setting that is usable at the aerodrome where the approach is to be conducted. Approach minima for aircraft categories are based on the airspeed at which the aircraft is to be maneuvered i.e. Category A = to 90 kts, B = 91 - 120 kts, C = 121 - 140 kts, D = 141 - 165 kts, E = over 165 kts.

Landing Minima
No PIC of an IFR aircraft shall conduct an instrument approach procedure except in accordance with the

minima specified in the CAP or the route inventory. No PIC of an IFR aircraft shall, unless the required visual reference necessary to continue the approach to land has been established, in the case of a CAT I or II precision approach, continue the final approach descent below the decision height or in the case of a non-precision approach, descend below the minimum descent altitude. Where the PIC of an IFR aircraft conducting an instrument approach does not establish the required visual reference, the PIC shall initiate a missed approach procedure at decision height (CAT I or II precision approach), or at the missed approach point (non precision approach). No PIC of an IFR aircraft shall conduct a precision approach to CAT II or CAT III minima unless the flight crew has received additionally specialized training specified in the *Manual of All Weather Operations (Categories I and II)* and the aircraft is operated in accordance with the procedures, the equipment requirements and the limitations specified in the manual.

Runway Visual Range (RVR)

Runway Visual Range is defined as, in respect of a runway, "the maximum horizontal distance, as measured by an automated visual landing distance system and reported by ATS for the direction of takeoff or landing, at which the runway, or lights delineating it, can be seen from a point above its centerline at a height corresponding to the average eye level of pilots at touchdown".

The concept of RVR is to produce a standard measurement for visibility in the runway environment.

Transmissometers are electronic devices, mounted at the edge of runways, that measure RVR. RVR A refers to RVR detection equipment that is located adjacent to the runway threshold, RVR B is located adjacent to the runway midpoint. RVR values can be compared to ground visibility report using the following scale:

Ground Visibility	RVR
1 mile	5,000 feet
3/4 mile	4,000 feet
1/2 mile	2,600 feet
1/4 mile	1,400 feet

RVR is not to be used for flight planning purposes. It is passed on to the pilot as a matter of routine and may only be used in the determination of visibility minima where the active runway is served by a transmissometer.

Approach Ban

Approaches are GOVERNED by RVR values only. Pilots non CARs Part VII (air operator) aircraft are PROHIBITED from completing an instrument approach past the outer marker or final approach fix to a runway served by an RVR if the RVR values are BELOW the following minima:

Measured RVR	Minimums
RVR A only	1200
RVR A and B	1200/600
RVR B only	1200

Exceptions to minimum RVR (when approach ban not in effect) for pilots conducting CAT III precision approach:

a) when the below-minimum RVR report is received, the aircraft is inbound on approach and has

passed the outer marker or the fix that serves as the outer marker of the instrument landing system in use for the runway;

b) the PIC has informed the appropriate ATC unit that the aircraft is on a training flight and that the PIC intends to initiate a missed approach procedure at or above the DH or MDA as appropriate;

c) the RVR is fluctuating above and below the minimum RVR or the ground visibility of the aerodrome as reported by an ATC unit or FSS is at least 1/4 mile.

With respect to approach restrictions, in the case of a local phenomenon or any fluctuations that affect RVR validity, where the ground visibility is reported by ATC or FSS to be at or above 1/4 mile, an approach may be completed.

EXAMPLE: An ILS CAT III approach is to be conducted to a runway by a non CARs Part VII aircraft. Transmissometers are located at position A and B. ATC reports "RVR A 1000 variable 800 - 1400, RVR B 800". Is an approach authorized?

ANSWER: Yes. An approach is authorized because RVR fluctuations are above RVR 1200. If RVR B had been below 600, an approach would not be authorized.

No PIC of an IFR aircraft shall commence a non-precision approach (e.g. GPS, LOC, NDB), an APV or a CAT I or CAT II precision approach where low visibility procedures are in effect.

Runway Visibility

When no reading from RVR A or RVR B for the runway of intended approach is available, the pilot or a qualified person may make an assessment of the runway visibility in accordance with CARs standards. This assessment is only valid for 20 minutes after it is made. Runway visibility is assessed while stationary at the runway threshold take-off point, at the taxiway holding position for the taxiway adjoining the runway threshold, or at a point adjacent to the runway threshold.

When assessing runway visibility, a pilot

a) assesses, in the runway direction, the furthest visible runway edge lights or landmarks within 10 degrees of the runway centreline that can be seen and recognized;

b) from the above assessment, determines the distance (in feet) based on a 200 foot runway edge light spacing or using the applicable Aerodrome Chart in the CAP;

c) immediately reports the distance assessed to ATS, if available, or to the inquiring party, as the runway visibility along the specified runway in the following format:

"RUNWAY VISIBILITY, RUNWAY [Runway number] ASSESSED AS [distance assessed] FEET AT [time] UTC"

If the runway visibility varies during the assessment, the pilot reports the lowest value observed. The lowest reported value is 200 feet, with lower values reported as "...LESS THAN 200 FEET ...". The highest reported value is 6,000 feet, with higher values reported as "...GREATER THAN 6,000 FEET...".

[Ref: CARs 602.129 - 130, 622.131; AIM RAC 9.19.2]

Continuous Listening Watch

Except in the case of a communications failure, where an aircraft is equipped with radiocommunication equipment, the PIC shall ensure that a listening watch is maintained on the appropriate frequency and where communications are required, communication is established with an ATC unit, FSS or CARS station, as applicable, on that appropriate frequency.

IFR Equipment Requirements

No person shall conduct a take-off in a power-driven aeroplane for the purpose of IFR flight unless it is equipped with the following:

a) airspeed indicator;
b) a magnetic compass or a magnetic direction indicator that operates independently of the aircraft electrical generating system;
c) a tachometer for each engine and for each propeller;
d) an oil pressure indicator for each engine employing an oil pressure system;
e) a coolant temperature indicator for each liquid-cooled engine;
f) an oil temperature indicator for each air-cooled engine having a separate oil system;
g) a manifold pressure gauge for each reciprocating engine equipped with a variable-pitch propeller, supercharged engine or turbocharged engine;
h) a means for the flight crew, when seated at the flight controls to determine the fuel quantity in each main fuel tank and if the aircraft employs retractable landing gear, the position of the landing gear;
i) a sensitive altimeter adjustable for barometric pressure;
j) a turn and slip indicator or turn coordinator, except where the aircraft is equipped with a standby attitude indicator that is usable through flight attitudes of 360° of pitch and roll, the aircraft may be equipped with a slip-skid indicator in lieu of a turn and slip indicator or turn coordinator;
k) an adequate source of electrical energy for all of the electrical and radio equipment;
l) in respect of every set of fuses of a particular rating that is installed on the aircraft and accessible to the PIC during flight, a number of spare fuses that is equal to at least 50 percent of the total number of installed fuses of that rating;
m) where the aircraft is operated so that an aerodrome is not visible from the aircraft, a stabilized magnetic direction indicator or a gyroscopic direction indicator;
n) where the aircraft is to be operated within the Northern Domestic Airspace, a means of establishing direction that is not dependent on a magnetic source;
o) an attitude indicator;
p) a vertical speed indicator;
q) an outside air temperature gauge;
r) a means of preventing malfunction caused by icing for each airspeed indicating system (ie pitot heat);
s) a power failure warning device or vacuum indicator that shows the power available to gyroscopic instruments from each power source;
t) an alternative source of static pressure for the altimeter, ASI and VSI;
u) sufficient radiocommunication equipment to permit the pilot to conduct two-way communications on the appropriate frequency;
v) sufficient radio navigation equipment to permit the pilot, in the event of the failure at any stage of the flight of any item of that equipment, including any associated flight instrument display, to proceed to the destination aerodrome or proceed to another aerodrome that is suitable for landing and where the aircraft is operated in IMC, to complete an instrument approach and, if necessary, conduct a missed approach procedure.

For night IFR operations, the following is also required:

a) a means of illumination for all of the instruments to operate the aircraft;
b) when carrying passengers, a landing light;
c) position and anti-collision lights the meet the required standards.

No person shall conduct a take-off or continue a flight in an aircraft where icing conditions are reported to exist or are forecast to be encountered along the route of flight unless the PIC determines that the aircraft is adequately equipped to operate in icing conditions in accordance with the standards of airworthiness under which the type certificate for that aircraft was issued, or current weather reports indicate that icing conditions no longer exist.

Holding Procedures

A holding pattern is used by ATC to keep an aircraft over a specified location, until a further clearance is issued. A holding clearance will include at least:

a) clearance to holding fix;
b) direction to hold from the holding fix;
c) specified radial, course, or inbound track;
d) DME distance specifications (if applicable);
e) Altitude or FL to maintain;
f) Time to expect further clearance (EFC) or approach clearance (EAC).

If a pilot reaches a clearance limit without receiving further clearance, the pilot is to hold on the inbound track to the clearance limit as published, and if not published, all turns right. If communications cannot be established with ATC, the pilot should proceed with the appropriate communications failure procedure.

Holding patterns must not exceed the following speeds:

a) propeller driven aircraft (including turboprop) to 30,000 feet: 175 kts IAS;
b) civil turbo-jet to 14,000 feet 230 knots, above 14,000 feet 265 kts IAS;
c) climbing while in a holding pattern: turboprop - normal climb speed, turbo-jet - 310 kts IAS or less.

For holds below 14,000 feet ASL, the inbound leg should be 1 minute. For holds above 14,000 feet ASL, the inbound leg should be 1.5 minutes. Adjust the timing of the outbound leg to achieve the proper inbound timing.

IFR Position Reports

IFR aircraft must make position reports over compulsory reporting points and as directed by ATC. A position report consists of:

a) identification;
b) position;
c) time over reporting point;
d) altitude;
e) type of flight plan/itinerary (VFR or IFR);
f) name of next reporting point and ETA;

g) name only of the next succeeding reporting point ("on request" reporting points are only named if specifically requested by ATC);

h) additional information.

Position reports should be made to IFR control units or other ATC/FSS unit. Position reports are unnecessary if ATS has told the pilot that the aircraft has been RADAR IDENTIFIED (unless the aircraft has been assigned an altitude not appropriate for the direction of flight). If the time estimated for the next applicable reporting point is in error by more than 3 minutes from that notified, the pilot must revise the next reporting point ETA.

IFR Communications Failure

If a 2-way communications failure occurs under IMC, complete the following:

a) Route: Use the route assigned by ATC in the last clearance. If being radar vectored, go direct to the fix, route or airway specified in the vector clearance. In the absence of an assigned route, use the route ATC has advised could be expected in a further clearance. Otherwise, use the flight planned route.

b) Altitude: Use the highest of the following altitudes for the route segment being flown - altitude last assigned in ATC clearance received and acknowledged, minimum IFR altitude, the altitude ATC has advised may be expected in a further clearance (climb to this altitude at the time ATC has specified to expect further clearance or new altitude).
Note: the intent is that the aircraft be operated at an altitude that provides the required obstruction clearance. If the comm failure occurs while being radar vectored, the pilot shall immediately climb to and maintain the appropriate minimum IFR altitude until arrival at the fix, route or airway specified in the clearance.

c) Descent for Approach: Upon reaching approach facility commence an appropriate descent procedure at whichever is later - the ETA as originally calculated or amended with ATC, the ETA last notified to and acknowledged by ATC, or the expected approach time last received and acknowledged.

If the comm failure occurs after the pilot has received and acknowledged holding instructions, the pilot should leave the holding point at the time specified in the clearance, EFC time or EAC time, whichever has been issued. If the holding fix is not a fix from which an approach begins, leave the fix at the EAC time if one has been received; if none has been received, upon arrival at the clearance limit, proceed to a fix from which an approach begins. Commence descent and/or approach as close as possible to the ETA as calculated from the filed ETE or as amended with ATC.

If the comm failure occurs on a transborder flight to the United States, communications failure procedures are essentially the same but it is the pilot's responsibility to comply with the US requirements. If a comm failure is experienced during a SID, comply with the comm failure procedure as outlined in the particular SID.

[Reference: CARs 602.121-130, 602.136, 602.137, 605.18, 605.30; AIM RAC 6.3.2, 8.1, 10.0]

1.20 Canadian Airspace Structure

Canadian Domestic Airspace (CDA)

CDA includes all airspace over the Canadian land mass, the Canadian Arctic, Canadian Archipelago and those areas of the high seas within the airspace boundaries.

CDA is geographically divided into the Southern Domestic Airspace and the Northern Domestic Airspace. In the SDA, magnetic track is used to determine cruising altitude for direction of flight, while true track is used in the NDA due to erratic indications from the magnetic compass.

The CDA is further divided vertically into the Low Level Airspace, which consists of all of the airspace below 18,000 feet ASL, and the High Level Airspace which consists of all airspace from 18,000 feet and above.

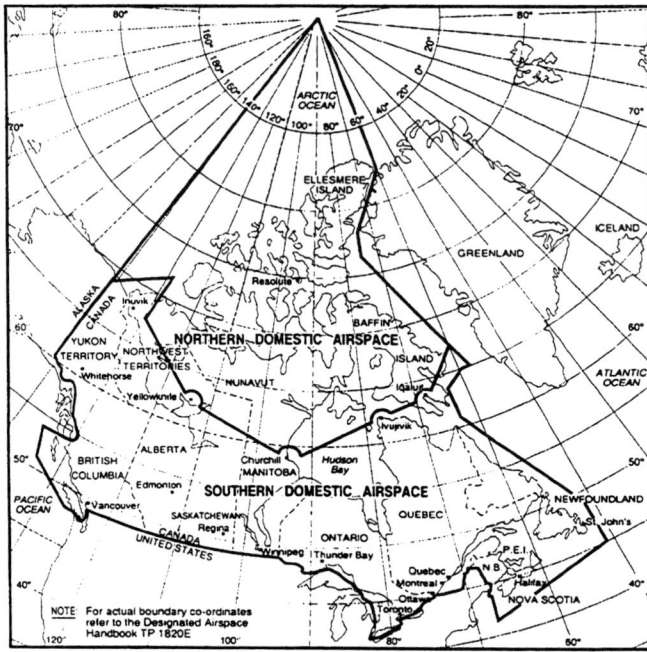

Fig. 1-5: Canadian Domestic Airspace.

Low Level Controlled Airspace

A CONTROL ZONE is airspace designated around certain airports to control IFR and VFR aircraft. The normal radius is 7 NM, although some control zones may have a 5 NM or 3 NM radius. The normal height is 3,000 feet AAE (Above Aerodrome Elevation).

A CONTROL AREA EXTENSION is designated around busy airports to provide additional controlled airspace for IFR aircraft. They are normally based at 2,200 feet AGL and extend up to 18,000 feet ASL. The outer portions may be based at higher levels.

A TERMINAL CONTROL AREA (TCA) is provided to service IFR aircraft. Aircraft operating within a TCA are subject to certain rules. A TCA is normally circular centered on the primary aerodrome with a 45 NM radius from 9,500 feet to 35 NM at 2,200 feet to 12 NM at 1,200 feet. A military TCA (MTCA) is the same as a TCA, except that it will be found at a military aerodrome.

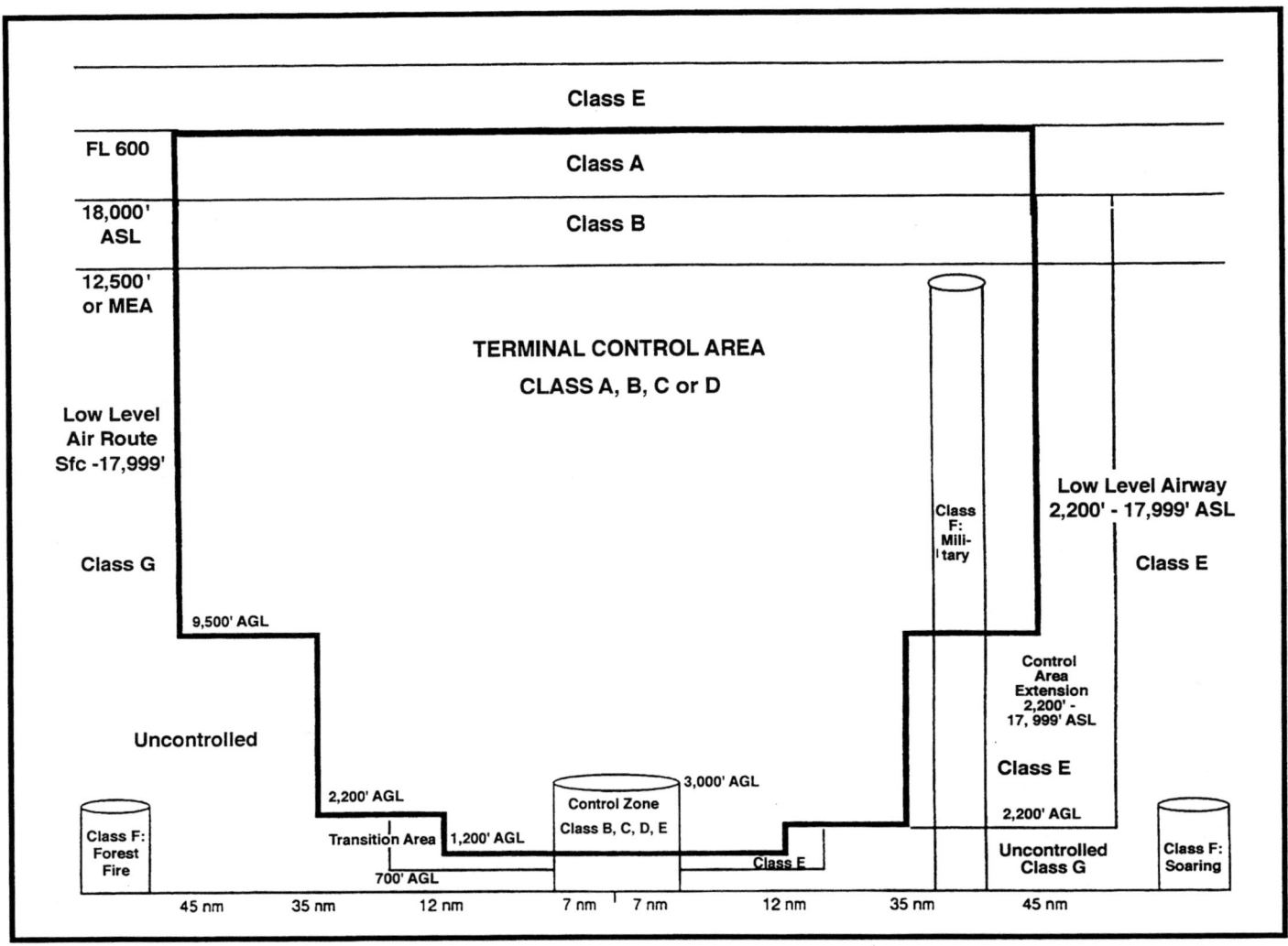

Fig. 1-6: Canadian Airspace Structure and Classifications.

A TRANSITION AREA provides additional airspace for IFR operations, and is normally based at 700 feet AGL, extending upwards to the base of overlying controlled airspace. The area around an aerodrome will normally be 15 NM radius, but shall be of sufficient size to contain all of the aerodrome published instrument approach procedures.

Airspace Classifications
Canadian Domestic Airspace is divided into the classifications Class A through G. The rules for operating in a particular type of airspace depends upon the classification of that airspace.

Class A
Consists of controlled, High Level Airspace from 18,000 feet to FL 600. Only IFR flight is permitted.

Class B
Consists of controlled, low level airspace above 12,500 feet, or the MEA (Minimum Enroute Altitude), whichever is higher, to, but not including 18,000 feet. May also include a control zone and associated terminal control area. Flights may be IFR or VFR. VFR aircraft must have a clearance from ATC prior to entering Class B airspace. Position reports are required. Aircraft must have a 2-way radio and appropriate navigation equipment to maintain airway/air route tracking. Where

weather falls below VFR, VFR aircraft must leave Class B airspace by the shortest route, by exiting horizontally or vertically, and inform ATC as soon as possible. If in a control zone, a request for Special VFR may be made. [NOTE: based on the latest LO charts, there is no designated Class B airspace currently within the NDA]

Class C
Controlled airspace. IFR and VFR permitted. VFR aircraft require a clearance to enter. 2-way radio required. Class C control zones and terminal Class C airspace. Traffic information and conflict resolution may be provided. Class C becomes Class E if appropriate ATC unit not in operation.

Class D
Controlled airspace. IFR and VFR permitted. VFR aircraft must establish contact with ATC to enter. 2-way radio required. Traffic information and conflict resolution may be provided. Class D control zones and terminal Class D airspace. Class D becomes Class E when appropriate ATC unit not in operation.

Class E
Class E airspace is designated where there is a need for controlled airspace, but it does not meet the criteria for Class A, B, C or D. VFR/IFR aircraft may enter. Includes low level airways, control area extensions,

transition areas or control zones established without an operating control tower.

Class F

Airspace where limitations must be put on aircraft operations. May be Class F ADVISORY (VFR aircraft should avoid unless participating in the activity) or Class F RESTRICTED (advance permission required to enter). Class F airspace is published on VFR and IFR charts. It includes alert areas, danger areas, rocket ranges, restricted areas, forest fire restrictions, military activity areas. May be controlled or uncontrolled, or a combination of both. Check NOTAMS for details on whether a particular Class F airspace is in operation. Identification code groups: CYA = Advisory (A = Acrobatics, T = training, F = aircraft test area, P = Parachute, S = Soaring, M = military operations, H = hang gliding). CYR = Restricted. CYD = Danger. B.C. = 100, Alberta = 200, Sask = 300, Man. = 400, Ont. = 500, Que. = 600, Atlantic = 700, Yukon = 800, NWT = 900. EXAMPLE: CYA 113 (A) means Class F Advisory airspace No. 113 in B.C. for acrobatics.

Class G

Airspace not designated A, B, C, D, E, or F. ATC has no authority. Flight information and alerting of SAR services provided by ATS. All uncontrolled domestic airspace. Includes all LOW LEVEL AIR ROUTES.

Airway and Air Route Structure

Low Level Airways are controlled airspace from 2,200 feet AGL to, but not including 18,000 feet ASL. VHF airways are 4 NM wide, either side of centerline, increasing where 4.5° lines intersect 4 NM width. LF/MF airways are 4.34 NM wide, either side of centerline, increasing where 5° lines intersect 4.34 NM width. Air routes have the same dimensions as airways, the differences being that air routes are in uncontrolled airspace and commence at the surface.

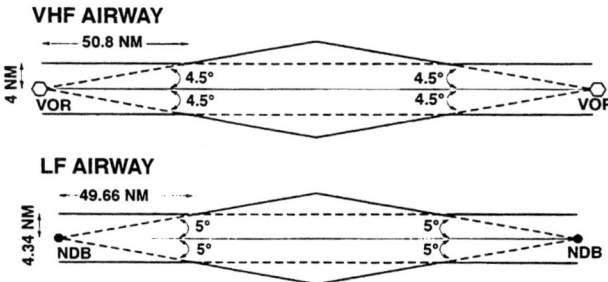

Fig. 1-7: Low Level Airways/Air Routes.

Flight Information Regions (FIR)

An FIR is airspace of defined dimensions extending upwards from the surface, in which flight information service (ie NOTAMS) and alerting services are provided. The FIRs are Vancouver, Edmonton, Winnipeg, Toronto, Montreal, Moncton and Gander.

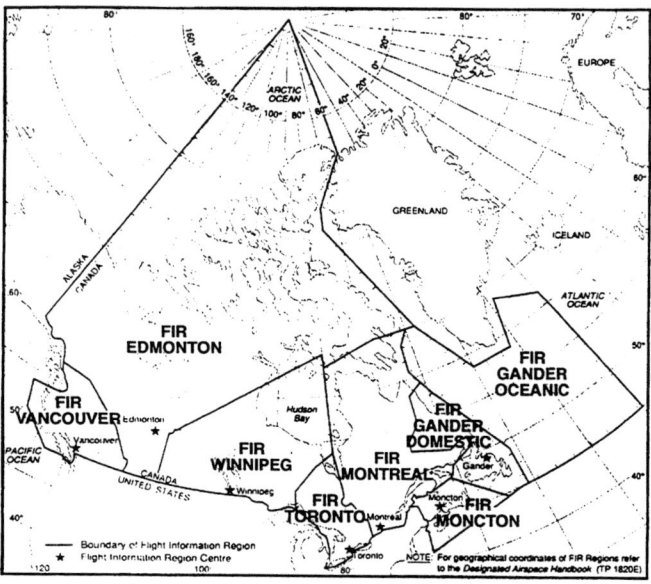

Fig. 1-8: Flight Information Regions.

1.21 NOTAM

A NOTAM is a notice containing information concerning conditions or changes in aeronautical facilities or hazards/procedures. NOTAMS are available at FSS stations or by voice advisory. Example of NOTAM format:

```
GG      CYZZLFNI                              (1)
141736 CYGLYFNI                               (2)
060007 NOTAMN CYOW OTTAWA INTL                (3)
CYOW ILS 07 U/S TIL APRX 0601191200           (4)
```

(1) Teletype priority and group address.
(2) Origination time (day, hour in UTC) and issuing location.
(3) 060007 = NOTAM continuity number (7th NOTAM of 2006 for that airport). NOTAMN = New NOTAM (R = Replacing, C = Cancelled, J = JBI/CRFI, Q = Query/response)
(4) 0601191200 = January 19, 2006 at 1200Z. If "aprx" used, cancelling NOTAM required.

[Reference: AIM MAP 5.0]

1.22 Air Traffic Services (ATS)

Air Traffic Control Services

The following air traffic control and information services are provided through area control centres, terminal control units and control towers:

a) Airport Control Service. Provided by airport control towers to aircraft and vehicles on the maneuvering area of an airport and to aircraft operating in the vicinity of an airport.
b) Area Control Service. Provided by ACCs to IFR and CVFR flights operating within specified control areas.
c) Terminal Control Service. Provided by IFR units

(ACCs) or TCUs to IFR and CVFR flights operating within specified control areas.

d) **Terminal Radar Service.** An additional service provided by IFR units to VFR aircraft operating within Class C airspace.

e) **Alerting Service.** To notify appropriate organizations for aircraft in need of search and rescue services, or to alert crash equipment, ambulances, doctors and any other safety services.

f) **Airspace Reservation Service.** Includes the services of the Airspace Reservation Unit (ARU) and ACCs in providing reserved airspace for specified air operations in controlled airspace, and in providing information concerning these reservations and military activity areas in controlled and uncontrolled airspace.

g) **Aircraft Movement Information Service.** Provided by ACCs for the collection, processing and dissemination of aircraft movement information for use by air defence units relative to flights operating into or within the ADIZ.

h) **Customs Notification Service (ADCUS).** Provided, on request, by ATC units for advance notification of customs officials for transborder flights at specified ports of entry.

i) **Flight Information Service.** Provided by ATC units to assist pilots of aircraft by supplying information concerning known hazardous flight conditions. This information will include data concerning unfavourable flight conditions and other known hazards; which may not have been available to the pilot prior to take off or which may have developed along the route of flight.

ATC service has been established primarily for the prevention of collisions and the expediting of traffic. ATC service takes precedence over flight information service, but every effort is made to provide flight information and assistance. Flight information is made available, whenever practicable, to any aircraft in communication with an ATC unit, prior to take-off or in flight, except where such service is provided by the aircraft operator. Many factors such as traffic volume, controller workload, communications frequency congestion and radar limitations may prevent a controller from providing this service.

IFR flights will be provided with information concerning

a) severe weather conditions;
b) weather conditions reported or forecast at destination or alternate aerodrome;
c) changes in the serviceability of navigation aids;
d) condition of airports and associated facilities;
e) other items considered pertinent to flight safety.

Flight Service Stations

FSS stations provide the following services:

a) **En Route Flight Information Service.** Continuous monitoring of assigned frequencies permits pilots the communications access to obtain and pass flight information, or to report emergencies should the need arise. In addition, the FSS relays IFR position reports and ATC clearances in areas where aircraft are beyond the communications range of the ATC facility responsible.

b) **Airport Advisory Service (AAS).** The FSS provides advisory information consisting of wind, preferred or active runway, time (departures only), altimeter, aircraft traffic, ground traffic, and other information to assist pilots to execute safe and expeditious departures and arrivals at uncontrolled airports.

c) **Remote Aerodrome Advisory Service (RAAS).** At aerodromes where the advisory is provided through RCO's, the service is referred to as a RAAS. RAAS consists of weather reports including wind and altimeter settings (METAR), the active or preferred runway (if known), field condition reports, NOTAM, PIREPs, and known aircraft and vehicle traffic. Since RAAS is a remote service provided by an FSS in another location, the pilot is responsible to comply with mandatory aerodrome traffic frequency procedures.

d) **Vehicle Control Service (VCS).** The FSS specialist controls vehicles operating on the maneuvering area of airports with a co-located tower and an FSS during the hours when the tower is closed. This service is also available at airports without a tower during the operating hours of the FSS. VCS is not available at sites served by RAAS.

e) **Flight Plan Service.** The FSS provides a preflight service which includes the provision of weather, NOTAM, RSC/JBI and other information. It accepts and processes flight plans and flight itineraries. An aviation information display is also maintained and is easily accessible to pilots. It assists pilots in compiling all the information essential to planning a safe flight.

f) **Surface Weather Observing Service.** The observation, recording, and dissemination of surface weather data, including specials, is performed by the FSS for aviation purposes.

g) **Aviation Weather Information Service (AWIS).** The FSS provides pertinent aviation weather information tailored to accommodate pilots at the preflight and en route stages. The service permits specialists to assist pilots in making decisions and calculations based on weather determinants.

h) **Aviation Weather Briefing Service (AWBS).** This is a fully interpretive preflight and en route weather briefing service provided toll-free by selected FSSs in each region. These sites are equipped with a full suite of weather products, including satellite and radar imagery. Briefers are trained to adapt meteorological information to fit the needs of all aviation users and to provide consultations and advice on special weather problems. Flight documentation for long range flights is also available, on request. This level of service is described as W1 in the CFS.

i) **VFR Alerting Service.** The FSS notifies SAR and conducts a communications search in the event that a VFR flight plan or flight itinerary is not closed within a specified time, or upon receiving an overdue report for an aircraft.

j) **Aeronautical Broadcast Service.** The FSS broadcasts weather and other information required by pilots to plan and complete flight safely.

k) **Navigation Assistance Service.** The FSS provides VDF assistance, at sites where the system is installed, to aircraft in emergency or potential emergency situations, or when requested by a pilot. Other navigation assistance may also be provided, depending on facilities available.

l) **Navigation Aids Monitoring Service.** The FSS monitors the status of navigation aids as assigned, and takes appropriate corrective and notification action should abnormal operation occur.

m) **NOTAM Service.** For designated locations and/or areas assigned to it, the FSS is the office responsible for co-ordination and dissemination of NOTAMs.

n) PIREPS. The FSS collects and distributes pilot reports of weather and other significant flight information.

o) Fixed Telecommunications Service. The FSS is connected to fixed telecommunications networks so that operational and administrative messages may be exchanged among FSS, other domestic and international aeronautical agencies, and aircraft in flight.

p) Paid Communications Service. On request from subscribing domestic and international air carriers, the FSS relays messages between an aircraft and its company for a cost recovery fee.

The majority of FSS stations provide flight information service 24 hours per day to the airports where they are located and to any number of assigned RCOs. Advisory services will also be provided from the FSS stations at controlled airports when the control towers are closed.

FSS stations give wind direction to the nearest 10° and wind speed/gusts to nearest 5 kts.

Remote Communications Outlets, Dial-Up RCOs

RCO's are remote VHF transmitters/receivers established as an extended communications capability to enable FSS specialists to provide flight information services in the vicinity of designated aerodromes. Flight Information Service En Route (FISE) is provided on the en route frequency and RAAS is provided on the MF.

An RCO can also be used to accept position reports and relay ATC clearances.

A Dial-Up Remote Communications Outlet (DRCO) is a standard RCO which has had a dial-up unit installed to connect the pilot with an ATS unit such as FSS via a normal telephone line. In this manner, the line is opened only after the communication has been initiated by the pilot or by ATS. The radio range of the RCO is unaffected by the conversion. To activate a DRCO, use the following procedure:

a) select the published RCO frequency on the aircraft radio;

b) key the radio microphone distinctly 4 times in a row, with not more than 1 second between keying. If successful, you will hear a dial tone, a series of touch tones, and a ringing signal;

c) if connected, there will be a recording that states "link established". Initiate the radio conversation as per standard radiotelephony practices, ie "Regina Radio, this is Cessna Golf Alfa Bravo Charlie, over". The ATS specialist may take a few moments to reply.

d) a "call terminated" message indicates that the telephone line has been disconnected.

ATC Clearances, Instructions and Information

Whenever an ATC clearance is received and accepted by the pilot, compliance shall be made with the clearance. If a clearance is not acceptable, the pilot should immediately inform ATC of this fact, since acknowledgement of the clearance alone will be taken by a controller as indicating acceptance.

A pilot shall comply with an ATC instruction which is directed to and received by the pilot, provided aircraft safety is not jeopardized.

A clearance will be identified by the use of some form of the word CLEAR in its content. An instruction will always be worded in such a manner as to be readily identified, although the word "instruct" will seldom be included. Pilots shall comply with and acknowledge receipt of all ATC instructions directed to and received by them.

Always remember that control is predicated on known air traffic only, and when complying with clearances or instructions, pilots are not relieved of the responsibility for practising safe procedures and good pilot judgement.

A clearance or instruction is only valid WHILE IN CONTROLLED AIRSPACE. Pilots crossing between controlled and uncontrolled airspace should pay close attention to the terrain and obstacle clearance requirements.

ATS personnel routinely inform pilots of conditions, observed by others or by themselves, which may affect flight safety and are beyond their control, such as observed airframe icing and bird activity. These are meant solely as assistance or reminders to pilots and are not intended in any way to absolve the pilot of the responsibility for the safety of the flight.

RADAR Services

General
The use of radar increases airspace utilization by allowing ATC to reduce the separation interval between aircraft. In addition, radar permits an expansion of flight information services, such as traffic information, radar navigation assistance and information on chaff drops, bird activity and severe weather information. Due to limitations inherent in all radar systems, it may not always be possible to detect aircraft, weather disturbances, etc. Where radar information is derived from secondary surveillance radar (SSR) only (without associated primary radar coverage) it is not possible to provide traffic information on aircraft that are not transponder equipped or to provide some of the other flight information.

Before providing radar services, ATC will establish identification of the aircraft concerned either through the use of position reports, identifying turns, or the use of transponders. Pilots will be notified whenever radar identification is established or lost.

Pilots are cautioned that radar identification of their flight does not relieve them of the responsibility for collision avoidance or terrain or obstacle clearance. ATC assumes responsibility for terrain or obstacle clearance when vectoring en route IFR and CVFR flights, and IFR aircraft being vectored for arrival until the aircraft is within the final approach area.

Radar Vectoring
Radar vectoring is used when necessary separation purposes, when required by noise abatement procedures, when requested by the pilot, or whenever vectoring will offer operation advantages to the pilot or the controller. When vectoring is initiated, the pilot will be informed of the location to which the aircraft is

being vectored. Pilots will be informed when radar vectoring is terminated, except when an arriving aircraft is vectored to the final approach course or to the traffic circuit.

When an aircraft is vectored to final approach or to the traffic circuit, the issuance of approach clearance indicates that normal navigation should be resumed. Normally, radar service will be continued until an aircraft leaves the area of radar coverage, enters uncontrolled airspace, or is transferred to an ATC unit not equipped with radar. The pilot will be informed when radar service is terminated.

MINIMUM RADAR VECTORING ALTITUDES (lowest altitude at which an aircraft may be vectored and still meet obstruction clearance criteria), which may be lower than minimum altitudes shown on navigation and approach charts, have been established at a number of locations to facilitate transitions to instrument approach aids. When an IFR flight is cleared to descend to the lower altitude, ATC will provide terrain and obstacle clearance until the aircraft is in a position from which an approved instrument approach or a visual approach can be commenced.

Visual Climb/Descent

When an aircraft is operated in VMC, an operational advantage may be gained for pilots and controllers by having the pilot request and ATC authorize a VISUAL CLIMB or a VISUAL DESCENT, as applicable, with respect to obstacles and terrain while on radar vectors. ATC authorization of a visual climb or descent under these circumstances constitutes acceptance by the pilot of the responsibility for terrain and obstacle avoidance. IFR separation normally provided between aircraft for the applicable classification of airspace will be maintained during the visual climb or visual descent phase of flight. Once the aircraft reaches (or passes) a minimum IFR altitude or an appropriate minimum vectoring altitude, responsibility for terrain and obstacle clearance reverts to ATC for as long as the flight is being radar vectored. Visual climbs or descents should only be requested when the pilot is assured of continuous visual reference with the terrain and obstacles throughout that phase of flight. To aid in the flow of air traffic, a controller may suggest a visual climb/descent to the pilot. In this case, the pilot has the option of accepting or not accepting the suggestion.

IFR Separation

Vertical Separation

The standard vertical separation minima is 1,000 feet when at or below FL 290, and 2,000 feet when above FL 290.

Lowest Usable Flight Levels

When the altimeter setting is less than 29.92, there will be less than 1,000 feet vertical separation between an aircraft flying at 17,000 feet ASL applying a local altimeter setting and another aircraft at FL 180 applying 29.92. Accordingly, if the altimeter setting is 29.92 or higher, the lowest usable flight level is FL 180. If the altimeter is from 29.91 to 28.92, the lowest usable flight level is FL 190, and if the altimeter is from 28.91 - 27.92, the lowest usable flight level is FL 200.

Lateral Separation

Lateral separation of IFR flights is provided by ATC in the form of "airspace to be protected" in relation to a holding procedure, instrument approach procedure or the approved track. The dimensions of protected airspace for a particular track take into account the accuracy of navigation that can be reasonably be expected. For track segments within signal coverage of NDB, VOR or TACAN stations and along bearings/courses/radials of such facilities, protected airspace takes into account the accuracy of available track guidance, accuracy of airborne receiver and indicator equipment, and a small pilotage tolerance. Separation is considered to exist provided the airspaces protected for each aircraft do not overlap. It is essential, therefore, that accuracy capability of navigation equipment be maintained.

Pilots of IFR or controlled VFR flights must adhere as closely as practicable to the centre line of their approved airway or track. If the aircraft inadvertently deviates from the approved track, immediate action must be taken to regain the centre line as soon as practicable. Pilots realizing that they are outside the airspace protected for their approved track must notify the appropriate ATC unit immediately.

In the low level airspace, the airspace to be protected is the full width of the airway.

In the high level airspace, all airspace is controlled within the Southern, Northern, and Arctic Control Areas. As a result, a high level is "a prescribed track between specified radio aids to navigation" and thus has no defined lateral dimensions. Therefore, the airspace to be protected for airways and or tracks in the high level airspace is the same as that for low level airways.

Along OFF AIRWAY TRACKS, the airspace to be protected is 45 NM each side of that portion of the track which is beyond navigational and signal coverage range.

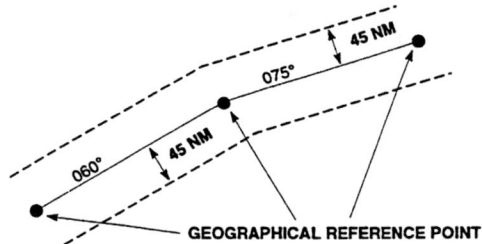

Fig. 1-9: Airspace to be protected along off airway tracks.

Additional airspace will be protected at and above FL 180 on the maneuvering side of tracks that change direction by more than 15° overhead navigation aids or intersections. It is expected that pilots of aircraft operating below FL 180 will make turns so as to remain within the normal width of airways or airspace protected for off airway tracks.

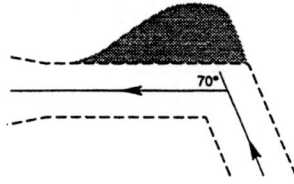

Fig. 1-10: Additional airspace to be protected for turns.

For lateral separation on an IAP, controllers consider the basic horizontal dimensions of intermediate approach areas, final approach areas and missed approach areas, for obstacle clearance purposes, as the airspace to be protected for aircraft conducting standard IAPs. Adequate horizontal separation is then deemed to exist when the airspace to be protected for such aircraft does not overlap the airspace to be protected for aircraft en route, holding or conducting simultaneous adjacent IFR approaches. Based on the airspace to be protected concept, it is the pilot's responsibility to remain within the limits or airspace to be protected. This can be accomplished by following the procedures published in the CAP or operator approved IAPs. If a pilot who is operating in controlled airspace anticipates being unable to conduct the approach as published, the pilot should inform ATC so that separation from other aircraft concerned can be increased as necessary.

Wake Turbulence Separation
Wake turbulence has its greatest impact on departure and arrival procedures, however, pilots should not assume that it will only be encountered in the vicinity of aerodromes. Caution should be exercised if a flight is conducted anywhere behind and less than 1,000 feet below a large aircraft.

Aircraft wake turbulence groups are as follows:

a) Group 1 (Heavy). All aircraft certified for a maximum take-off weight of 300,000 lbs or more.
b) Group 2 (Medium). Aircraft certified for a maximum take-off weight of between 12,500 and 300,000 lbs.
c) Group 3 (Light). Aircraft certified for a takeoff weight up to 12,500 lbs inclusive.

For radar departures, controllers generally apply the following radar separation minimum between a preceding IFR/VFR aircraft and an aircraft vectored directly behind it and at less than 1,000 feet below:

Heavy behind a Heavy . . . . 4 miles
Medium behind a Heavy . . . . 5 miles
Light behind a Heavy . . . . 6 miles
Light behind a Medium . . . . 4 miles

For non-radar departures, controllers will apply a two minute separation interval to any aircraft that takes off into the wake of a known heavy aircraft if the aircraft concerned commences the take-off from the threshold of the same runway, or any following aircraft departs from the threshold of a parallel runway that is located less than 2,500 feet away from the runway used by the preceding heavy aircraft. ATC does not apply this two minute spacing interval between a light following a medium aircraft in the above circumstances, but will issue wake turbulence advisories to light aircraft.

Controllers will apply a three minute separation interval to any aircraft that takes off into the wake of a known heavy aircraft, or a light aircraft that takes off into the wake of a known medium aircraft if the following aircraft starts it take-off roll from an intersection or from a point further along the runway than the preceding aircraft, or if the controller has reason to believe that the following aircraft will require more runway length for take-off than the preceding aircraft.

ATC will also apply separation intervals of up to three minutes when the projected flight paths of any following aircraft will cross that of a preceding heavy aircraft.

Regardless of the above, ATC will not guarantee that wake turbulence will not be encountered.

If a pilot is advised to wait before being given clearance for take-off due to wake turbulence, the pilot may request a PILOT WAIVER of wake turbulence separation. Although controllers are not permitted to initiate waivers to wake turbulence separation minimums, they will issue take-off clearance to pilots who have waived wake turbulence requirements on their own initiative. In some situations a pilot waiver may not be granted such as with intersection take-offs of a light or medium behind a heavy, or a light or medium take-off following a heavy missed approach.

[Reference: AIM RAC 1.1, 1.2, 1.5, 4.2]

1.23 Cruising Altitudes

The appropriate cruising altitude is determined on the basis of the height of obstructions, aircraft performance and weather conditions. Cruising altitudes are based on the aircraft TRACK (path flow over the ground). The MAGNETIC TRACK is utilized in the SDA, and the TRUE TRACK is utilized in the NDA.

VFR Cruising Altitudes (Below 18,000 ft ASL):

Aircraft Track	
000° - 179°	180° - 359°
3,500	4,500
5,500	6,500
7,500	8,500
9.500	10,500
11,500	12,500
etc.	etc.

IFR/CVFR Cruising Altitudes (below 18,000 ft ASL):

Aircraft Track	
000° - 179°	180° - 359°
3,000	4,000
5,000	6,000
7,000	8,000
9,000	10,000
11,000	12,000
etc.	etc.

Some airways or air routes use cruising altitudes that are an exception to the above requirements: refer to the CFS or IFR charts for specific requirements. The exceptions do not apply to VFR flights.

At or above 18,000 feet ASL but below FL 290:

Aircraft Track	
000° - 179°	180° - 359°
FL 190	FL 180
FL 210	FL 200
FL 230	FL 220
etc.	etc.

At or above FL 290 (use 4,000 ft intervals):

Aircraft Track

000° - 179°	180° - 359°
FL 290	FL 310
FL 330	FL 350
FL 370	FL 390
etc.	etc.

[Reference: CARs 602.34; AIM RAC 2.3]

1.24 Flight Within Designated Mountainous Regions

When operating IFR within the designated mountainous regions, but outside areas for which minimum IFR altitudes have been established (e.g. MEA, MOCA, MSA, minimum vectoring altitudes, 100 NM safe etc.).

a) Aircraft shall be flown at least 2,000 feet above the highest obstacle within 5 NM, for areas 1 and 5.
b) Aircraft shall be flown at least 1,500 feet above the highest obstacle within 5 NM, for areas 2, 3 and 4.

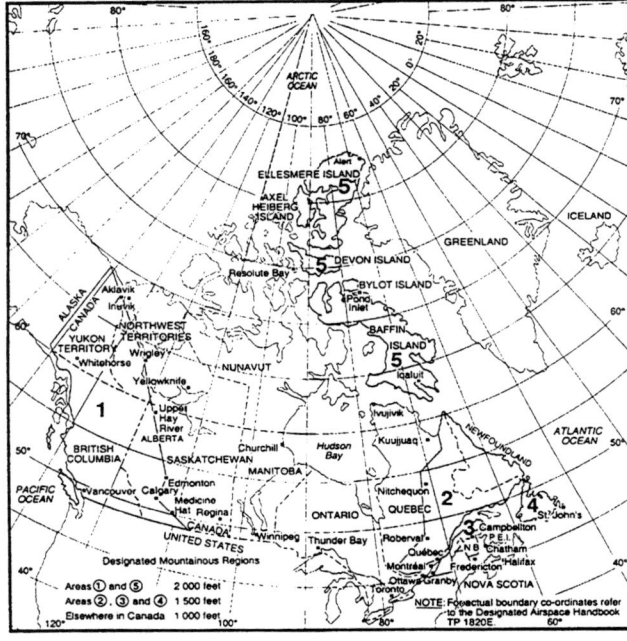

Fig. 1-11: Designated Mountainous Areas.

NOTE: In extremely cold temperatures or high variations in pressure, aircraft should be operated at least 1,000 feet above the MEA within designated mountainous regions.

[Reference: CARs 602.125; AIM RAC 2.12; CAP Gen 4]

1.25 Major Errors of the Pressure Altimeter

The altimeter is calibrated to indicate true altitude only under ISA conditions. In extremely cold temperatures, the true altitude will be lower than the indicated altitude. Therefore, pilots must add correction factors to all relevant altitudes on any approaches in conditions of extreme cold weather. Fig. 1-11 (b) below provides an example of this kind of chart.

An IFR cruising altitude or vectoring altitude, once accepted by the pilot, must be maintained. For example, if a pilot accepts "maintain 3000", the pilot should not apply an altitude correction. Radar vectors altitudes assigned by ATC are temperature compensated and require no corrections.

EXAMPLE: Assume an airport has an elevation of 2,000 feet ASL and surface temperature of -20° C. Based on the cold air correction factor table below, determine the corrections to be applied and indicated altitudes for a published Procedure Turn altitude of 3,500 feet, and published MDA straight-in approach minimums of 2,600 feet.

ANSWER: Correction for procedure turn will be 210 feet, which will yield an indicated altitude of 3,710 feet. Correction for MDA straight-in approach minimums will be 90 feet, which will yield an indicated altitude of 2,690 feet.

[Reference: AIM RAC 9.17; CAP Gen 21]

1.26 Speed Adjustment - Radar Controlled Aircraft

ATC may ask for a speed adjustment during radar vectoring. Such requests will be expressed in 10 kt units. Pilots should maintain +/- 10 KIAS of that speed requested, and inform ATC should a specified speed be unsafe. Issuance of an approach clearance normally cancels a speed adjustment. If the controller requires that a pilot maintain a speed adjustment after issuing

Airport Temp.	Altitude Correction Chart Height above the elevation of the altimeter setting source (feet)													
° C	200	300	400	500	600	700	800	900	1,000	1,500	2,000	3,000	4,000	5,000
0	20	20	30	30	40	40	50	50	60	90	120	170	230	290
-10	20	30	40	50	60	70	80	90	100	150	200	290	390	490
-20	30	50	60	70	90	100	120	130	140	210	280	430	570	710
-30	40	60	80	100	120	130	150	170	190	280	380	570	760	950
-40	50	80	100	120	150	170	190	220	240	360	480	720	970	1,210
-50	60	90	120	150	180	210	240	270	300	450	600	890	1,190	1,500

Fig. 1-11 (b): Altitude Correction Chart.

an approach clearance, the controller will restate it.

The minimum speeds used by ATC when issuing speed adjustments are:

a) All aircraft, when 20 NM or more from destination - when above 10,000 feet ASL - 250 kts, when below 10,000 feet ASL - 210 kts;
b) Turbo-jet aircraft, when less than 20 NM from destination - 160 kts.
c) Propeller aircraft, less than 20 NM from destination - 120 kts.

[Reference: AIM RAC 1.8.3]

1.27 Aircraft General Operating Rules

Reckless or Negligent Operation of Aircraft
No person shall operate an aircraft in such a reckless or negligent manner as to endanger or be likely to endanger any person's life or property.

Fitness of Flight Crew Members
No operator of an aircraft shall require any person to act as a flight crew member and no person shall act as a flight crew member, if either the person or the operator has any reason to believe, having regard to the circumstances of the flight to be made, that the person is suffering or is likely to suffer from fatigue, or is otherwise unfit to perform properly the person's duties as a flight crew member.

Alcohol or Drugs - Crew Members
No person shall act as a crew member of an aircraft within 8 hours after consuming an alcoholic beverage, while under the influence of alcohol or while using any drug that impairs the person's faculties to the extent that the safety of the aircraft or of persons on board the aircraft is endangered in any way.

Alcohol or Drugs - Passengers
INTOXICATING LIQUOR means a beverage that contains more than 2.5% proof spirits. No person shall consume on board an aircraft an intoxicating liquor unless the intoxicating liquor has been served to that person by the operator of the aircraft or, where no flight attendant is on board, has been provided by the operator of the aircraft. No operator of an aircraft shall provide or serve any intoxicating liquor to a person on board the aircraft, where there are reasonable grounds to believe that the person's faculties are impaired by alcohol or a drug to an extent that may present a hazard to the aircraft or to persons on board the aircraft.

No operator of an aircraft shall allow a person to board the aircraft, where there are reasonable grounds to believe that the person's faculties are impaired by alcohol or a drug to an extent that may present a hazard to the aircraft or to persons on board the aircraft. The operator of an aircraft my allow a person whose faculties are impaired by a drug to board an aircraft, where the drug was administered in accordance with a medical authorization and the person is under the supervision of an attendant.

Smoking
No person shall smoke on board an aircraft during take-off or landing or when directed not to smoke by the PIC. No person shall smoke in an aircraft lavatory. No person shall tamper with or disable a smoke detector installed in an aircraft lavatory without permission from a crew member or the operator of the aircraft.

Aircraft Operating Limitations
No person shall operate an aircraft unless it is operated in accordance with the operating limitations set out in the AFM or other authorized document.

Portable Electronic Devices
No operator of an aircraft shall permit the use of a portable electronic device on board an aircraft, where the device may impair the functioning of the aircraft's systems or equipment. No person shall use a portable electronic device on board an aircraft except with the permission of the operator of the aircraft.

Fuelling with Engines Running
No persons operating an aircraft shall permit the fueling of the aircraft while an engine used for the propulsion of the aircraft is running and passengers are on board the aircraft or are embarking or disembarking unless it is done so in accordance with detailed requirements specified in the Private Operator, Commuter Operator, Airline Operator standards as appropriate (see s. 1.09 for Air Taxi - fuelling with passengers on board for general rules).

Starting and Ground Running of Aircraft Engines
No person shall start an engine of an aircraft unless:

a) a pilot's seat is occupied by a person who is competent to control the aircraft;
b) precautions have been taken to prevent the aircraft from moving;
c) in the case of a seaplane, the aircraft is in a location from which any movement of the aircraft will not endanger persons or property.

No person shall leave an engine of an aircraft running unless

a) a pilot's seat is occupied by a person who is competent to control the aircraft;
b) where no persons are on board the aircraft, precautions have been taken to prevent the aircraft from moving, and the aircraft is not left unattended.

Aircraft Icing
CRITICAL SURFACES means the wings, control surfaces, propellers, horizontal stabilizers, vertical stabilizers or any other stabilizing surface of an aircraft and, in the case of an aircraft that has rear mounted engines, includes the upper surface of its fuselage.

No person shall conduct or attempt to conduct a take-off in an aircraft that has frost, ice or snow adhering to any of its critical surfaces. A person may conduct a take-off in an aircraft that has frost adhering to the underside of its wings that is caused by cold-soaked fuel, if the take-off is conducted in accordance with the aircraft manufacturer's instructions for take-off under those conditions.

Where conditions are such that frost, ice or snow may reasonably be expected to adhere to the aircraft, no person shall conduct or attempt to conduct a take-off unless, for non Airline operators, the aircraft has been inspected immediately prior to take-off (by PIC, flight crew member or trained person designated by operator) to determine whether any frost, ice or snow is adhering to any of its critical surfaces or the operator has established an approved aircraft inspection program and the dispatch and take-off of the aircraft are in accordance with that program. For Airline operators, the aircraft must be operated under an approved aircraft inspection program and the dispatch and take-off of the aircraft must be in accordance with that program.

Where, before commencing take-off, a crew member of an aircraft observes that there is frost, ice or snow adhering to the wings of the aircraft, the crew member shall immediately report that observation to the PIC, and the PIC or a flight crew member designated by the PIC shall inspect the wings of the aircraft before take-off. Before an aircraft is de-iced or anti-iced, the PIC of the aircraft shall ensure that crew members and passengers are informed of the decision to do so.

Overflight of Built-Up Areas or Open Air Assemblies of Persons during Take-offs, Approaches and Landings

Except if conducting a take-off, approach or landing at an airport or military aerodrome, no person shall conduct a take-off, approach or landing in an aircraft during which the aircraft will overfly a built-up area or an open-air assembly of persons, unless the aircraft is operated at an altitude from which, in the event of an engine failure or any other emergency necessitating an immediate landing, it would be possible to land the aircraft without creating a hazard to persons or property on the surface.

Take-Offs, Approaches and Landings within Built-up Areas of Cities and Towns

Unless otherwise permitted (ie aerial work), no person shall conduct a take-off, approach or landing in an aircraft within a built-up area of a city or town, unless that take-off, approach or landing is conducted at an airport or a military aerodrome.

A person may conduct a take-off or landing in an aircraft within a built-up area of a city or town at a place that is not located at an airport or a military aerodrome where the place is not set apart for the operation of aircraft, the flight is conducted without creating a hazard to persons or property on the surface and the aircraft is operated for the purpose of a police operation that is conducted in the service of a police authority or for the purposes of saving human life.

Minimum Altitudes and Distances

Except where conducting a take-off, approach or landing, no person shall operate an aeroplane over a built-up area or over an open air assembly of persons unless the aircraft is operated at an altitude from which, in the event of an emergency necessitating an immediate landing, it would be possible to land the aircraft without creating a hazard to persons or property on the surface, and, in any case, at an altitude that is not lower than 1,000 feet above the highest obstacle located within a horizontal distance of 2,000 feet from the aeroplane. In other circumstances, at a distance less than 500 feet from any person, vessel, vehicle or structure.

Permissible Low Altitude Flight

A person may operate an aircraft at altitudes and distances less than those minimums specified above where the aircraft is operated at altitudes and distances that are no less than necessary for the purposes of the operation in which the aircraft is engaged, the aircraft is operated without creating a hazard to persons or property on the surface and the aircraft is operated

a) for a police operation that is conducted in the service of a police authority;
b) for saving human life;
c) for fire-fighting or air ambulance operations;
d) for administration of the fisheries legislation;
e) for the administration of national or provincial parks;
f) for flight inspection.

A person may operate an aeroplane, to the extent necessary for aerial work in an aerial work zone, less than the 1,000 feet minimum. As well, an aeroplane can be operated at less than the 500 feet minimum, where the aircraft is operated without creating a hazard to persons or property on the surface and the aircraft is operated for the purpose of aerial application or aerial inspection, aerial photography conducted by the holder of an air operator certificate, or flight training conducted by or under the supervision of a qualified flight instructor.

Carriage of Persons during Low Altitude Flight

No person engaged in aerial application or aerial inspection at altitudes less than 500 feet AGL may carry on board any person other than a flight crew member, unless that person's presence on board is essential to the purposes of the flight.

Fuel Dumping

No person shall jettison fuel from an aircraft in flight unless it is necessary to do so in order to ensure aviation safety and all appropriate measures are taken to minimize danger to human life and damage to the environment, as far as the circumstances permit.

Supersonic Flight

No person shall operate an aircraft at a true Mach number of 1.00 or greater.

Flight Over the High Seas

The PIC of a Canadian aircraft that is in flight over the high seas shall comply with the applicable ICAO rules.

Transoceanic Flight

No PIC of a single engine aircraft, or of a multi-engined aircraft that would be unable to maintain flight in the event of the failure of any engine, shall commence a flight that will leave CDA and enter airspace over the high seas unless the PIC holds an IFR rating and the aircraft is IFR equipped. The aircraft must also have an HF radio capable of transmitting and receiving on a minimum of two appropriate international air-ground general purpose frequencies and hypothermia protection for each person on board. The aircraft must also carry sufficient fuel to meet VFR/IFR fuel requirements and in addition, carry contingency fuel equal to at least 10% extra to complete the flight to the aerodrome of destination.

1.28 Aircraft Airworthiness

Flight Authority

FLIGHT AUTHORITY means a certificate of airworthiness, special certificate of airworthiness, flight permit or validation of a foreign document attesting to an aircraft's fitness for flight, or a foreign certificate of airworthiness.

No person shall operate an aircraft in flight unless a flight authority is in effect in respect of the aircraft, the aircraft is operated in accordance with the conditions set out in the flight authority and the flight authority is carried on board the aircraft.

Availability of Aircraft Flight Manual (AFM)

No person shall conduct a take-off in an aircraft for which an AFM is required by airworthiness standards, unless the AFM (or AOM where authorized) is available to the flight crew members at their duty stations. The AFM or where an AOM has been established, shall include all of the amendments and supplementary material that are applicable to the aircraft type.

Markings and Placards

No person shall conduct a take-off in an aircraft in respect of which markings or placards are required by the applicable standards of airworthiness unless the markings or placards are affixed to the aircraft or attached to a component of the aircraft in accordance with those standards.

Minimum Equipment Lists

Air operators require a formal system for the deferral of defects. MINIMUM EQUIPMENT LISTS (MELs) are lists of systems and equipment installed in an aircraft that show the degree to which defects may be allowed for a limited period.

The Minister may, in accordance with the *MMEL/MEL Policy and Procedures Manual*, establish a Master Minimum Equipment List (MMEL) for each type of aircraft, and the Minister shall approve a MEL in respect of each operator of that type of aircraft.

Unserviceable and Removed Equipment - General

No person shall conduct a take-off in an aircraft that has equipment that is not serviceable or from which equipment has been removed if, in the opinion of the PIC, aviation safety is affected.

A person may conduct a take-off in an aircraft that has equipment that is not serviceable or from which equipment has been removed where the aircraft is operated in accordance with the conditions of a flight permit that has been issued specifically for that purpose.

Unserviceable and Removed Equipment - Aircraft with a Minimum Equipment List

Where a MEL has been approved for an operator using an aircraft, no person shall conduct a take-off in the aircraft with equipment that is not serviceable or that has been removed unless the aircraft is operated in accordance with any conditions or limitations specified in the MEL and a copy of the MEL is carried on board the aircraft. Where the conditions or limitations specified in a MEL are in conflict with the requirements of an AD, the AD prevails.

Unserviceable and Removed Equipment - Aircraft without a Minimum Equipment List

Where a MEL has not been approved in respect of the operator of an aircraft, no person shall conduct a take-off in the aircraft with equipment that is not serviceable or that has been removed, where that equipment is required by

a) the standards of airworthiness that apply to day or night VFR or IFR flight, as applicable;
b) any equipment list published by the aircraft manufacturer respecting aircraft equipment that is required for the intended flight;
c) an air operator certificate, a private operator certificate, a special flight operations certificate or

d) an AD;
e) the CARs.

Where a MEL has not been approved in respect of the operator of an aircraft, and the aircraft has equipment other than the equipment required by the above that is not serviceable or that has been removed, no person shall conduct a take-off in the aircraft unless

a) where the unserviceable equipment is not removed from the aircraft, it is isolated or secured so as not to constitute a hazard to any other aircraft system or to any person on board the aircraft;
b) the appropriate placards are properly installed;
c) an entry recording these actions is made in the journey log.

Requirement to Keep Technical Records

Every owner of an aircraft shall keep the following technical records in respect of the aircraft:

a) a journey log;
b) a separate technical log for the airframe, each installed engine and each variable-pitch propeller;
c) an approved empty weight and balance report.

Technical Records - General

Every person who makes an entry in a technical record shall make the entry accurately, legibly and in a permanent manner, and enter the person's name and signature or employee identifier and date the entry.

The owner of an aircraft shall ensure that all necessary measures are taken to protect the technical records for the aircraft from damage and loss.

Every person who brings into use a new volume of an existing technical record shall make the entries relating to the preceding volume that are necessary to ensure that an unbroken chronological record is maintained.

Where a person alters an entry on a technical record for the purpose of correcting the entry, the person shall do so by striking out the incorrect entry in such a manner that the underlying information remains legible, and inserting the correct entry together with the date of the alteration, the reason for the alteration if necessary, the persons's name and signature or employer identifier or electronic user code.

Journey Log Requirements

The particulars required to be entered in a journey log are set out in the CARs (ie PIC, date, air time, flight time, defects etc.).

No person shall make a single entry in a journey log in respect of a series of flights unless the aircraft is operated by the same PIC throughout, the series, or an approved daily flight record is used in a flight training operation.

The owner of an aircraft shall retain every entry in a journey log for a period of not less than one year. A three year retention period applies where the aircraft is registered to a Canadian and the journey log is used for the purposes of recording particulars of aircraft flight time.

Unless recorded in the operational flight plan or operational flight data sheet, the PIC of an aircraft engaged in a commercial air service and operating in international flight shall record in the journey log the following particulars in respect of each flight:

a) the names of all of the crew members and their duty assignments;
b) the places and times of departure and arrival;
c) the flight time;
d) the nature of the flight, such as private, aerial work, scheduled or non-scheduled;
e) any incidents or observations relating to the flight.

Journey Log - Carrying on Board

No person shall conduct a take-off in an aircraft unless the journey log is on board the aircraft. A person may conduct a take-off in an aircraft without carrying the journey log on board where it is not planned that the aircraft will land and shut down at any location other than the point of departure.

Requirements for Technical Records other than the Journey Log

Engine parts, propeller and airframe parts and components installed and maintenance work must be recorded in the appropriate technical record. The particulars of the maintenance work including the name and identification number of the person who made the original entry must be recorded.

Transfer of Records

Every owner of an aircraft who transfers title of an aircraft, airframe, engine, propeller or appliance to another person shall, at the time of transfer, also deliver to that person all of the technical records that relate to that aeronautical product.

[Reference: CARs 605.03 - 605.10, 605.92, 605.93, 605.94, 605.95, 605.96, 605.97]

1.29 Aircraft Speed Limit

No person shall operate an aircraft in excess of an IAS of 200 kts, when below 3,000 feet AGL and within 10 NM of a controlled airport. It is also prohibited to operate an aircraft at speeds in excess of 250 kts IAS when below 10,000 feet ASL. This requirement does not apply to aircraft climbing to a cruising altitude above 10,000 feet or in cases where the minimum safe flight speed is greater than the above, or where a special flight operations certificate has been obtained (ie airshow).

[Reference: CARs 602.32, 603.02]

1.30 Security Control of Air Traffic

To permit identification and control of aircraft in the interest of national security, the Air Defence Identification Zone (ADIZ) has been established.

All aircraft operating in the ADIZ must have a 2-way radio and maintain a listening watch on the appropriate frequency. Pilots must file a Defence Flight Plan or Defence Flight Itinerary. The estimated time of penetration of the ADIZ boundary (or arrival at a reporting point) must be provided and revised if more than +/- 5 minutes or +/- 20 NM of estimate.

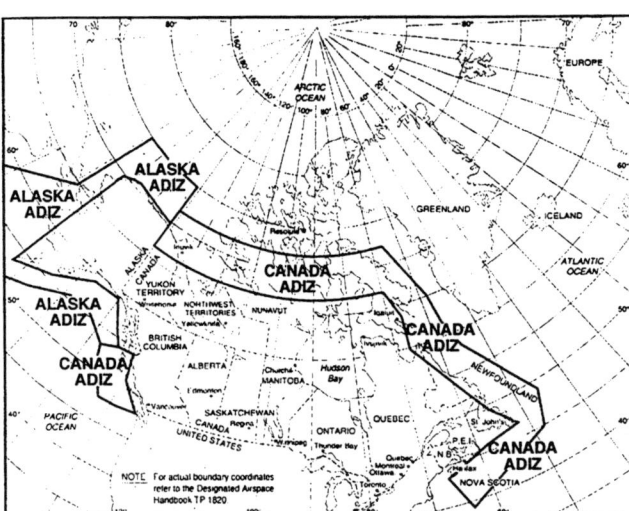

Fig. 1-12: ADIZ.

In times of war or national emergency, the ESCAT (Emergency Security Control of Air Traffic) plan may come into effect. The ESCAT plan involves a system for the control or civilian aircraft with respect to diversion, landing, grounding, dispersal and control of air navigation aids.

To ensure the effectiveness of the ESCAT plan, periodic tests will be conducted without any prior notice. The test message will read as follows:
ATTENTION; THIS IS AN ESCAT TEST. THIS IS AN ESCAT TEST. ALL AIRCRAFT WILL ACKNOWLEDGE THIS MESSAGE AND CONTINUE NORMAL OPERATIONS. Since these tests are considered essential to national security, cooperation of all pilots and agencies is necessary.

In times of an actual emergency, the following message will be broadcast by ATS: ATTENTION: THIS IS ESCAT. THIS IS ESCAT. ALL AIRCRAFT WILL ACKNOWLEDGE.

[Reference: CARs s. 602.145, 602.146; AIM RAC 2.13, 3.9, 12.8]

1.31 Transportation Safety Board and Aircraft Accidents

The purpose of accident safety investigation is to prevent recurrence of accidents or incidents. The Canadian Transportation Accident Investigation and Safety Board (TSB) is responsible for accident investigation in Canada. A team of investigators is on 24 hr stand-by duty.

An AVIATION OCCURRENCE means any accident or incident associated with the operation of aircraft and any situation or condition that the TSB has reasonable grounds to believe could, if left unattended, induce an accident or incident.

A REPORTABLE AVIATION ACCIDENT means an accident resulting directly from the operation of an aircraft where

a) a person sustains a serious injury or is killed as a result of being on board the aircraft, coming into contact with any part of the aircraft or its contents, or being directly exposed to the jet blast or rotor downwash of the aircraft;

b) the aircraft sustains damage or failure that adversely affects the structural strength, performance or flight characteristics of the aircraft and that requires major repair or replacement of any affected component part;

c) the aircraft is missing or inaccessible.

A REPORTABLE AVIATION INCIDENT means an incident resulting directly from the operation of an aeroplane having a maximum certificated takeoff weight greater than 5,700 kgs where a serious situation develops such as an engine failure or precautionary shutdown, smoke or fire occurs, control difficulties are encountered, uncontrolled vibration, depressurization requiring emergency descent, gear up landing, crew member physically incapacitated posing safety threat, risk of collision or near miss etc.

The TSB should be notified as soon as possible of any reportable aviation accidents or reportable aviation incidents (see AIP, GEN section for detailed reporting procedures).

Following an accident, the site must not be disturbed, and the PIC, crew and others must preserve and protect the accident site as best as possible, except to rescue people, prevent fire or avoid other dangers.

[Reference: AIM GEN 3.0]

1.32 High Level Airspace

CDA is divided into High Level Airspace (18,000 ft ASL and above) and Low Level Airspace (below 18,000 ft).

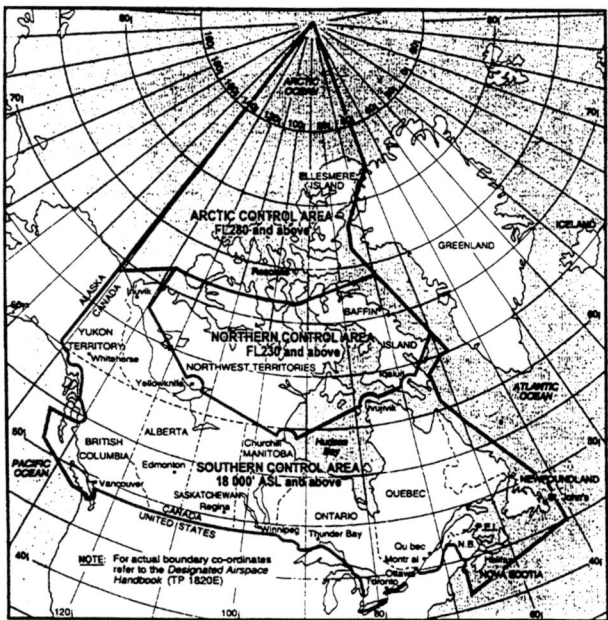

Fig. 1-13: High Level Controlled Airspace.

Controlled airspace within the High Level Airspace is divided into three separate areas: the Southern Control Area (SCA), the Northern Control Area (NCA) and the Arctic Control Area (ACA).

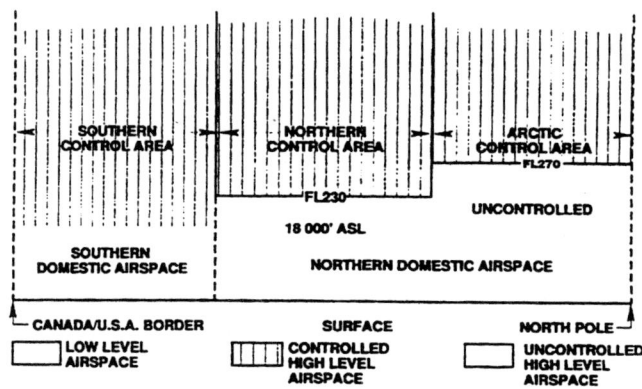

Fig. 1-14: Cross section of high level airspace. Note areas of uncontrolled high level airspace.

A PROFILE DESCENT permits an arriving aircraft to conduct an uninterrupted descent from cruising altitude or flight level toward the airport. Once issued, a profile descent is an ATC clearance, but is not considered to be an approach clearance. Busier airports such as Toronto-Pearson have published profile descent procedures in the CAP.

ATC will use the phrase WHILE IN CONTROLLED AIRSPACE in conjunction with altitude if an aircraft will be entering or leaving controlled airspace. In addition, ATC will specify the point at which an aircraft is to leave or enter controlled airspace laterally if the instruction is required for separation purposes. The altitude assigned by ATC need only reflect the minimum safe IFR altitude within controlled airspace: the pilot should be alert for the possibility of a higher minimum safe IFR altitude outside of controlled airspace and should request higher if uncertain.

When operating enroute in uncontrolled airspace, pilots should continuously monitor 126.7 MHz whenever practicable. Position reports should be made over navaids along the route flown to appropriate stations. Immediately before changing altitude, commencing an IAP or departing IFR, pilots should broadcast their intentions over 126.7 MHz whenever practicable.

Position Reporting within the NCA and ACA

Within the NCA, position reports are to be made when over compulsory reporting points, when requested by ATC, and in accordance with the following:

a) For flights on north or south tracks (315°T clockwise through 045°T or the reciprocals), report over fixed reporting lines coincident with each 5° of latitude north or south of, and including, 65°N latitude.

b) For flights on predominantly east or west tracks (046°T through 134°T or the reciprocals), report over fixed reporting lines coincident with each 10° of longitude east and west of, and including, 100°W longitude, except that where 20° of longitude will be traversed in less than 60 minutes the flight may report over such reporting lines spaced at 20° intervals. Longitude will be expressed in degrees only. Latitude will be expressed in degrees and minutes.

c) Flights operating within the Northern Track System shall report over 100°W, 90°W, 80°W, 70°W and

60°W longitude, over designated compulsory reporting points or as requested by ATC. The code name of the track and reporting meridian can be used in the position report.

EXAMPLE: Aircraft is over Bravo Track: "Bravo 80 at (time)".

When within the ACA, pilots should report when over the 141°W, 115°W and 60°W meridians. If the flight will not cross these meridians, the pilot should make the position report at the time of entry/exit from the ACA.

[Reference: AIM RAC 2.6, RAC 12.7]

1.33 Operational & Emergency Equipment

General

No person shall operate a Canadian registered aircraft unless the operational and emergency equipment required is carried on board.

Requirements for Power Driven Aircraft

No person shall conduct a take-off in a power-driven aircraft, other than an ultra-light aeroplane, unless the following operational and emergency equipment is carried on board:

a) a checklist or placards that enable the aircraft to be operated in accordance with the limitations in the AFM, AOM, POH or any equivalent document provided by the manufacturer. The checklist or placards shall enable the aircraft to be operated in normal, abnormal and emergency conditions and shall include a pre-start check, pre-take-off check, post take-off check, pre-landing check, and emergency procedures. The emergency procedures shall include emergency operation of fuel, hydraulic, electrical and mechanical systems (where applicable), emergency operation of instruments and controls (where applicable), engine inoperative procedures and any other procedure that is necessary for aviation safety;
b) where the aircraft is operated in VFR OTT, night VFR flight or IFR flight, all of the necessary current aeronautical charts and publications covering the route of the proposed flight and any probable diversionary route;
c) a hand held fire extinguisher in the cockpit that is of a type suitable for extinguishing the fires that are likely to occur, designed to minimize the hazard of toxic gas concentrations and readily available in flight to each flight crew member;
d) a timepiece that is readily available to each flight crew member;
e) where the aircraft is operated at night, a flashlight that is readily available to each crew member;
f) a first aid kit.

The above checks and emergency procedures shall be performed and followed where applicable.

Survival Equipment - Flights over Land

No person shall operate an aeroplane over land unless there is carried on board survival equipment, sufficient for the survival on the ground of each person on board, given the geographical area, the season of the year and anticipated seasonal climatic variations, that provides the means for

a) starting a fire;
b) providing shelter;
c) providing or purifying water;
d) visually signalling distress.

These requirements do not apply in the following cases:

a) aeroplanes operated within 25 NM of the aerodrome of departure and that have the capability of radiocommunication with a surface-based radio station for the duration of the flight;
b) multi-engined aircraft operated south 66° 30' north latitude in IFR flight within controlled airspace or along designated air routes;
c) aircraft operated by an air operator, where the aircraft is equipped with equipment specified in the air operator's COM;
d) aircraft that are in a geographical area where and at a time of year when the survival of the persons on board is not jeopardized.

Life Preservers and Flotation Devices

No person shall conduct a take-off or a landing on water in an aircraft or operate an aircraft over water beyond a point where the aircraft could reach shore in the event of an engine failure, unless a life preserver, individual flotation device or personal flotation device is carried for each person on board.

No person shall operate a land aeroplane at more than 50 NM from shore unless a life preserver is carried for each person on board.

Every life preserver, individual flotation device and personal flotation device shall be stowed in a position that is easily accessible to the person for whose use it is provided, when that person is seated.

Life Rafts and Survival Equipment - Flights over Water

No person shall operate over water a single-engined aeroplane, or a multi-engined aeroplane that is unable to maintain flight with any engine failed, at more than 100 NM, or the distance that can be covered in 30 minutes of flight at the cruising speed filed in the flight plan or flight itinerary, whichever distance is the lesser, from a suitable emergency landing site unless LIFE RAFTS are carried on board and are sufficient in total rated capacity to accommodate all of the persons on board.

No person shall operate over water a multi-engined aeroplane that is able to maintain flight with any engine failed at more than 200 NM, or the distance that can be covered in 60 minutes of flight at the cruising speed filed in the flight plan or flight itinerary, whichever distance is the lesser, from a suitable emergency landing site unless life rafts are carried on board and are sufficient in total rated capacity to accommodate all of the persons on board. A person may operate over water a transport category aircraft that is an aeroplane, at up to 400 NM, or the distance that can be covered in 120 minutes of flight at the cruising speed filed in the flight plan or flight itinerary, whichever distance is the lesser, from a suitable emergency site without the liferafts required being carried on board.

Life rafts where required shall be stowed so that they are easily accessible for use in the event of a ditching, installed in conspicuously marked locations near an exit and equipped with an attached survival kit, sufficient for the survival on water of each person on board the aircraft, given the geographical area, the season of the year and anticipated seasonal climatic variations, that provides a means for providing shelter, providing or purifying water, and visually signalling distress.

[Reference: CARs 602.58, 602.59, 602.60, 602.61, 602.62, 602.63]

1.34 Aircraft Seats, Belts & Harnesses

Generally, a seat and an individual safety belt is required for every person in an aircraft except infants (anyone less than two years of age). Infants must be held securely in the arms of an adult. Aircraft dropping parachutists need not have seats, or seat belts, but must have safety harnesses secured to the primary structure of the aircraft.

Hang gliders and ultra light aircraft do not require seats and seat belts, but must have a harness or other means to secure a person to the aircraft.

A harness or safety belt is to be fastened while the aircraft is taking off or landing and at any other time when so directed by a member of the crew or by a sign displayed in the aircraft. The PIC is to remain at his/her seat while the aircraft is in flight with his/her safety belt or harness fastened.

A PIC shall direct all persons on board the aircraft to fasten safety belts during movement of the aircraft on the surface, during take off and landing, and at any time during the flight that the PIC considers it necessary.

Shoulder harnesses are mandatory for flight crew on aeroplanes manufactured after July 18, 1978.

[Reference: CARs 605.16, 605.22 - 605.25]

1.35 Oxygen Equipment Requirements

Oxygen Supply: Unpressurized Aircraft

No person shall operate an unpressurized aircraft unless it is equipped with sufficient oxygen dispensing units and oxygen supply to comply with these requirements:

a) an oxygen supply must be available for all crew members and 10% of passengers (not less than one passenger), for the entire period of flight exceeding 30 minutes at cabin pressure altitudes above 10,000 feet ASL but not exceeding 13,000 feet ASL;

b) an oxygen supply must be available for all persons on board the aircraft for the entire period of flight at cabin pressure altitudes above 13,000 feet ASL.

For aircraft operated in an air transport service, the period of flight oxygen availability must be at least one hour.

Oxygen Supply: Pressurized Aircraft

No person shall operate a pressurized aircraft unless it is equipped with sufficient oxygen dispensing units and oxygen supply to provide, in the event of cabin pressurization failure at the most critical point during the flight, sufficient oxygen to continue the flight to an aerodrome suitable for landing while complying with these requirements:

a) an oxygen supply must be available for all FLIGHT CREW MEMBERS and 10% of passengers (not less than one passenger), for the entire period of flight exceeding 30 minutes at cabin pressure altitudes above 10,000 feet ASL but not exceeding 13,000 feet ASL, and for the entire period of flight a cabin pressure altitudes above 13,000 feet ASL. For aircraft operated in an air transport service in these conditions, a period of flight of not less than 30 minutes, and for flight crew members , two hours for aircraft the type certificate of which authorizes flight at altitudes exceeding FL 250;

b) an oxygen supply must be available for all PASSENGERS for the entire period of flight at cabin pressure altitudes exceeding 13,000 feet ASL. For aircraft operated in an air transport service under these conditions, a period of flight of not less than 10 minutes.

Use of Oxygen

Where an aircraft is operated at cabin pressure altitudes above 10,000 feet ASL but not exceeding 13,000 feet ASL, each crew member shall wear an oxygen mask and use supplemental oxygen for any part of the flight at those altitudes that is more than 30 minutes in duration.

Where an aircraft is operated at cabin pressure altitudes above 13,000 feet ASL, each person on board the aircraft shall wear an oxygen mask and use supplemental oxygen for the duration of the flight at those altitudes.

A QUICK DONNING MASK means an oxygen mask that can be secured by a person using one hand on the person's face within 5 seconds, and that provides an immediate supply of oxygen. The pilot at the flight controls of an aircraft shall use an oxygen mask if the aircraft is not equipped with quick donning masks and is operated at or above FL 250 or the aircraft is equipped with quick donning masks and is operated above FL 410.

[Reference: CARs 605.31, 605.32]

1.36 Emergency Radio Frequency and Visual Interception Signals

The emergency frequency is 121.5 MHz (243.0 MHz military). All aviation 2-way radios must be capable of communicating on 121.5. When operating within sparsely settled areas, 121.5 must be continuously monitored or when operating a Canadian aircraft over water more than 50 NM from shore (unless carrying out

Signals Initiated by Intercepting Aircraft and Responses by Intercepted Aircraft

Series	Intercepting Aircraft Signal	Meaning	Intercepted Aircraft Response	Meaning
1.	DAY - Rocking wings from a position in front and, normally, to the left of intercepted aircraft and, after acknowledgement, a slow level turn, normally to the left, on to the desired heading. NIGHT - Same and, in addition flashing navigational lights at irregular intervals. DAY OR NIGHT - Meteorological conditions or terrain may require the intercepting aircraft to take up a position in front and to the right of the intercepted aircraft and to make the subsequent turn to the right. If the intercepted aircraft is not able to keep pace with the intercepting aircraft, the latter is expected to fly a series of race-track patterns and to rock its wings each time it passes the intercepted aircraft.	You have been intercepted. Follow me.	AEROPLANES: DAY - Rocking wings and following. NIGHT - Same and, in addition flashing navigational lights at irregular intervals. HELICOPTERS: DAY or NIGHT - Rocking aircraft, flashing navigational lights at irregular intervals and following.	Understood, will comply.
2.	DAY or NIGHT - An abrupt breakaway manoeuvre from the intercepted aircraft consisting of a climbing turn of 90 degrees or more without crossing the line of flight of the intercepted aircraft.	You may proceed.	AEROPLANES: DAY or NIGHT - Rocking wings. HELICOPTERS: DAY or NIGHT - Rocking aircraft.	Understood; will comply.
3.	DAY - Circling aerodrome, lowering landing gear and overflying runway in direction of landing or, if the intercepted aircraft is a helicopter, overflying helicopter landing area. NIGHT - Same and, in addition, showing steady landing lights.	Land at this aerodrome.	AEROPLANES: DAY - Lowering landing gear, following the intercepting aircraft and, if after over-flying the runway landing is considered safe, proceeding to land. NIGHT - Same and, in addition showing steady landing lights (if carried). HELICOPTERS: DAY or NIGHT - Following the intercepting aircraft and proceeding to land, showing a steady landing light (if carried).	Understood, will comply.

Signals Initiated by Intercepted Aircraft and Responses by Intercepting Aircraft

Series	Intercepted Aircraft Signal	Meaning	Intercepting Aircraft Response	Meaning
4.	AEROPLANES: DAY - Raising landing gear while passing over landing runway at a height exceeding 300 m (1,000 feet) but not exceeding 600 m (2,000 feet) above AAE, and continuing to circle the aerodrome. NIGHT - Flashing landing lights while passing over landing runway at a height exceeding 300 m (1,000 feet) but not exceeding 600 m (2,000 feet) AAE, and continuing to circle the aerodrome. If unable to flash landing lights, flash any other lights available.	Aerodrome you have designated is inadequate.	DAY or NIGHT - If it is desired that the intercepted aircraft follow the intercepting aircraft to an alternate aerodrome, the intercepting aircraft raises its landing gear and uses the Series 1 signals prescribed for intercepting aircraft. If it is decided to release the intercepted aircraft, the intercepting aircraft uses the Series 2 signals prescribed for intercepting aircraft.	Understood, follow me. Understood, you may proceed.
5.	AEROPLANES: DAY or NIGHT - Regular switching on and off of all available lights but in such a manner as to be distinct from flashing lights.	Cannot comply.	DAY or NIGHT - An abrupt breakaway manoeuvre from the intercepted aircraft consisting of a climbing turn of 90° or more without crossing the line of flight of the intercepted aircraft.	Understood.
6.	AEROPLANES: DAY or NIGHT - Irregular flashing of all available lights. HELICOPTERS: DAY or NIGHT - Irregular flashing of all available lights.	In distress.	DAY or NIGHT - An abrupt breakaway manoeuvre from the intercepted aircraft consisting of a climbing turn of 90° or more without crossing the line of flight of the intercepted aircraft.	Understood.

Fig. 1-15: Intercept Signals.

communications on other frequencies or cockpit duties/equipment limitations do not permit monitoring two VHF frequencies). When an aircraft is intercepted by a military aircraft, the pilot should advise ATC of the interception, attempt to contact the intercepting aircraft on 121.5 MHz, squawk 7700 and follow visual instructions from the intercepting aircraft.

[Reference: CARs 602.143, 602.144; AIM COM 5.10, SAR 4.6]

1.37 Canadian Minimum Navigation Performance Specifications (CMNPS) Airspace

General

CMNPS airspace is that controlled airspace within the CDA between FL 330 and FL 410 as indicated. This airspace is contained for the most part in the ACA and NCA with a small portion in the SCA.

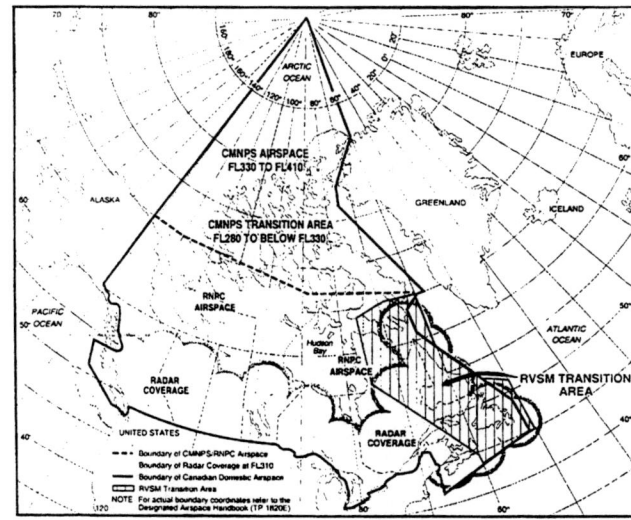

Fig. 1-16: CMNPS Airspace.

To conduct RNAV operations in CMNPS airspace, in which reduced ATC separation criteria can be applied, aircraft must be certified as being capable of navigating within specified tolerances.

CMNPS Transition Airspace

In order to permit both CMNPS certified and non certified aircraft to operate above FL 280, a transition area exists from FL 270 to below FL 330 underlying the lateral limits of CMNPS airspace.

CMNPS Aircraft Certification

Only aircraft certified by the state of aircraft registry or the state of the operator as meeting the MNPS of either the North Atlantic or Canada are permitted to operate within the CMNPS airspace unless the ATC unit concerned indicates that the non-certified aircraft may be accommodated without penalizing certified aircraft.

Certification is dependent upon crew training and navigation performance capability such that:

a) the standard deviation of lateral track errors is less than 6.3 NM;
b) the proportion of total flight time spent by aircraft 30 NM or more off the cleared track is less than 1 hour in 2,000 flight hours;
c) the proportion of total flight time spent by aircraft between 50 and 70 NM off the cleared track is less than 1 hour in 8,000 flight hours.

Such navigation performance capability shall be verified by the state of registry or the state of operator, as appropriate. Aircraft certified to operate within designated airways and company approved routes, that are completely in signal coverage of ground based navaids, satisfy CMNPS requirements when operating within the protected airspace for airways and company approved routes.

The following minimum navigation systems may be deemed to satisfy the CMNPS:

a) aircraft transitioning CDA to/from another continent must be equipped with two long range navigation systems (ie OMEGA/VLF, INS, GNSS) or one navigation system using the inputs from one or more sensor systems plus one short range navigation system (ADF, VOR/DME);
b) aircraft operating within North America on routes that lie within reception of ground based navigation aids must be equipped with a single long range navigation system plus a short range navigation system (ADF, VOR/DME);
c) aircraft operating on high level airways or company approved routes must be equipped with dual short range navigation systems (ADF, VOR/DME).

Flight Planning

The navigation equipment suffix Y shall be used on flight plans to indicate that the aircraft is CMNPS certified. The equipment suffix X (NAT MNPS certification) is acceptable in lieu of CMNPS certification.

Partial or Complete Loss of Navigation Capability while Operating within CMNPS Airspace

CMNPS operations and the associated ATC separation minima depend upon the accuracy of the navigation systems. ATC is to be advised immediately at any time that a pilot is uncertain of the aircraft position, or of an on board navigation systems failure or degradation.

Upon entry into CMNPS airspace, or as soon as practical thereafter, flight crews are to cross check the accuracy of their long range navigation system with information obtained from station referenced aids. Navigation systems shall be updated if the cross check indicates such action is considered necessary.

Air to Ground Communications

Aircraft operating in CMNPS airspace are to communicate with ATS facilities as published on the Enroute High Altitude charts. VHF coverage is limited and is supplemented by HF facilities. During periods of HF unreliability, aircraft are to make position reports immediately upon coming within range (200 NM) of any published VHF facility. When possible, ATS facilities are to be selected in the following priority:

a) directly to an ACC (peripheral stations);
b) any FSS or RCO on the published frequency;
c) Baffin (Cambridge Bay), Montreal (Iqulauit) or Churchill radio on HF;
d) Alert (CYLT) may be contacted on VHF 126.6 MHz or 5680 KHz, 6705 KHz or any DND North Warning Radar Station on 126.7 MHz. These stations have limited means of forwarding information and should only be used as a last resort.

[Ref: AIM RAC 12.5]

1.38 Aeronautical Information Publications

The Transport Canada Aeronautical Information Manual (AIM) is a pre-flight reference document providing summarized rules and procedures of a lasting nature in a single primary document. It is intended for operation of aircraft in CDA. It also contains some important CARs in original format that are of interest to pilots.

The term SHOULD, when used in the AIM (or any regulation) means that pilots should conform to the applicable procedure, while the word SHALL means that the applicable procedure is mandatory and is supported by regulatory authority.

Remember that the final authority for any rule or procedure is not the AIM but the original regulation or rule in its legislative version in the *Aeronautics Act*, CARs or other enactment. If any doubt exists or for the final legal authority on a rule or regulation, refer to these regulatory sources and not the AIM.
The AIP Canada (ICAO) contains aeronautical information of a lasting nature and is intended to satisfy international ICAO requirements. Various government publications and charts are considered to form part of the AIP Canada (ICAO) such as VFR and IFR charts, Canada Flight Supplement, DAH etc.

An AERONAUTICAL INFORMATION CIRCULAR (AIC) provides advance notification of major changes to legislation, regulations, procedures or purely administrative matters where the text is not part of the AIM. An AVIATION NOTICE is used to disseminate information which is mainly of interest to specific TC regions or when the information is of a one- time nature and the notice will have served its purpose on first reading.

AIRAC CANADA is a notice which is issued weekly by NAV CANADA to provide advance notification to chart makers and producers of aeronautical information concerning changes within the CDA. This notice ensures that all users of the CDA have the same information on the same date.

[Ref: AIM MAP 6.0]

1.39 Flight Data Recorders and Cockpit Voice Recorders

Flight Data Recorder (FDR)

An approved FDR is required in the following cases:

a)   a multi-engined turbine powered pressurized aeroplane having a MCTOW of more than 5,700 kgs (12,566 lbs) and that has a passenger seating configuration, excluding pilot seats, of 10 or more, manufactured after October 11, 1991 and is operated by a Commuter operator;
b)   an multi-engined turbine powered aeroplane operated by an Airline operator;
c)   after July 31, 1997 or for an aircraft operated commercially, after February 28, 1997, any other multi-engined turbine powered aircraft that has a passenger seating configuration excluding any pilots seats, of 10 or more and that was manufactured after October 11, 1991.

Cockpit Voice Recorder (CVR)

An approved CVR is required in the following cases:

a)   a multi-engined, turbine powered pressurized aeroplane that has a MCTOW of more than 5,700 kgs (12,500 lbs) and that has a passenger seating configuration, excluding any pilots seats, of 6 or more, is required to be operated by two pilots pursuant to the type certificate or air operator requirements and is operated by a Commuter operator;
b)   any multi engined turbine powered aeroplane operated by an Airline operator;
c)   after July 31, 1997 or for an aircraft operated commercially, after February 28, 1997, any other multi engined turbine powered aircraft that has a passenger seating configuration, excluding any pilot seats, of six or more, and for which two pilots are required by the type certificate or air operator requirements.

Use of FDR/CVR

Unless otherwise permitted, no person shall operate an aircraft for which a FDR or CVR is required unless

a)   in the case where a FDR is required, the FDR is operated continuously from the start of the take-off until the completion of the landing;
b)   in the case where a CVR is required, the CVR is operated continuously from the time at which electrical power is first provided to the recorder before the flight to the time at which electrical power is removed from the recorder after the flight.

No person shall erase any communications pertaining to the flight being undertaken where such communications have been recorded by a CVR.

Where a MEL has been approved for an operator, the operator may operate the aircraft without a serviceable FDR or CVR if the aircraft is operated in accordance with the MEL.

Where a MEL has not been approved for the operator, the operator may operate the aircraft without a serviceable FDR for a maximum period of 90 days after the date of failure of the FDR if, where a CVR is also required, the CVR is serviceable and the aircraft technical records show the date of the failure of the FDR.

Where a MEL has not been approved for the operator, the operator may operate the aircraft without a serviceable CVR for a maximum period of 90 days after the date of failure of the CVR, where a FDR is required, the FDR is serviceable and the aircraft technical records show the date of the failure of the CVR.

[Reference: CARs 605.33, 605.34]

1.40 Altitude Alerting Systems and Ground Proximity Warning System (GPWS)

Altitude Alerting System or Device

No person shall conduct a take-off in a turbo-jet powered aeroplane unless it is equipped with an approved altitude alerting system or device.

An aeroplane above may be operated without a serviceable altitude alerting system or device:

a)   where a MEL has been approved for the operator and the aeroplane is operated in accordance with the MEL;
b)   where a MEL has not been approved for the operator of the aeroplane, the aeroplane may be operated from the place where the operator or PIC takes possession of the aeroplane to a place where the aeroplane can be equipped with such a system or device, for the sole purpose of conducting a flight test, PPC or flight crew training, or where the system or device becomes unserviceable after take-off, until it reaches an aerodrome at which the system or device can be repaired or replaced.

Ground Proximity Warning System (GPWS)

GPWS is an aircraft based, radar warning system which activates when prescribed minimum ground clearance is not met. No Commuter or Airline operator shall conduct a take-off in a turbo-jet powered aeroplane that has a MCTOW of more than 15,000 kgs (33,069 lbs) and for which a type certificate has been issued authorizing the transport of 10 or more passengers, unless the aeroplane is equipped with a GPWS.

An aeroplane above may be operated without a serviceable GPWS if an MEL has been approved for the operator and the aeroplane is operated in accordance with the MEL.

Where in the interests of safety, it is necessary during a flight to deactivate any mode of a GPWS, the PIC may deactivate that mode if the deactivation is performed in accordance with the AFM, AOM, flight manual supplement or MEL.

[Reference: CARs 605.36, 605.37]

1.41 De-Icing and Anti-Icing Equipment

No person shall conduct a take-off or continue a flight in an aircraft where icing conditions are reported to exist or are forecast to be encountered along the route of flight unless the PIC determines that the aircraft is adequately equipped to operate in icing conditions in accordance with the standards of airworthiness under which the type certificate for that aircraft was issued, or current weather reports or pilot reports indicate that icing conditions no longer exist.

[Reference: CARs 605.30]

1.42 Standby Attitude Indicator

No person shall conduct a take-off in a turbo-jet powered aeroplane as a commercial air operator unless that aeroplane is equipped with a standby attitude indicator. After July 31, 1997, no person shall conduct a take-off in a transport category aircraft unless the aircraft is equipped with a standby attitude indicator.

The standby attitude indicator shall be powered from a source independent of the electrical generating system, and continue to operate for a minimum of 30 minutes (without selection) after total failure of the electrical generating system.

[Reference: CARs 605.41, 625.41]

1.43 Emergency Locator Transmitter (ELT)

General

No person shall operate an aircraft unless it is equipped with one or more ELT's as required by the aircraft type and area of operation. An ELT required to be carried shall be armed as specified in the AFM, AOM, POH or equivalent document provided by the manufacturer.

ELT's help search crews locate downed aircraft. They send out a signal on 121.5 MHz and/or 243.0 MHz. The signal can be detected by SARSAT and COSPAS satellites, or by any aircraft listening over the emergency frequencies.

ELT Categories

There are 5 types of ELT:

a) TYPE A or AD - Automatic ejectable or Automatic Deployable. Automatically ejects from the aircraft and activates during crash force.
b) TYPE F or AF - Fixed or Automatic Fixed. Activates based on inertia switch when aircraft subject to crash force. Most GA aircraft use this type.
c) TYPE AP - Automatic Portable. This type is similar to Type F or AF except that the antenna is integral to the unit for portable operation.
d) TYPE P - Personnel. This type has no fixed mounting and does not transmit automatically.
e) TYPE W or S - Water Activated or Survival. Transmits automatically when immersed in water. It is waterproof and floats and should be tethered to life rafts or survivors.

Schedule of Requirements

Review Fig. 1-17 for a schedule that outlines the requirements to carry an ELT.

Exceptions

An aircraft may be operated without an ELT on board where the aircraft is

a) a glider, balloon, airship, ultra-light aeroplane or gyroplane;
b) a multi-engined turbo-jet aeroplane of more than 5,700 kgs (12,500 lbs) MCTOW that is being operated in IFR within controlled airspace over land and south of 66° 30' N latitude;
c) registered under the laws of another state, equipped with a serviceable radio transmitter that is approved by that state for search and rescue purposes, having a distinctive siren like sound capable of communication on frequencies 121.5/243.0 MHz;
d) operated by the holder of a flight training unit operating certificate, engaged in flight training and operated within 25 NM of the aerodrome of departure;
e) engaged in a flight test;
f) a new aircraft engaged in flight operations incidental to manufacture, preparation or delivery of the aircraft.

As well, where an aircraft is required to carry an ELT, the aircraft may be operated without a serviceable ELT if

a) where a MEL has been approved for the aircraft operator, the aircraft is operated in accordance with the MEL;
b) where an MEL has not been approved for the aircraft operator, the operator repairs or removes the ELT at the first aerodrome at which repairs or removal can be accomplished, on removal of the ELT from the aircraft, sends the ELT to a maintenance facility, displays on a readily visible placard within the aircraft cockpit, for the period of ELT removal from the aircraft, a notice stating that the ELT has been removed and setting out the date of removal and re-equips the aircraft with a serviceable ELT within 90 days after the date of the removal.

Column I	Column II	Column III
Aircraft	**Area of Operation**	**Minimum Equipment**
1. All aircraft except those exempted.	Over land	One ELT of type AD, AF, AP, A, or F.
2. Large multi-engine turbojet aeroplanes engaged in an air transport service carrying passengers.	Over water at a distance from land that requires the carriage of life raft pursuant to CAR 602.63.	Two ELTs of type W or S or one of each.
3. All aircraft that require an ELT other than those set out in item 2.	Over water at a distance from land that requires the carriage of life raft pursuant to CAR 602.63.	One ELT of type W or S.

Fig. 1-17: ELT Schedule of requirements.

ELT Activation

No person shall activate an ELT except in an emergency. A person may activate an ELT during the first 5 minutes of any hour for a duration of not more than 5 seconds for the purpose of testing it. Where an ELT has been inadvertently activated during flight, the PIC of the aircraft shall ensure that the nearest ATC unit, FSS or CARS is so informed as soon as possible and the ELT is switched off.

[Reference: CARs 605.38, 605.39, 605.40]

1.44 ATC Special Procedures

Adherence to Mach Number

Within CDA, aircraft shall adhere to the Mach number assigned by ATC unless approval is obtained from ATC to make a change or until the pilot receives the initial descent clearance approaching destination. If it is essential to make an immediate temporary change in the Mach number (such as due to turbulence), ATC shall be notified as soon as possible that such a change has been made.

If it is not possible, because of aircraft performance, to maintain the last assigned Mach number during en route climbs and descents, pilots shall advise ATC at the time of the climb/descent request.

Parallel Offset Procedures

ATC may request that an aircraft fly a parallel offset from an assigned route. This maneuver and subsequent navigation is the responsibility of the pilot. When requested to offset or regain the assigned route, the pilot should change heading by 30° to 45° and report when the offset or assigned route is attained.

In a radar environment, ATC will provide radar monitoring and the required separation.

In a non-radar environment, ATC will apply parallel offsets to RNPS certified aircraft operating within high level RNPC airspace (see Fig. 1-17 for boundaries of RNPC airspace) in order to accomplish an altitude change with respect to same direction aircraft.

The following phraseology is normally used for parallel offset procedures:

"(Flight identification) PROCEED OFFSET (number) MILES (right/left) OF CENTRELINE (track/route) AT (point/now) UNTIL (point/time)".

Structured Airspace

During specific periods, certain portions of domestic high level airspace may be structured for one-way traffic in which flight levels inappropriate to the direction of flight may be assigned. Aircraft operating in a direction contrary to the traffic flow will be assigned those flight levels appropriate to the direction of flight except in specific instances, such as turbulence. When the airspace is not structured for one-way traffic, flight levels appropriate for direction of flight will be used. ATC will transition aircraft to the appropriate altitude for the direction of flight before aircraft exit the defined areas or before termination of the indicated times.

EXAMPLE: Within the Montreal FIR/CTA, FL 350 is structured and available as an eastbound altitude between 0100 UTC and 0900 UTC daily (DT 0000 UTC and 0800 UTC), within an area bounded by a line commencing at 53°28'N 80°00'W to 56°00'N 75°00'W to 63°50'N 75°00'W to 62°45'N 80°00'W to the point of beginning.

Required Navigation Performance Capability Airspace (RNPC)

General
RNPC airspace is that controlled airspace within the CDA as depicted at Fig. 1-17. RNPC airspace is established to accommodate area navigation (RNAV) operations and is contained within the SDA and NCA.

RNPC Aircraft Certification
Only aircraft certified by the State of Registry or the State of the Operator as meeting the required navigation performance capability are permitted to conduct RNAV operations.

Certification is dependent upon crew training and navigation equipment that permits position determination within +/- 4 NM. Such navigation performance capability shall be verified by the State of Registry or the State of Operator, as appropriate. Aircraft that are certified for operations in CMNPS and NAT MNPS airspace satisfy all requirements for RNPC certification.

The minimum navigation equipment to satisfy the RNPC certification is one long-range area navigation, plus a short-range navigation system (VOR/DME or ADF).

Flight Planning
The navigation equipment suffix "R" shall be used on flight plans to indicate that the aircraft is RNPC certified. The use of the equipment suffix "Y" (CMNPS certification) or "X" (NAT MNPS certification) is acceptable in lieu of RNPC certification.

RNAV/DME Distance
ATC requests for distance information from RNAV certified aircraft shall be based on RNAV distances. DME based on TACAN or VOR/DME shall be used only if ATC indicates such information in the request.

RNAV Equipment Failure Procedures

RNAV operations and the associated ATC separation minima depend upon the accuracy of the RNAV systems. ATC is to be advised immediately at any time that a pilot is uncertain of the aircraft position or of an on-board navigation system failure or degradation.

Canadian Minimum Navigation Performance Specifications Airspace (CMNPS)

Review s. 1.37 in this manual for complete information on CMNPS Airspace and procedures.

Canadian Domestic Routes

General

Within North American Airspace, various route and track systems exist in order to provide effective management of airspace and traffic. Under specified conditions, random routes may be flight planned or requested.

North American Route Program (NARP)

The NARP is a joint FAA and Nav Canada program which integrates the FAA National Route Program (NRP) and the Canadian Domestic Random Route Program. The objective of the NARP is to harmonize and adopt common procedures, to the extent possible, applicable to random route flight operations at and above FL 290 within the conterminous U.S. and Canada. Flights may participate in the NARP under specific guidelines and filing requirements, provided the flight originates and terminates within conterminous U.S. and Canada or for North Atlantic international flights operating within the North American Route (NAR) system. NARP procedures and specific Nav Canada and FAA requirements are contained in the CFS.

Preferred IFR Routes (including RNAV)

Preferred IFR routes provide guidance in planning routes, minimize route changes and allow for an efficient and orderly management of traffic. The flight planning of preferred routes is not mandatory, but will result in fewer traffic delays. Preferred routes procedures and description are published in the CFS.

Fixed RNAV Routes

In order to accommodate RNAV operations, all fixed RNAV routes are published in the CFS. Fixed RNAV routes are referred to as "T" routes.

Northern Control Area Random Routes

Within the NCA, flights operating on random routes shall flight plan and make position reports as specified in s. 1.32 in this manual.

Canadian Track Structures

Northern Control Area Track Structure

The NCA Track System allows for reduced lateral separation, and facilitates the application of Mach number techniques. The tracks are contained within the SCA and NCA and extend upward from FL 280. The system is primarily used by international flights operating North America/Europe (NAT) and North America/Alaska-Orient (PAC). The tracks are depicted on Enroute High Altitude Charts.

The mandatory use of NCA tracks and the availability to random route is different for NAT and PAC traffic. For flight planning an NCA or Lateral track, the flight plan routing is indicated by using the abbreviation "NCA" or "LAT" as appropriate, followed by the letter or number of the track.

EXAMPLE: Track Bravo - NCA B. Lateral 3 - LAT 3.

For flights operating within the NCA Track System, position reports are to be indicated by the compulsory reporting point designator, or the code name of the track and the reporting line meridian (the parallel of latitude need not be included in the position report).

EXAMPLE: Bravo Track at 80°W - "BRAVO 80 AT (time)".

Aircraft operating on NCA/NAT tracks and NCA/PAC are also subject to Eastbound and Westbound restrictions, depending on the time of day and date of year - refer to the AIP/CFS for specific restrictions.

Southern Control Area Track System

The SCA Track System is primarily used by international traffic operating between the mid-west and western U.S. and Europe via NAT. The tracks are within the SCA and extend upwards from FL 180. The tracks are depicted on Enroute High Altitude Charts.

The SCA tracks are completely optional for flight planning. Entry or exit from the SCA tracks may be at designated reporting points or at the reporting points coincident with the longitudes 80°W and 90°W. Lateral transitions between tracks may be flight planned or requested between significant reporting points. For flight planning an SCA track, the route is indicated by using the abbreviation "SCA" followed by the letter of the track.

EXAMPLE: SCA Hotel Track - SCAH.

Flights operating within the SCA Track System shall report over reporting lines coincident with the longitudes 80°W and 90°W, designated reporting points or as requested by ATS.

ACA Track Structure

The ACA Track Structure (ACATS) is a system of published tracks in the ACA serving international flights operating between Europe and Alaska/Orient. The routes are depicted on Enroute High Altitude charts.

The ACATS is established to enhance the utilization of the airspace and thus facilitate more efficient use of optimum flight levels and ATC separation minimums. The use of named waypoints along the ACATS route will assist in the application of data link technologies through ADS reporting and controller pilot data link communications.

The ACA tracks are identified by letters O, P, Q, R and S. These tracks are laterally separated throughout the Edmonton FIR and complement the Polar Track Structure (PTS) in the Reykjavik CTA and the fixed route system in the Anchorage FIR.

The use of these tracks is not mandatory, and have been published to facilitate flight planning. If the flight is planned along the complete length of one of the ACA tracks or a portion thereof, the track shall be defined in Item 15 of the flight plan using the abbreviation "ACA" followed by the track letter.

EXAMPLE: ADREW DCT TAVRI **ACAQ** LT

Flights may leave or join the ACATS routes in the Edmonton FIR at the identified waypoints. Flights operating on ACATS routes shall report at designated compulsory reporting points or as requested by ATS.

Flights that will enter the ADIZ while in the ACA may forward the required estimated time and place of ADIZ entry as part of their 115°W longitude position report.

Traffic Alert and Collision Avoidance Systems and Airborne Collision Avoidance Systems

General

TCAS is the acronym for the Traffic Alert and Collision Avoidance System developed in the U.S. by the FAA, while the Airborne Collision Avoidance System (ACAS) is the name applied by ICAO for similar systems.

TCAS/ACAS is designed to operate independently of ATC and, depending on the type of TCAS/ACAS, will display proximate traffic, providing Traffic Alerts (TAs) and Resolution Advisories (RAs). TAs provide information on proximate traffic and are intended to assist the flight crew in visual acquisition of conflicting traffic and to alert pilots to the possibility of an RA. RAs are divided into two categories: preventative advisories, which instruct the pilot to maintain or avoid certain vertical speeds; and corrective advisories, which instruct the pilot to deviate from the current flight path (i.e. "CLIMB" when the aircraft is in level flight). In an encounter between two TCAS/ACAS equipped aircraft, their computers will communicate using the Mode S transponder data link which has the capability to provide complementary RAs (i.e. one climbing and one descending). Aircraft without transponders are invisible to TCAS/ACAS equipped aircraft; thus, TAs or RAs are not provided.

These are three types of TCAS/ACAS:

a) TCAS/ACAS I is a less sophisticated system which will provide a warning of proximate traffic (TA) without guidance to avoid potential collisions;
b) TCAS/ACAS II consists of a computer, pilot displays, a Mode S transponder, modified instantaneous vertical speed indicators, controls, wiring and antennas which provide both TAs and vertical plane RAs;
c) TCAS/ACAS III is a more advanced system (still under development) which will provide TAs and both horizontal and vertical plane RAs (i.e. turns) as well as climbs and descents.

Use of TCAS/ACAS

The U.S. is the only state in the world which mandates the use of TCAS/ACAS. The following TCAS/ACAS requirements must be complied with in order to operate in U.S. airspace:

a) TCAS/ACAS I. All aircraft with 10 to 30 passenger seats were required to be equipped with TCAS/ACAS I by December 31, 1995.
b) TCAS/ACAS II. All aircraft with more than 30 passenger seats were required to be equipped with TCAS/ACAS II by December 30, 1993. Since its development, TCAS/ACAS has undergone a number of modifications, particularly in the system's computer logic.

Transport Canada TCAS/ACAS Policy

While TC encourages the installation of TCAS/ACAS, the equipment will not be made mandatory in Canadian airspace for the foreseeable future.

Operational Approval

For Canadian operators, TCAS/ACAS II operational approval is accomplished through TC approval of pertinent training programs, checklists, operations manuals or training manuals, maintenance programs, MELs or other documents applicable to the operator.

In order for Canadian operators to meet the regulatory requirements, they must address the following training, checking, and currency issues for TCAS/ACAS flight crew qualification:

a) initial ground training;
b) initial flight training;
c) initial checking;
d) recurrent training;
e) recurrent checking;
f) currency.

Pilot Immunity for Enforcement Action for Deviating from Clearances

In accordance with CARs, pilots are permitted to deviate from a clearance in order to follow a resolution advisory in Canadian airspace. After responding to the resolution advisory, the pilot shall, as soon as possible, advise ATC of the deviation and return to the altitude in the previous clearance, or obtain another one. The policy outlined below is the same as applies in U.S. airspace.

The TC policy for Canadian airspace with respect to enforcement investigation into a deviation from an assigned altitude in response to a TCAS/ACAS RA, and the use of TCAS/ACAS recorded data is provided below.

The use of TCAS/ACAS II may result in a flight crew deviating from an assigned altitude for a short period of time. During the investigation of the incident, all factors will be considered, including factors that are TCAS/ACAS related, before a final determination is made. Specifically, enforcement action will not be taken against flight crew who deviate from a clearance issued by ATC when that deviation is in response to a TCAS/ACAS generated RA and the response is in accordance with the operator's approved flight procedures. Likewise, enforcement action will not be taken if the operator's procedures allow a crew not to follow a displayed RA because of other information that may be available to the pilot.

Mode S Transponder Approval and Unique Codes

Along with performing all the functions of Mode A and C transponders, Mode S transponders also have a data link capability. Mode S transponders are an integral component of all TCAS/ACAS II installations.

There is no requirement to replace existing Mode A or C transponders with Mode S transponders until it becomes impossible to maintain presently installed Mode A and C transponders.

Airworthiness approval must be obtained by Canadian aircraft operators who install Mode S transponders. Each Canadian registered aircraft with a Mode S transponder must receive a unique code assignment, which must be loaded in the transponder, through their regional TC licencing office.

Pilot/Controller Actions
In order to use TCAS/ACAS in the most effective and safest manner, the following pilot and controller actions are necessary:

a) pilots shall not maneuver their aircraft in response to TAs only;

b) pilots shall notify the appropriate ATC unit, as soon as possible, of the deviation, including its direction, and when the deviation has ended;

c) in the event of an RA to alter the flight path, the alteration of the flight path should be limited to the minimum extent necessary to comply with the RA;

d) when a pilot reports an maneuver induced by an RA, the controller should not attempt to modify the aircraft flight path until the pilot reports returning to the terms of the existing ATC instruction or clearance, but should provide traffic information as appropriate;

e) pilots who deviate from an ATC instruction or clearance in response to an RA shall promptly return to the terms of that instruction or clearance when the conflict is resolved.

Pilot and Controller Interchange
ICAO is currently developing pilot/controller phraseologies. For phonetic clarity purposes, the term TCAS is used. Review Fig. 1-18 for a list of TCAS/ACAS phraseology.

Circumstances	Phraseologies	
After modifying vertical speed to comply with a TCAS/ACAS RA	Pilot:	(Call Sign) TCAS CLIMB (or DESCENT);
	Controller:	(acknowledgement);
After TCAS/ACAS "Clear of conflict" is annunciated in the cockpit	Pilot:	(Call Sign) RETURNING TO (assigned clearance);
	Controller:	(acknowledgement) (or alternate instructions);
After the response to a TCAS/ACAS RA is completed	Pilot:	(Call Sign) TCAS CLIMB (or DESCENT), RETURNING TO (assigned clearance);
	Controller:	(acknowledgement) (or alternate instructions);
After returning to clearance after responding to a TCAS/ACAS RA	Pilot:	(Call Sign) TCAS CLIMB (or DESCENT), COMPLETED (assigned clearance) RESUMED;
	Controller:	(acknowledgement) (or alternate instructions);
When unable to comply with a clearance because of a TCAS/ACAS RA	Pilot:	(Call Sign) UNABLE TO COMPLY, TCAS RA;
	Controller:	ROGER.

Fig. 1-18: Pilot - Controller TCAS Phraseologies.

[Ref: AIM RAC 12.1 - 12.17]

1.45 North Atlantic Operations

General Aviation Aircraft

No PIC of a single-engined aircraft, or of a multi-engined aircraft that would be unable to maintain flight in the event of the failure of any engine, shall commence a flight that will leave CDA and enter airspace over the high seas unless

a) the PIC holds a pilot licence with an instrument rating;

b) the aircraft is equipped as per CARs requirements

for power driven IFR aircraft;

c) the aircraft has an HF radio capable of transmitting and receiving on a minimum to two appropriate international air-ground general purpose frequencies;

d) the aircraft is equipped with hypothermia protection for each person on board;

e) the aircraft carries sufficient fuel to meet the normal CARs fuel reserve requirements plus, in addition, carries contingency fuel equal to at least 10% of the fuel required by CARs to complete the flight to the aerodrome of destination.

North American Routes (NAR)

The NAR System interfaces with the NAT oceanic and domestic airspace, and is used by air traffic transiting the North Atlantic. NARs extend to/from established oceanic coastal fixes to major airports throughout Canada and the U.S. NAR procedures and routes are published in the CFS.

NAT Organized Track System

For subsonic traffic, organized tracks are formulated and published in a NAT Track Message via the Aeronautical Fixed Telecommunication Network (AFTN) to all interested operators. The day-time structure is published by Shanwick ACC and the night-time structure by Gander ACC. The hours of validity of the two Organized Track Systems (OTS) are normally 1130 - 1800 UTS at 30°W for daytime OTS, and 0100 - 0800 UTC at 30°W for nightime OTS (the hours of validity are specified in the track message).

The most northerly track of a day OTS is designated as NAT Track Alpha, the adjacent track to the south, as NAT Track Bravo; etc. For the night OTS, the most southerly track is designated as Track Zulu; the adjacent track to the north, as Track Yankee; etc. Flight levels are allocated for use within the OTS and, in most cases, details of domestic entry and exit routings associated with individual tracks are provided in the NAT Track Message.

To permit an orderly change-over between successive OTS, a period of several hours is interposed between the termination of one system and the commencement of the next. During these periods, operators are expected to file random routes or use the co-ordinates of a track in the system about to come into effect.

Eastbound traffic crossing 30°W at 1030 UTC or later and westbound traffic crossing 30°W at 0000 UTC or later should plan to avoid the OTS.

Flight Rules

Over the high seas, the lower limit of all NAT OCA's is FL 55 with no upper limit. Throughout the NAT Region, airspace at and above FL 55 is Class A controlled airspace, and below FL 55 is Class G uncontrolled airspace.

Flights shall be conducted in accordance with IFR (even when not operating in IMC when operated at or above FL 60).

ATC clearances to climb or descend maintaining one's

own separation while operating in VMC shall not be issued.

Flight Planning Procedures

Routes

Flights conducted wholly or partially outside the OTS shall be planned along great circle tracks joining successive significant points.

For flights operating predominately in an east-west direction:

a) south of 70°N, the planned tracks shall be defined by significant points formed by the intersection of half or whole degrees of latitude at each 10° of longitude (60°W, 50°W, 40°W). For flights operating north of 70°N, significant points are defined by the parallels of latitude expressed in degrees and minutes with longitudes at 20° intervals;

b) the distance between significant points shall, as far as possible, not exceed one hour of flight time. Additional significant points should be established when required because of aircraft speed or the angle at which meridians are crossed. When the flight time between successive significant points is less than 30 minutes, one of the points may be omitted.

For flights operating predominately in a north-south direction, the planned tracks shall be defined by significant points formed by the intersection of whole degrees of longitude with parallels of latitude spaced at 5° (65°N, 60°N, 55°N).

For flights planning to operate within the OTS from the entry point into oceanic airspace to the exit point as detailed in the daily NAT track message, the track shall be defined in Item 15 of the flight plan by the abbreviation "NAT" followed by the code letter assigned to the track.

For eastbound NAT flights planning to operate on the OTS, the second and third route options should be indicated at the end of Item 18 of the flight plan.

EXAMPLE: RMKS/ .... O2.X370 O3. V350. (Option 2 is Track X at FL 370; Option 3 is Track V at FL 350).

Pilots of potential non-stop westbound flights may submit a flight plan to any suitable aeronautical radio facility or designated intersection east of 70°W. The route and altitude to any of the approved regular or alternate aerodromes may be specified. Prior to reaching the flight planned fix or clearance limit, the pilot, after assessing the onward flight conditions, will advise ATC of the intended destination and request an ATC clearance accordingly. If flight to the airport of destination is undesirable, the pilot will request an appropriate ATC clearance to the alternate airport. If an onward ATC clearance from the fix designated in the flight is not obtained by the time the fix is reached, the pilot must proceed towards the alternate airport in accordance with the flight plan and amendments thereto.

Airspeed

TAS or Mach number is to be entered in Item 15 of the flight plan.

Altitude

The planned cruising level(s) for the oceanic portion of the flight is to be included in Item 15 of the flight plan. Flight level allocations may be requested for Reduced Vertical Separation Minimums (RVSM) - see below. Requests for a suitable alternative flight level may be indicated in Item 18 of the flight plan.

Estimated Times

For flights operating on the OTS, the accumulated elapsed time only to the first oceanic FIR boundary are to be entered in Item 18 of the flight plan. For flights operating wholly or partly on the OTS, accumulated estimated times to significant points en route are to be entered in Item 18 of the flight plan.

Aircraft Approval Status and Registration

For flights certified as being in compliance with MNPS and intending to operate wholly or partly in MNPS airspace, the approval status (MNPS) shall be indicated in Item 10 by entering the letter "X". It is the pilots' responsibility to ensure that specific approval has been given for MNPS operations.

For flights certified as being in compliance with RVSM Minimum Aircraft System Performance Specification (MASPS) and intending to operate wholly or partly at RVSM designated altitudes, the approval status (RVSM) shall be indicated in Item 10 by entering the letter "W". It is the pilots' responsibility to ensure that specific approval has been given for RVSM operations.

For those aircraft being in compliance with both MNPS and RVSM, the letters "X" and "W" shall be entered in Item 10.

Filing

NAT operators are to forward all flight plans for eastbound NAT flights to those Canadian ACCs in which the flight will traverse their FIR/CTSs. These flight plans are to include the EET for each CTA boundary in Item 18 of the flight plan. The AFTN address for Canadian ACCs are: Gander - CZQXZQZX; Moncton - CZQMZQZX; Montreal - CZULZQZX; Toronto - CZYZZQZX; Winnipeg - CZWQZQZX; Edmonton - CZEGZQZX; Vancouver - CZVRZQZX.

Flight plans for flights departing from points within adjacent regions and entering the NAT region without intermediate stops should be submitted at least 3 hours prior to departure.

Where possible, operators are to file eastbound NAT flight plans at least 4 hours prior to the ETA at the coast-out fix specified in the flight plan.

Preferred Routes Messages

NAT operators are to send Preferred Routes Messages (PRM) for eastbound and westbound flights to the following: EGGXZOZX (Shanwick ACC); EGTTZZE (London Flow Management Unit); KCFCZDZX (FAA Air Traffic Control System Command Centre); KZNYZRZX (New York ARTCC); BIRDZQZX (Reykjavik ACC); LPPOZOZX (Santa Maria ACC); CZHQZDZX (Canadian Air Traffic Management Unit); CZQXZQZX (Gander ACC); CZQMZQZX (Moncton ACC); CZULZQZX (Montreal ACC).

The following format is to be used for westbound PRMs ("EB" is substituted for "WB" on eastbound PRMs):

[PRIORITY] [DEST ADDRESS] [DEST ADDRESS] ----
[DATE TIME OF ORIGIN] [ORIGIN ADDRESS]
[MESSAGE TYPE]-[COMPANY]-[WB]-[YYMMDD AT30W]-

[(DEP/DEST) (FIRST UK POINT) (ANCHOR POINT) (OCA RPS)
(LANDFALL) (INLAND FIX) (NUMBER OF FLT 01-99)]

EXAMPLE (eastbound):
FF EGGXZOZX EGTTZDZE CZQXZQZX CZQMZQZX
CZULZQZX CYHQZDZX KCFCZDZX KZNYZRZX
BIRDZQZK LPPOZOZX
120936 EHAMKLMW
PRM-KLM-E-930213-
KJFK/EHAM TOPPS YAY 53/50 53/40 54/30 54/20
54/15 BABAN BLUFA 03
CYMX/EHAM YML FOXXE 57/50 58/40 58/30 57/20
56/10 MAC BLUFA 01

PRMs are to be sent no later than 1000 UTC for east-bound flights, and no later than 1900 UTC for west-bound flights.

Clearances

Oceanic Clearances
Pilots intending to operate in the Gander OCA should note the following:

a) Clearances for VFR climb or descent will not be granted.
b) The Mach number to be maintained will be speci-fied for turbojet aircraft.
c) ATC will specify the full route details for aircraft cleared on a route other than an organized track or flight plan route. The pilot is to read back the full details of the clearance, including the cleared track.
d) ATC will issue an abbreviated oceanic clearance to aircraft that will operate along a NAT organized track. The abbreviated clearance will include the track letter, the flight level and the Mach number to be maintained (for turbojet aircraft). The pilot is to read back the clearance, including the Track Message Identification (TMI) number. ATC will confirm the accuracy of the read-back and the TMI number.
e) Pilots who receive an oceanic clearance which specified "via flight plan route" are to read back the flight plan route from the oceanic entry point to the exit point.
f) If the aircraft is designated to report meteorological information, the pilot will be advised by the inclu-sion of the phrase "SEND MET REPORTS" in the clearance.

Domestic Clearances - NAT Westbound Traffic
Pilots proceeding westbound across the North Atlantic and entering CDA within the Gander, Moncton and Montreal FIRs should comply with the following proce-dures:

a) Flights that have been cleared by ATC via the flight planned route prior to reaching CDA will not be issued en route clearances upon entering Domestic airspace and are to follow the flight planned route as cleared. Domestic en route clearances will be issued for flights that have been rerouted and exit oceanic airspace at other than the flight planned exit fix, at a pilot's request for another routing, or if a flight plan has not been received by the ACC.
b) Flights that have been rerouted from the flight planned route and enter CDA within 120 NM of the flight planned oceanic exit point can anticipate a clearance to regain the flight planned route by the Inland Navigation Fix (INF) unless the pilot re-quests a different routing. For flights beyond 120 NM form the flight planned oceanic exit point, a clearance will be issued following consultation with the pilot.
c) ATC will use the latest flight plan received before a flight departs. Subsequent changes to the flight plan route after departure, including any changes received by the pilot from operations/dispatch, must be requested directly by the pilot on initial contact with the appropriate domestic ACC. Direct requests from flight operations/dispatchers to ATC to reclear aircraft will only be considered under exceptional circumstances and are not an accept-able alternative to a pilot initiated request for a reclearance.
d) Domestic reclearances by ATC may contain either the route specified in full detail or a NAR.

If entering CDA within the Edmonton FIR, the onward domestic routing will have been established in co-ordination between the Reykjavik and Edmonton ACCs, and additional domestic clearance is not required. If there has been a change in route from the filed flight plan, clarification of the onward routing may be ob-tained form Edmonton ACC on request.

Westbound turbojet aircraft which have proceeded across the North Atlantic and have entered CDA shall maintain the last Mach number assigned by ATC, unless approval is obtained from ATC to make a change, or until the pilot receives an initial descent clearance approaching destination.

Oceanic Clearance Delivery
Unless otherwise advised by ATC, the following oceanic clearance delivery procedures are in effect daily be-tween 2320 and 0530 UTC for all eastbound oceanic flights operating above FL 280 that transit the Gander Domestic FIR/CTA:

a) Clearance delivery frequencies are published daily in the "Remarks" section on the eastbound NAT track message. Pilots are to contact Gander Clear-ance Delivery on the frequency for the track/route as per the NAT track message to which the aircraft is proceeding. Contact with clearance delivery should be made when within 200 NM of the speci-fied clearance delivery frequency location. In the even that contact cannot be established, pilots are to advise ATC on the assigned sector control frequency. The following frequencies and frequency locations will normally be used: Natash (YNA) at 50°11'N 61°47'W — 135.45 MHz, Allen's Island at 46°50'N 55°47'W - 128.45 MHz, and Churchill Falls (UM) at 53°35'N 64°14'W - 128.7 MHz.
b) For those operators that do not receive the NAT track message, pilots are to contact Gander Clear-ance Delivery on one of the frequencies listed above when within 200 NM of the frequency location. In the event that contact cannot be established, pilots are to advise ATC on the assigned sector control frequency.

Pilots are to maintain a continuous listening watch on the assigned control sector frequency while obtaining the oceanic clearance.

Data link oceanic clearances must be verified during the times indicated above with Gander Clearance Delivery. Aircraft with authorization may provide verification by stating the message sequence number. Outside the indicated hours, oceanic clearances are to be verified on the appropriate control frequency.

ATC will not normally advise pilots to contact Gander Clearance Delivery. There is no requirement for pilots to confirm receipt of an oceanic clearance from Gander Clearance Delivery with the assigned ATC sector frequency.

Due to frequency congestion on both the clearance delivery and sector control frequencies, pilots should refrain from unnecessary lengthy discussions with respect to oceanic clearances and procedures. Constructive comments and complaints should be processed postflight through the company operations.

Position Reports

Requirements
Unless otherwise requested by ATC, flights shall make position reports at the significant points listed in the flight plan.

The contents of a position report at geographical coordinates are to be expressed as follows:

a) for generally eastbound or westbound aircraft, latitude shall be expressed in degrees and minutes, longitude in degrees only;
b) for generally northbound or southbound aircraft, latitude shall be expressed in degrees only, longitude in degrees and minutes.

Position reports shall include the reported position, the next reporting point and estimated time and the succeeding reporting point as per the cleared route. If the estimated time over the next reporting point is found to be in error by three minutes or more, a revised estimated time shall be transmitted as soon as possible to the appropriate ATC unit.

Position information shall be based on the best obtainable navigation fix. The time of fixing aircraft position shall be arranged so as to provide the most accurate position information and estimates possible.

When making position reports, all times shall be expressed in UTC, giving both the hour and minutes.

Communications
All flights operating in the Gander OCA should report on the appropriate international air-to-ground frequencies.

In addition to maintaining a listening watch on the appropriate en route frequency, flights are to establish and maintain communication with Gander, Moncton, or Montreal as soon as possible on the appropriate frequency (see AIP RAC paragraph 11.9.2 for a detailed list of these frequencies).

Eastbound flights that traverse the Gander Domestic FIR are required to establish contact with Gander Clearance Delivery as specified under "Oceanic Clearance Delivery" frequencies above.

If an aircraft in the Gander OCA is unable to communicate with Gander Oceanic, pilots are to endeavour to pass position reports by relay through

a) another oceanic centre with which communication has been established;

b) another aircraft: in the NAT region, when out of range of VHF ground stations, 123.45 MHz may be used for air-to-air communications, including the relaying of position reports; or
c) another aircraft on frequency 121.5 or 243.0 MHz, if no other means is available.

Communications Failure - NAT Traffic

The following procedures are intended to provide general guidance for NAT aircraft experiencing a communications failure, and do not supersede the standard communications failure procedures as required by CARs.

General
If the aircraft is so equipped, a pilot experiencing a two-way radio communications failure shall operate the transponder on Code 7600 and Mode C.

The pilot shall attempt to contact any ATC facility, inform them of the difficulty, and request that information be relayed to the ATC facility with whom communications are intended.

Communications Failure Prior to Entering NAT Oceanic Airspace
If operating WITH a received and acknowledged oceanic clearance, the pilot shall enter oceanic airspace at the cleared oceanic entry point, level and speed, and proceed in accordance with the received and acknowledged oceanic clearance. Any level or speed changes required to comply with the oceanic clearance shall be completed within the vicinity of the oceanic entry point.

If operating WITHOUT a received and acknowledged oceanic clearance, the pilot shall enter oceanic airspace at the first oceanic entry point, level and speed, as contained in the filed flight plan and proceed via the filed flight plan route to landfall. The first oceanic level and speed shall be maintained to landfall.

Communications Failure Prior to Exiting NAT Oceanic Airspace
IF CLEARED ON FLIGHT PLAN ROUTE, the pilot shall proceed in accordance with the last received and acknowledge oceanic clearance, including level and speed, to the last specified oceanic route point, normally landfall, then continue on the flight plan route. Maintain the last assigned oceanic level and speed to landfall. After passing the last specified oceanic route point conform with the relevant State procedures and regulations.

IF CLEARED ON OTHER THAN FLIGHT PLAN ROUTE, the pilot shall proceed in accordance with the last received and acknowledged oceanic clearance, including level and speed, of the last specified oceanic route point, normally landfall. After passing this point, conform with the relevant State procedures and regulations, rejoining the filed flight plan route by proceeding, via published ATS routes where possible, to the next significant point ahead as contained in the filed flight plan.

Transponder Operation

The pilot shall operate the transponder at all times on Mode A and C, Code 2000 during flight in the NAT Region, except the last ATC assigned code must be retained for a period of 30 minutes after entry into NAT airspace unless otherwise directed by ATC. This procedure does not affect the use of the special purpose codes 7500, 7600 and 7700.

North Atlantic Minimum Navigation Performance Specification Airspace

Compliance with MNPS is required by all aircraft operating within the following defined airspace boundaries:

a) between FL 285 and FL 420;
b) between latitudes 27°N and the North Pole;
c) bounded in the east, by the eastern boundaries of CTAs Santa-Maria, Shanwick Oceanic and Reykjavik;
d) bounded in the west, by the western boundaries of CTAs Reykjavik and Gander and New York Oceanic, excluding the area west of 60°W and south of 38°30'N.

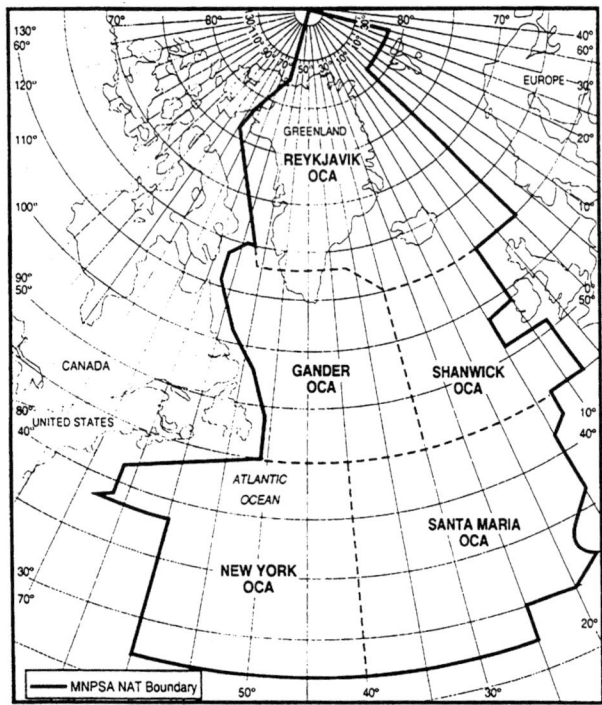

Fig. 1-19: NAT MNPS Airspace.

Aircraft used to conduct flights within NAT MNPS airspace shall have navigation performance capability such that, among other factors, the standard deviation of lateral track errors shall be less than 6.3 NM.

The required navigation performance capability shall be verified by the State of Registry or the State of the Operator as appropriate. TC is responsible for authorizing all Canadian civilian registered aircraft to fly within NAT MNPS.

In order to justify consideration for State approval for unrestricted operations in the NAT MNPS airspace, an aircraft is required to be equipped with two fully serviceable Long-Range Navigation Systems (LRNSs) which may consist of

a) two Inertial Navigation Systems (INS);
b) two Flight Management Systems (FMS) with two Inertial Reference Systems (IRS);
c) two primary-means GPS;
d) one INS and one FMS/IRS;
e) one INS and one GPS;
f) one FMS/IRS and one GPS.

Each LRNS must be capable of providing a continuous indication to the flight crew of the aircraft position relative to track. It is desirable that the navigation system employed for the provision of steering guidance should be capable of being coupled to the automatic pilot.

North Atlantic Reduced Vertical Separation Minimum (RVSM)

RVSM designated altitudes are currently from FL 290 to FL 410.

In RVSM, specific flight levels are assigned so as to allow 1,000 ft vertical separation, thereby permitting greater traffic flow. These designated altitudes depend on whether the flight is on random tracks eastbound or westbound, and whether the flight is on the organized track system. Refer to the AIM at paragraph RAC 11.22 for the specific flight level assignments available and for complete details on the RVSM approval process.

In Flight Procedures
Operators must meet minimum equipment specifications to operate at RVSM designated altitudes. The following equipment should operate normally:

a) two primary altitude measurement systems;
b) one automatic altitude-keeping device;
c) one altitude alerting device.

In the event that any of the required equipment fails prior to entering RVSM altitudes, a new clearance should be requested in order to avoid RVSM designated altitudes.

In level cruise flight, for RVSM operations, it is essential that the aircraft maintains the cleared flight level. Except in contingency situations, aircraft should not deviate from the cleared flight level without an ATC clearance. If the pilot is notified by ATC of an Assigned Altitude Deviation (AAD) error greater than 300 ft, the pilot should return to the cleared flight level as soon as possible.

During altitude changes, aircraft should not overshoot or undershoot the cleared flight level by more than 150 ft.

Meteorological Effects
Turbulence can be detrimental to accurate height keeping required for RVSM. If an aircraft reports greater than moderate turbulence and is within 5 minutes of another aircraft at 1,000 ft separation, ATC will endeavour to establish 2,000 ft separation.

ATC may temporarily suspend RVSM in selected areas of MNPS airspace and the RVSM transition area because of adverse weather conditions.

[Ref: AIM RAC 11.1 - 11.23]

1.46 Reduced Vertical Separation Minimum (RVSM)

General
RVSM refers to the application of 1,000 ft vertical separation at and above FL 290 between RVSM certified aircraft operating in designated RVSM airspace and RVSM Transition airspace. The ability to reduce vertical separation requirements will greatly increase the numbers of aircraft that can be flown along existing airways.

RVSM Airspace
RVSM Airspace is depicted below at Fig. 1-20. It is controlled airspace extending from FL 290 up to and including FL 410 as outlined below. RVSM Transition Airspace is also from FL 290 up to and including FL 410, as outlined below.

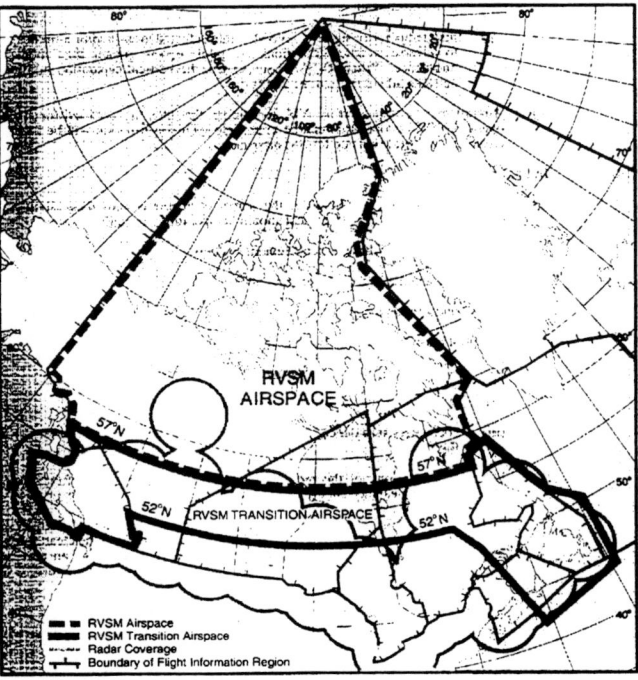

Fig. 1-20: RVSM Airspace.

Air Traffic Control Procedures
When in RVSM airspace, ATC will reduce vertical separation minimums from 2,000 feet to 1,000 feet. Altitudes will be assigned on the basis of the aircraft track. ATC may temporarily suspend RVSM within selected areas and/or altitudes due to adverse weather conditions. When RVSM is suspended, the vertical separation minimum between all aircraft will be 2,000 feet.

Aircraft Requirements
Before entering RVSM airspace, the status of required equipment should be reviewed. The following equipment should be operating normally:

a)   two primary altitude measurement systems;
b)   one automatic altitude keeping device (autopilot);
c)   one altitude alerting device.

ATC should be notified in the event any of the required equipment malfunctions.

When in level cruise flight, it is essential that the aircraft maintains the cleared flight level. During cleared transition between flight levels, the aircraft should not overshoot or undershoot the assigned level by more than 150 feet.

Pilots shall report "reaching" any altitude assigned within RVSM airspace.

Only aircraft having received RVSM approval may enter RVSM airspace.

[Ref: AIM RAC 12.16]

Notes:

Notes:

Notes:

Airframes, Engines, Systems

2.01 Airframes

Wing Design

WING PLANFORM refers to the shape of a wing when viewed from above (i.e. delta, rectangular). WINGLETS are small vertical rectangles attached to the wingtip that reduce induced drag by reducing the area of the wing tip disturbed by vortices and making a net increase in effective wing span (but marginally increasing form drag). VORTEX GENERATORS are small vertical pieces on the upper surface of a wing that delay boundary layer separation, improving take-off and landing performance. Vortex generators are also employed in design of high speed aerofoils to delay onset of shock waves - see s. 4.02. A CANARD is a horizontal stabilizer assembly mounted at the front of the aircraft, that has the same basic function as a standard horizontal stabilizer and that improves stall characteristics. WING SLATS are miniature aerofoils that open forward of the main aero-foil to smooth out the airflow over the wing. At high angles of attack, slats automatically move forward ahead of the wing. WING SLOTS are openings in the leading edge of aerofoils which allow air to pass through forward portion of the wing at high angles of attack. FLAPS are high lift devices that produce the same amount of lift at lower speed, by increasing aerofoil camber. Flaps can be fitted both to the leading edge and to the trailing edge of the aerofoil. WING FENCES are small vertical fins on the upper aerofoil that smoothen airflow and improve handling.

SPOILERS disrupt lift by interfering with the smooth airflow on the upper or lower surface of the wing. Spoil-ers are used either to assist in aileron control (known as "roll spoilers) or are linked to brakes to disrupt lift on landing and enhance braking action (by increasing weight on wheels).

A TRIMMING DEVICE is a small control surface used to assist the pilot in maintaining the proper position of a primary control surface. A SERVO TAB is used primarily on large aeroplanes and functions similar to a trim tab. However, only the servo tab moves in response to the movement of the pilot's flight control, and the force of the airflow on the servo tab then moves the primary control surface.

2.02 Reciprocating Engines

Basic Principles

The engine is often referred to as the POWERPLANT because, in addition to giving the aircraft power for forward motion, it also provides energy for electrical, hydraulic and pneumatic systems.

RECIPROCATING ENGINES involve pressures from burning and expanding gases that cause a piston to move up and down in an enclosed cylinder. This reciprocating motion of the pistons is transferred through a connecting rod into rotary motion by a crankshaft, geared to a propeller.

The basic parts of a reciprocating engine are the crankcase, cylinders, pistons, connecting rods, valves, spark plugs and crankshaft.

The main types of aircraft reciprocating engines are defined on the basis of the cylinder arrangement. The main classifications are HORIZONTALLY OPPOSED, IN-LINE, and RADIAL.

COMBUSTION is achieved by igniting fuel-air within the cylinder. Most aircraft piston engines operate on a 4-stroke system. The valves are timed so that they open slightly early and close slightly late, termed VALVE LEAD or VALVE LAG. This timing improves efficiency.

TOP DEAD CENTRE is the highest position of travel of the piston. BOTTOM DEAD CENTRE is the lowest position of travel of the piston. COOLING is generally achieved by air coming in contact with baffles or fins added to the engine cylinders.

Supercharging and Turbocharging

A SUPERCHARGER is an internally driven compressor, powered directly by the engine. The purpose of super-charging is to maintain sea level rated horsepower as altitude increases. The supercharger is installed down-stream from the carburetor, and it compresses the fuel-air mixture after it leaves the carburetor. Feeding the higher pressure fuel-air mixture into the intake manifold is FORCED INDUCTION, which results in a super-charged engine.

Forced induction can be used to increase normal sea level horsepower at low altitudes, or to make up for the loss of power at high altitudes to obtain sea level horse-power. For a supercharger to operate effectively, the temperature of the fuel-air mixture must be maintained within a specified range.

In a TURBOCHARGER, an impeller is powered by re-direction of exhaust gases to increase the pressure of the intake air. The advantage of a turbocharger over a supercharger is that engine horsepower is not lost to drive the impeller. The turbocharger compresses intake air alone, not fuel-air as in the supercharger. The redirection of exhaust gases is effected by the WASTEGATE, which is a valve controlled by a governor. When the wastegate is open, the engine is normally aspirated. When the wastegate is closed, the engine is turbocharged.

The Carburetor and Fuel Injection

The function of the carburetor is to mix the correct ratio of fuel to air, to vaporize it and to deliver the mixture to the cylinders. The ratio of fuel to air is governed by weight (not volume). The ratio for best power is 14 pounds air to 1 pound fuel.

As the incoming air passes through the venturi, its speed increases and a low pressure area is created. When the nozzle sprays fuel into the low pressure area, the fuel is vaporized. The throttle controls the butterfly valve, which regulates the amount of fuel/air mixture entering the engine and thereby controls the power output.

The MIXTURE control governs the amount of fuel that is sent through the discharge nozzle. A high proportion of fuel to air is a RICH mixture, and a low proportion of fuel to air is a LEAN mixture. Adjust the mixture control as specified in the operating handbook. When the mixture is too lean, the engine will run hot and the excessively hot engine temperatures may cause detonation. When the mixture is too rich, the engine will be burning more fuel than is necessary, and spark plug fouling may result. Climbing without leaning the mixture may lead to an excessively rich mixture, because the air becomes less dense with altitude. BEST POWER MIXTURE is that fuel/air ratio at which the most power can be obtained for any given throttle setting.

CARBURETOR HEAT is used to prevent or melt ice which may form in the carburetor. Carburetor ice forms when the outside air is humid and the temperature drop within the venturi is enough to form ice. Carburetor ice is possible in temperatures from -6° C to +32° C. An indication of carburetor ice is a drop in manifold pressure. The carburetor heat control is used to melt the ice. Using carburetor heat results in a reduction of power and a more rich mixture, because the warm, heated air being sent to the carburetor is less dense. Carburetor heat should only be used when ice is suspected or is likely to form (such as at reduced power settings), because it results in hot, unfiltered air going into the engine, reducing efficiency and supplying dirt.

When using carburetor heat, remember that the engine loses an average of 9 percent of its power when heat is applied. This is due to the reduced volumetric efficiency of heated air and the loss of the ram air feature.

As the ambient temperature decreases, the effect of carburettor heat on the engine efficiency also decreases.

In a FUEL INJECTED engine, a fuel injection nozzle supplies a fine spray of fuel which is discharged into the cylinders. The result is that a carburetor is unnecessary, and most of the problems associated with carburetor ice do not exist. Fuel injection systems also provide better fuel distribution to the engine cylinders. The FUEL CONTROL UNIT of a fuel injection system is that part of the system where the throttle and mixture controls effect a resulting fuel flow to the cylinders. Negative aspects of fuel injection systems are vapour lock during ground operations on hot days and difficulty starting a hot engine.

ALTERNATE INDUCTION AIR is used in fuel injected engines to provide an alternate source of intake air should the primary intake experience a blockage (for example, caused by buildup of airframe ice during flight into freezing rain).

2.03 Gas Turbine Engines

There are five sections to a jet engine: the INLET, COMPRESSOR, COMBUSTOR, TURBINE and NOZZLE. The inlet, considered part of the airframe, converts the ram-air pressure to static pressure. The compressor takes large volumes of air, increases its pressure, and delivers it to the combustor. The combustor separates the air and divides it into two streams: one flows into the combustor where it burns, and the other is used for combustor cooling. The hot gases then flow through the turbine, where part of the energy is used to drive the compressor. The other gases flow to the nozzle, which increases the momentum of the gas. The change of momentum from inlet to nozzle produces the majority of the thrust force.

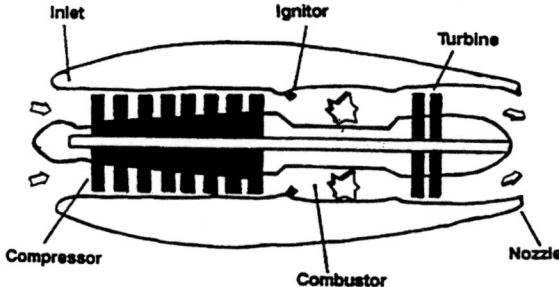

Fig. 2-1: Cutaway of basic gas turbine engine.

SPARK IGNITORS are used to light gas turbines. Starting systems generally consist of a pneumatic type starter where air from a high pressure bottle or from an Auxiliary Power Unit (APU) is forced through the system while the spark ignitors are activated. AUTO-RELIGHT ignition systems are designed to provide instantaneous re-ignition if an engine begins to lose power.

Early jet engines used a CENTRIFUGAL COMPRESSOR, which is still in use today in the design of many modern jet engines. The centrifugal compressor uses the principle of a centrifuge to compress the intake air by rotating it at high speeds, such that the air is thrown outwards and collected by a DIFFUSER and fed to the combustors at high pressure. Some advantages to the centrifugal compressor design are that it provides a relatively high pressure for each stage, it provides good compression efficiency over a high range of rotational speeds, it has relatively low weight, and given the relative simplicity of its design, it is less costly to manufacture and is more durable. Disadvantages to the centrifugal design are that given the rapid loss of energy it can only utilize a maximum of two stages, and it takes up a large frontal area leading to increased aerodynamic drag.

The more modern compressor design is that of the AXIAL FLOW COMPRESSOR, which has a much smaller diameter than the centrifugal compressor and hence less associated aerodynamic drag. With axial flow compressors, many more compressor stages can be added without increasing the diameter of the engine, as airflow is directed back parallel to the rotational axis of the compressor.

Compressor blades are like miniature aerofoils: they have an angle of attack, and the lift generated is airflow. The compressor blades duct air into STATOR VANES (immobile), which gradually increases air pressure as it is forced into smaller constrictions as it hits further compressor blades and stator vanes. COMPRESSION RATIO is the ratio between compressor discharge pressure and engine inlet pressure. The compression ratio must not get lower than the back pressure from the hot section, otherwise the airflow reverses.

ENGINE PRESSURE RATIO (EPR) is the ratio of turbine discharge total pressure to compressor inlet total pressure. For example, an EPR of 2.09 means the pressure at the turbine discharge is 2.09 times as great as at the compressor inlet.

Due to turbulence, excessively high angles of attack, or sudden reduction of airflow, the compressor blades may stall, resulting in reduced engine airflow. A COMPRESSOR STALL may be limited to a few blades only (often unnoticeable), or a partial or total area of the compressor face. A COLD STALL is when only a few blades stall or small area of the compressor is affected. A HOT STALL is when all the compressor blades stall, causing severe engine damage and possible reversal from increased pressure from burner, causing fire to blast out the inlet with or without a loud noise. INLET GUIDE VANES regulate air to the compressor (at idle they are closed down, and full open at about 70% RPM). COMPRESSOR BLEEDS dump air overboard from the compressor when there is an excessive build-up of air pressure (such as from rapid change of throttle to idle/full thrust), preventing compressor stalling.

A HUNG START is any start on a turbine engine where the engine fails to accelerate to normal idle RPM. Temperatures increase rapidly and can exceed temperature limitations.

A HOT START occurs when fuel has been introduced to the combustors with too low compressor RPM, and fuel then burns at excessively hot temperatures without any rearward pressure flow of exhaust gases into the turbine section. A hot start can lead to serious engine damage.

COMPRESSOR BLEED AIR from the compressor is used to aid in cooling of the turbine and to provide pressurization for the internal air and oil systems. Due to very high temperatures associated with bleed air, temperature sensors are employed and bleed overtemp sensor warnings are displayed in the cockpit.

The Fuel Control Unit (FCU) in a gas turbine engine collects such inputs as power or thrust lever position, air pressure and engine temperature, and then meters fuel as needed to the engine. Some FCU's are mechanical, while others are electrical, known as Electrical

Control Units (ECU). ECU's may also be termed FADEC's: Full Authority Digital Engine Controls. Some turbine engine aircraft are fitted with both ECU's and FCU's for back-up purposes.

TURBOFAN JET ENGINES use one more turbine wheels to drive a large, low speed FAN. The fan acts like a propeller: it moves a large volume of air back and provides a significant amount of total thrust. Total thrust is combined fan thrust and engine thrust. Turbofan engines are more efficient because a larger volume of air is accelerated and exhaust velocities are reduced. BYPASS RATIO is the volume of air that bypasses the compressor, burner and turbine sections (accelerated only by fan) compared to the volume of air passing through the engine.

TURBOPROP ENGINES use exhaust gas energy to power turbines. In a FREE TURBINE engine, the POWER (N2) TURBINE is mechanically free from the COMPRESSOR (N1) TURBINE. The power turbine drives the propeller shaft. A GEAR REDUCTION BOX reduces the propeller shaft RPM. The compressor turbine drives the compressor and accessories gear box.

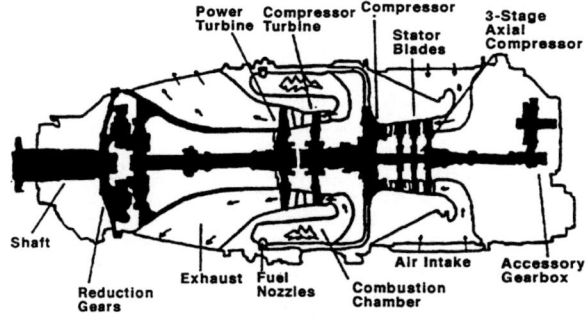

Fig. 2-2: Pratt & Whitney PT-6 Free Turbine engine.

BETA RANGE is the region of propeller position for taxi operations. In beta range, the propeller pitch and fuel flow are controlled by the power levers. LOW PITCH STOPS prevent the propeller from entering beta range or reverse in cruise flight.

In a PT6 turboprop, the function of the POWER TURBINE GOVERNOR is to prevent power turbine overspeed by reducing fuel flow.

A TORQUEMETER is an instrument that provides a means of accurately measuring shaft horsepower in the reduction-gear assembly. Engine power is determined from reading the torquemeter.

The temperature of turbine exhaust gases is monitored on a TURBINE INLET TEMPERATURE (TIT) gauge, especially during start when overheating damage is most likely. The AFM will provide various temperature operating limits under various conditions.

COMPRESSOR INLET SCREENS are used in some turbine engine systems to minimize damage caused by ingestion of foreign objects such as stones and other debris. PARTICLE SEPARATORS utilize centrifugal pressure from inlet suction to direct sand and debris into a sediment trap.

Gas turbine engines performance is reduced in conditions of reduced air density, such as high ambient temperatures and higher altitudes. Humidity does not generally adversely affect gas turbine engine performance because 65% - 70% of the airflow going through the engine is used for cooling the combustion mixture and the remaining relative amount of suspended moisture has no measurable effect.

2.04 Propellers

The function of the propeller is to convert the turning movement of the crankshaft into thrust. As a propeller rotates, it moves the aircraft forward by moving a large mass of air backward. A propeller is similar to an airfoil, because it has an angle of attack and creates lift. Lift, in the case of a propeller, is thrust.

A propeller blade is twisted. The blade angle changes from the hub to the tip with the greatest angle of incidence at the hub and the smallest at the tip. The reason for this twist is to produce uniform lift from the hub to the tip. As the blade rotates there is a difference in the actual speed of the various portions of the blade. The tip of the blade travels faster than the part near the hub because the tip travels a greater distance than the hub in the same length of time.

When the propeller rotates, it creates various forces resulting in yaw. TORQUE is caused by the propeller forcing the aircraft to rotate in the opposite direction, and results in a left turning tendency. ASYMMETRIC THRUST results in the downgoing portion of the propeller producing more pull than the upgoing portion, at high angles of attack. This is also known as "P-Factor". PRECESSION refers to the tendency of a propeller to act like a gyroscope: if the aircraft tail is raised, the nose yaws to the left (this effect is most apparent with tail-wheeled aircraft during the take-off roll). SLIPSTREAM is rotating air pushed back from the propeller, which strikes the fin from the left and causes a yaw to the left.

Generally, the adverse yaw caused by the propeller can be corrected for by design, such as off-setting the fin to reduce propeller slipstream. In flight, apply rudder to correct adverse yaw.

BLADE ANGLE is the angle of the propeller blade. PROPELLER EFFICIENCY is the ratio of thrust horsepower to brake horsepower. GEOMETRIC PITCH is the theoretical distance a propeller should advance in one rotation. EFFECTIVE PITCH is the actual distance a propeller advances in one revolution. PROPELLER SLIP is the difference between geometric pitch and effective pitch.

Most single engine training aircraft have FIXED PITCH propellers (constant blade angle). Other aircraft have VARIABLE PITCH propellers, which utilize a system in the propeller hub using oil pressure to vary the propeller blade angle. With CONSTANT SPEED propellers, the pilot selects a propeller RPM, and an oil regulated governor adjusts blade angle to maintain the selected RPM. FINE PITCH means small blade angles, and is used for take-off and climb. With COARSE PITCH, the propeller blade travels a greater forward distance per revolution. Coarse pitch is used for cruise.

In a COUNTERWEIGHT CONTROLLABLE SYSTEM, counterweights force the prop blades to increase their angle of attack. Oil pressure reduces the blade angle. Where there is a loss of oil pressure, the blade angle increases. Some models are automatic feathering when all oil pressure is bled from the system.

In a CONSTANT SPEED CONTROLLABLE SYSTEM, the twisting moment of the propeller blade results from centrifugal force, moving the blades into fine pitch. Oil pressure in the propeller hub counteracts this tendency. Gears from the crankshaft drive the governor oil pump: if the propeller RPM gets too high, the governor opens an oil passage, and greater oil pressure increases the propeller blade angle.

To summarize, in an oil-counterweight, feathering prop, a loss of oil pressure results in the propeller going into feather position. In a non-counterweight propeller system, a loss of oil pressure results in full fine pitch.

Earlier models of light twins (such as Piper Twin Commanche) have propellers that rotate in the same direction. The result, at higher angles of attack, is asymmetric thrust: the down-going blade (right side in clockwise rotating props) produces more thrust than the left side. Should an engine fail, a greater turning moment and more adverse yaw results from a failure of the left engine, because the right engine's thrust is furthest away from the centerline of the aircraft longitudinal axis. This is why the left engine is termed the CRITICAL ENGINE: its loss results in the most adverse handling qualities.

Later models of light twins (such as Piper Seminole) have COUNTER-ROTATING propellers to reduce adverse yaw associated with asymmetric thrust: the right engine has a left rotating propeller. The result is more stable single engine handling following an engine failure of the left engine. A left rotating crankshaft is designated with an "L" prefix, such as "LT10-540-F2BD".

2.05 Fuels

OCTANE RATING refers to the ability of a fuel to burn slowly and expand evenly, rather than explode suddenly. Fuel grades are coloured according to their octane rating:

Colour	Fuel
Red	80/87
Blue	100 LL (low lead)
Green	100/130 (high lead)
Straw/undyed	Turbine

The aircraft operating manual specifies the correct grade of fuel for the aircraft. Insure that the proper fuel grade is utilized. If the proper grade of fuel is not available, use the next higher grade, but NEVER use a lower grade. Using an octane rating fuel that is too low for the aircraft causes DETONATION, which is when the fuel combusts too rapidly and will cause serious damage to the engine. If a higher octane fuel is used, insure the engine is run hot so as to burn off deposits accumulated on the spark plugs.

PRE-IGNITION is caused by hot spots or carbon particles inside the piston igniting the fuel/air mixture before the spark plug fires, with an un-controlled firing of one or a few cylinders. VAPOUR LOCK occurs in conditions of high temperature when fuel is vaporized in the fuel line, resulting in fuel blockage. FUEL DENSITY changes with temperature: the colder the temperature, the heavier the fuel. Generally, conventional grades weigh 6.0 lbs per U.S. gallon, turbine Jet A 6.8 lbs/USG, turbine Jet B 6.5 lbs/USG.

Fuel can become CONTAMINATED by water and other foreign substances. Always top up the aircraft fuel tanks after flight, to prevent water accumulation in the air-space inside the fuel tanks. To ensure fuel is not contaminated, drain a fuel sample and inspect it for proper colour, water, sediment and brilliance. Warm fuel has a greater capacity to hold water.

If straining fuel to remove sediments or water, it should be strained from the lowest point in the fuel system, such as fuel sumps, if equipped.

The aircraft and fuelling equipment through which fuel passes require BONDING. Bonding prevents sparks by equalizing electrical potentials. The hose nozzle must be bonded to the aircraft before the tank cap is removed. Grounding of the fuel service vehicle and bonding of the service vehicle and hose nozzle to the aircraft before fuelling will dissipate static electricity.

2.06 Fuel Systems

Simple fuel systems are normally GRAVITY FEED, which means that the force of gravity alone pushes the fuel to the engine, and the fuel cells are located above the engine. More complex aircraft or aircraft having low fuel cells require an engine driven FUEL PUMP. Such systems often also employ a standby ELECTRIC PUMP as a back-up. A FUEL PRIMER is a small hand pump which is used to pump fuel into the engine prior to starting in cold weather. Care should be taken not to over-prime, because the engine will be flooded, resulting in a risk of backfire or engine fire.

FUEL VENTS prevent a vacuum forming inside the fuel tanks. Venting is accomplished with either a fuel vent designed in the fuel cap, or an over-flow vent placed under the fuel tank. Venting insures proper air pressure within the tanks for a steady fuel flow.

Large transport aeroplanes have multiple fuel tanks such that fuel is distributed from many sources. HIGH PRESSURE fuel pumps provide fuel delivery to the engines directly, while LOW PRESSURE fuel pumps deliver fuel from the tanks to the high pressure fuel pumps. AUXILIARY FUEL PUMPS are used to transfer fuel between tanks and act as emergency back-up. CROSSFEED systems permit draining fuel from tanks on opposite engines so as to be able to utilize fuel supply from an engine that has been shut down.

FUEL QUANTITY MEASURING STICKS are used to obtain accurate measurement of fuel in fuel tanks. DRIPSTICKS are older fuel measuring sticks that indicate fuel quantity based on unlocking them from beneath the fuel tank, then lowering them until fuel starts to drip and this will indicate the fuel quantity. The more modern MAGNASTICKS slide freely once unlocked from beneath the fuel tank, stopping automatically at the fuel tank quantity based on alignment of the top of the magnastick within the fuel tank with a magnetized fuel float.

A FUEL JETTISON SYSTEM or fuel "dumping" system is required for transport category aircraft where the maximum take-off weight exceeds the maximum landing weight (this is a type certificate requirement). A fuel jettison system must be able to jettison engine fuel within 15 minutes and must be operable during all operating conditions. As well, fuel jettisoning systems must be designed so as to provide for a means to prevent jettisoning the fuel (in the tanks used for take-off and landing) below the level allowing climb from sea level to 10,000 feet and thereafter allowing 45 minutes cruise at a speed for maximum range.

2.07 Ignition Systems

Magnetos supply electrical current to the spark plugs. The magnetos are driven by the engine crankshaft, and are independent of the aircraft battery or electrical system. Two magnetos supply DUAL IGNITION (two spark plugs per cylinder) for more efficient combustion, better engine performance and safety (should one magneto fail, the other will continue to deliver current, with a slight loss of RPM). If the magneto grounding wire is broken, the mags can continue to supply electrical current when the mags are in the OFF position.

2.08 Oil Systems

The functions of oil in an aircraft engine are for cooling, sealing, lubrication and cleaning. Oils are graded according to viscosity. For engines of moderate horse-power, use the following grades:

Grade	Season
120	Summer
100	Fall or Spring
80	Winter
65	Arctic

Hot engine oil is pumped through an OIL COOLER having baffles which reduces oil temperature and thereby aids in engine cooling. An OIL FILTER is used to trap sediments and impurities. The PRESSURE RELIEF VALVE provides constant oil pressure. There are two methods of lubrication: FORCE FEED (dry sump) and SPLASH (wet sump). In force feed lubrication, oil is sprayed under pressure. In splash lubrication, oil is contained in a sump or reservoir and splashes over the engine as the crankshaft rotates.

OIL DILUTION is adding a small amount of fuel to the oil within the engine to aid start-up in winter: the engine should be hot before departure to burn off any fuel remaining.

CHIP DETECTORS are small magnets that are placed within the oil system to detect the presence of ferrous metal particles. Should the chip detectors pick up metallic chips or flakes, serious internal wear has occurred and engine malfunction is likely.

MINERAL OIL has no additives and is used to break-in new engines. DETERGENT OIL has various additives for cleaning, dispersion and suspension of particles. Never add detergent oil to mineral oil (drain the mineral oil first). Never use automotive oil in an aircraft engine. Most oil weighs 2 lbs/litre.

When flying, monitor the oil pressure gauge and oil temperature gauge. If the oil pressure is too low, oil is not being delivered to all parts of the engine, and the flight should be terminated as soon as possible. If the oil temperature is too high, the oil may not be viscous enough to lubricate all parts of the engine, or the oil level may be too low. The owner's manual provides the correct range of oil temperature and oil pressure.

2.09 Electrical Systems

Aircraft electrical systems are either 12 or 24 volts, direct current (DC). Either a GENERATOR or ALTERNATOR is incorporated to supply electrical current to the system and re-charge the battery. The advantage of an alternator is that it supplies electrical current at low RPM, while the older generator requires high RPM to function adequately. INVERTERS convert DC power to AC power. TRANSFORMER RECTIFIER UNITS (TRU) convert AC power to DC.

Power is supplied through one or more BUS BARS, and the MASTER SWITCH controls this power to all circuits. All electrical circuits are protected by fuses or circuit breakers. An AMMETER indicates the flow of current, in amperes, from the alternator to the battery or from the battery to the electrical system. When the engine is operating and the master switch is on, the ammeter shows the charging rate applied to the battery. If the alternator is not functioning or the electrical load exceeds the output of the alternator, the ammeter shows the battery discharge rate. The ammeter should always show on the "+" side of the "0" mark. If the needle is in the "-" side, electrical energy is being drawn from the battery instead of the alternator.

An OVER-VOLTAGE SENSOR is a warning light which, when illuminated, indicates that the alternator has been shut down due to an over-voltage condition, and the battery is supplying all electrical power. The over-voltage sensor may be reset by turning the master switch off and then back on again. If the warning light does not illuminate, normal alternator charging has resumed; however, if the light again illuminates, a malfunction has occurred and the flight should be terminated as soon as possible. The over-voltage warning light may be tested by momentarily turning off the ALT side of the master switch and leaving the BAT side on.

A VOLTMETER indicates the voltage in the electrical system. The VOLTAGE REGULATOR prevents the battery from being over-charged or the generator or alternator from overloading the system.

Prior to start, all electricals (radios, instruments, lighting etc) are powered by the battery. After start, all electricals are powered by the generator or alternator, as the case may be. When starting, the starter switch activates a SOLENOID, which in turn activates a flow of current to the STARTER MOTOR, which causes the engine to turn over.

An APU provides back-up and auxiliary electrical generation for ground operations, engine starting and emergency back up.

If an electrical system becomes overloaded, LOAD SHEDDING is required to reduce electrical demand, an non-essential electrics are shut down.

Large aircraft are generally AC powered.

Since turbine aircraft generally put very high demand on batteries for start, often a Ground Power Unit (GPU) will be used for starting whenever available.

CIRCUIT BREAKERS are employed in electronic systems to protect circuits. Some circuit breakers can be re-set and "pop out" when activated. Typically, where a circuit breaker has popped, it can only be re-set once: refer to the AFM for the particular aircraft for re-set procedures. GENERATOR CONTROL UNITS protect generators and associated buses from faults.

BUS BARs provide a source of electrical power from which various circuits can be sourced. Advanced aircraft utilize a number of inter-connected bus bars so as to provide redundancy such that should one bus bar fail, another can be utilized to maintain power.

2.10 Hydraulic Systems

A hydraulic system consists of liquid under pressure which provides a convenient method of remote control for operation of components such as landing gear, flaps and speed brakes. Liquids are practically incompressible, so that when a force is applied at any point in a liquid, a corresponding pressure will be transmitted equally throughout the liquid.

A simple hydraulic system consists of a jack, control, pump and reservoir. Fluid is supplied from the reservoir to an engine-driven pump which generates pressure. A manually operated control directs the fluid to the desired end of the hydraulic jack, forcing the jack piston to move. When the ram reaches the end of its travel, the pump increases its pressure due to the build up, causing a relief valve to open, allowing the fluid to return to the reservoir.

To keep hydraulic lines clean, filters are employed in the lines.

HYDRAULIC PUMPS are either VARIABLE DELIVERY, where hydraulic pressure is provided at different pressures depending upon system requirements, or CONSTANT DELIVERY, where hydraulic pumps provide hydraulic pressure at a fixed, unchanging rate.

Various types of hydraulic ACTUATORS convert hydraulic pressure to desired resulting pressured movements.

PRESSURE RELIEF VALVES prevent excessive hydraulic pressures from developing and damaging components.

A HYDRAULIC ACCUMULATOR consists of a large cylinder in which a piston separates hydraulic fluid from air under pressure. The cylinder can be charged to the correct pressure with air from an external source. A gauge will indicate the pressure available. Pilots should ensure the correct pressure is available before start up, because low pressure may result in sluggish operation of services. A hydraulic accumulator has the following functions:

a) to smooth out the fluctuations in the flow of oil when there is a restriction in the supply pipe (which may be intentionally designed into the system, such as with the flap circuit, to prevent too rapid operations);
b) to act as an emergency pressurized oil supply for the operation of hydraulic brakes and smaller services such as air intake shutters;
c) to give an initial impetus to other hydraulic components.

The accumulator of the hydraulic system of a B-727 is pre-charged to about 1,000 PSI. When the system is depressurized, the accumulator should read 1,000 PSI. When the hydraulic system is on, both the hydraulic and accumulator gauge will indicate the same pressure.

Think of the accumulator primarily as a shock absorber: when a valve opens or closes on a 3200 PSI system, there is quite a shock to the lines.

2.11 Exhaust Systems

Aircraft heating systems often obtain heat energy from waste heat from the exhaust manifold. Therefore, pilots should carefully inspect the exhaust tubes for indications of leakage so as to prevent the possibility of carbon monoxide poisoning.

2.12 Undercarriage and Brake Systems

The function of a landing gear system is to take the shock of landing and allow maneuvering on the ground. Landing gear can be fixed or retractable. Retractible landing gear is operated by mechanical, hydraulic, electrical or pneumatic means. The main advantage of a retractable landing gear system is that it reduces airframe drag and thereby improves performance.

BRAKES are used for slowing, stopping, holding, or steering while on the ground. Brake systems are normally independent, having their own hydraulic fluid reservoir and are entirely independent of the main hydraulic system. Brake systems are powered by master cylinders similar to those used in cars. The system consists of a reservoir, two master cylinders, mechanical linkage which connects each master cylinder with its corresponding brake pedal, connecting fluid lines, and a brake assembly in each main landing gear wheel. Each master cylinder is actuated by toe pressure on the respective brake pedal. The master cylinder builds up pressure by the movement of a piston inside a sealed, fluid filled cylinder. The resulting hydraulic pressure is transmitted through the fluid line which is connected to the brake assembly in the wheel.

Prior to flight, brake discs should be checked for even wear with no signs of cracking or pitting, and there should be no leaks or discontinuities in any visible portions of hydraulic lines.

In transport aircraft, braking systems are normally hydraulically powered. Secondary accumulator systems are available as back-up upon loss of hydraulic pressure. BRAKE TEMPERATURE MONITORS provide an indication of brake overheat.

THRUST REVERSERS are fitted into exhaust systems which effectively reverse exhaust gas flow. One typical jet reverser system utilizes "clamshells" (or MECHANICAL BLOCKAGE reversers) at the rear of the engines. Once the reversers are in position, power may be applied to assist in reducing forward kinetic energy. Reverser systems are a great advantage to landing on slippery surfaces.

Another type of thrust reverser system is known as an AERODYNAMIC BLOCKAGE or CASCADE type thrust reverser. This type of thrust reverser is used in turbofan engines and is based on a series of cascading, movable vanes located within either the fan exhaust or hot exhaust, and these vanes are deflected so as to re-direct the escaping gases into a forward direction.

SPOILERS also assist in braking.

ANTI SKID SYSTEMS monitor wheel deceleration and provide brake release to achieve strongest braking. Anti skid is based on anti skid valves which are controlled by an anti skid control unit; brake release is provided if excessive wheel deceleration is detected by transducers. Wheel speed measurements for each individual wheel determine whether the system is activated.

Anti-skid control units also provide touchdown brake lock protection and hydroplaning protection based on ground speed inputs.

Generally, for turbine powered aircraft, stopping distance following slightly high approach speeds are minimized by setting the aircraft down firmly on main wheels, initiating reverse thrust, setting nosewheel down, apply full reverse thrust followed by braking. Aerodynamic braking by holding off longer is much less efficient at reducing stopping distance.

2.13 Anti Icing and De-Icing Systems

ANTI ICING SYSTEMS prevent the formation of ice on the airframe, while DE-ICE systems remove ice that has already formed. The anti-icing systems on large aircraft generally use hot bleed air and electrical heating elements to anti-ice certain areas. Ice detectors are often employed. Details regarding the correct operation of the system used will be found in the AFM.

Aircraft ground de-icing is important to insure an aircraft will remain ice free for departure. It is also a regulatory requirement that snow/frost/ice accumulation be removed before flight. Refer to s. 4.12 in this manual for further details on critical surface contamination.

Aircraft ground anti-ice de-ice equipment used by airline operators is generally truck-mounted, mobile units that have de-ice/anti-ice fluids contained in tanks, a heater to worm the fluid to its best application temperature, an aerial boom/basket device and fluid dispersal system including pumps and a spray nozzle.

HOLDOVER TIME refers to how long the beneficial effect of the anti-ice fluid lasts. Many factors affect holdover time, such as temperature, water amount, relative humidity, winds, type of anti-ice fluid used etc. Anti-ice fluid itself is a contaminant and must be removed before flight. On larger, transport aircraft, the higher speed at take-off will tend to clear itself. Holder time is calculated as beginning at the start of the final application of an approved de-icing/anti-icing fluid and expiring when the fluid is no longer effective. The fluid is no longer effective when its ability to absorb more precipitation has been exceeded.

TYPE I de-icing fluids contain various percentages of glycol mixed with water. The maximum amount of glycol to water should be about 50 - 50. Type I fluids provide some protection against re-freezing but don't provide much protection against further contamination.

TYPE II anti-icing fluids are more viscous. They tend to stick to the surface and absorb precipitation. These fluids are effective anti-icers because of their high viscosity and pseudoplastic behaviour. They are designed to remain on the wings of an aircraft during ground operations or short term storage, thereby providing some anti-icing protection and will readily flow off the wings during take-off. When these fluids are subjected to shear stress (such as during take-off), their viscosity decreases drastically, allowing the fluids to flow off the wings and causing little adverse effect on the aircraft's aerodynamic performance.

TYPE III fluids have properties between Types I and II, so provide a longer holdover time than Type I, but less than Type II. Its shearing and flow-off characteristics are designed for aircraft that have a shorter time to the rotation point, which makes it acceptable for aircraft that have a Vr of less than 100 kts.

TYPE IV fluids meet the Type II standard and in addition have a significantly longer holdover time, and accordingly, holdover time tables are available for Type IV. The product is dyed green as it is believed that the green product will provide for application of a more consistent layer of fluid to the aircraft and will reduce the likelihood that fluid will be mistaken for ice. However, as these fluids do not flow as readily as conventional Type II fluid, caution should be exercised to ensure that enough fluid is used to give uniform coverage. Research indicates that the effectiveness of a Type IV fluid can be seriously diminished if proper procedures are not followed when applying it over a Type I fluid.

All fluid users are advised to ensure that these fluids are applied evenly and thoroughly and that an adequate thickness has been applied in accordance with the manufacture's recommendations. Particular attention should be paid to the leading edge area of the wing and horizontal stabilizer.

Holdover tables are provided for use in de-ice and anti-ice operations and all notes and cautions should be carefully reviewed (see Fig. 2-4).

The purpose of a Snowfall Visibility vs. Intensity Chart is to obtain a HOT holdover table value that should be used in a given anti-ice or de-ice procedure (see Fig. 2-3).

EXAMPLE: Refer to the Snow Visibility vs Snowfall Intensity Chart at Fig. 2-3 below. Assume that the daytime visibility in snowfall is less than 1/2 of a statute mile and the temperature is -5° C. What is the snowfall rate that will be used to determine which HOT holdover table value is appropriate for the fluid?

ANSWER: For these conditions the snowfall rate that would apply is "heavy".

Once the weather conditions and effective intensities are known a specific holdover time can be calculated based on holdover tables.

EXAMPLE: Refer to the SAE Type II Fluid Holdover Times table at Fig. 2-4. Assume freezing fog is present, and the outside air temperature is -1° C. Your company uses the Ground Icing Operations Standard and your aircraft is anti-iced with an undiluted ("neat") SAE Type II fluid. What is the acceptable Decision Criteria Time for these weather conditions?

ANSWER: 35 minutes. Since fluid is undiluted, we use a concentration of 100%. For freezing fog, a range of from 35 minutes to 1 hour 30 minutes is provided, however, according to the cautions provided, the only acceptable decision criteria time is the SHORTEST time within the applicable holdover time table cell, therefore, holdover time is 35 minutes based on the data provided.

SNOW VISIBILITY VS SNOWFALL INTENSITY CHART
(For Sample Questions Only)

Lighting	Temperature Range		Visibility in statute miles		
	°C	°F	Heavy*	Moderate*	Light*
Daylight	Above -1	Above 30	<1	1 - 2	>2
	-1 to -7	30 to 19	<½	½ - 1¼	>1¼
	Below -7	Below 19	<3\8	3\8 - 5\8	>5\8
Darkness	Above -1	Above 30	<2	2 - 4	>4
	-1 to -7	30 to 19	<1	1 - 2½	>2½
	Below -7	Below 19	<¾	¾ - 1¼	>1¼

* Light snow intensity is defined as less then 1mm/hr equivalent liquid water, moderate intensity as 1mm/hr to 2.5mm/hr equivalent liquid water, and heavy as greater than 2.5mm/hr equivalent liquid water.

Fig. 2-3: Snow visibility vs snowfall intensity chart.

SAE TYPE II FLUID HOLDOVER TABLE[1]

Guideline for Holdover Times Anticipated for Type II Fluid Concentrations as a Function of Weather Conditions and OAT

THE RESPONSIBILITY FOR THE APPLICATION OF THESE DATA REMAINS WITH THE USER

OAT		Type II Fluid Concentration Neat Fluid/Water (Vol% / Vol%)	Approximate Holdover Times Under Various Weather Conditions (hours:minutes)						
°C	°F		FROST[2]	FREEZING FOG	MODERATE SNOW	FREEZING DRIZZLE[4]	LIGHT FREEZING RAIN	RAIN ON COLD SOAKED WING	OTHER[5]
above 0°	above 32°	100/0	12:00	0:35-1:30	0:20-0:55	0:30-0:55	0:15-0:30	0:05-0:40	
		75/25	6:00	0:25-1:00	0:15-0:40	0:20-0:45	0:10-0:25	0:05-0:25	
		50/50	4:00	0:15-0:30	0:05-0:15	0:05-0:15	0:05-0:10		
0 to -3	32 to 27	100/0	8:00	0:35-1:30	0:20-0:45	0:30-0:55	0:15-0:30	**CAUTION** No holdover time guidelines exist	
		75/25	5:00	0:25-1:00	0:15-0:30	0:20-0:45	0:10-0:25		
		50/50	3:00	0:15-0:30	0:05-0:15	0:05-0:15	0:05-0:10		
below -3 to -14	below 27 to 7	100/0	8:00	0:20-1:05	0:15-0:35	0:15-0:45[3]	0:10-0:25[3]		
		75/25	5:00	0:20-0:55	0:15-0:25	0:15-0:30[3]	0:10-0:20[3]		
below -14 to -25	below 7 to -13	100/0	8:00	0:15-0:20	0:15-0:30				
below -25	below -13	100/0	Type II fluid may be used below -25°C (-13°F) provided the freezing point of the fluid is at least 7°C (13°F) below the OAT and the aerodynamic acceptance criteria are met. Consider use of SAE Type I when Type II fluid cannot be used.						

°C = Degrees Celsius   OAT = Outside Air Temperature
°F = Degrees Fahrenheit   VOL = Volume

NOTES

1. Based on test of neat fluids with the lowest viscosity deliverable on the aircraft, yet meeting Type II WSET and HHET.
2. During conditions that apply to aircraft protection for ACTIVE FROST.
3. The lowest use temperature is limited to -10° C (14° F).
4. Use light freezing rain holdover times if positive identification of freezing drizzle is not possible.
5. Heavy snow, snow pellets, ice pellets, moderate and heavy freezing rain and hail.

CAUTIONS

THE TIME OF PROTECTION WILL BE SHORTENED IN HEAVY WEATHER CONDITIONS, HEAVY PRECIPITATION RATES OR HIGH MOISTURE CONTENT. HIGH WIND VELOCITY OR JET BLAST MAY REDUCE HOLDOVER TIME BELOW THE LOWEST TIME STATED IN THE RANGE. HOLDOVER TIME MAY ALSO BE REDUCED WHEN AIRCRAFT SKIN TEMPERATURE IS LOWER THAN OAT.

THE ONLY ACCEPTABLE DECISION CRITERIA TIME IS THE SHORTEST TIME WITHIN THE APPLICABLE HOLDOVER TIME TABLE CELL.

FLUIDS USED DURING GROUND DE-ICING ARE NOT INTENDED FOR AND DO NOT PROVIDE ICE PROTECTION DURING FLIGHT.

A LIST OF THE APPROVED HOT TABLES IS INCLUDED IN THE CURRENT COMMERCIAL AND BUSINESS AVIATION ADVISORY CIRCULAR (CBAAC) "GROUND ICING UPDATE".

Fig. 2-4: Sample holdover table.

2.14 Fire Detection and Extinguishing Systems

In advanced aircraft, there are a variety of methods to achieve fire detection. HEAT SENSORS can be located in various parts of the engine compartment. INFRARED LIGHT SENSORS are not sensitive to heat, but can detect the light from a flame or glowing hot metal. SMOKE DETECTORS can be located in the baggage compartment or engine compartments. Once activated, the fire detection system will illuminate a cockpit warning indicator light.

More advanced aircraft will have a means of FIREWALL SHUTOFF, whereby a cockpit control will cut off fuel and hydraulic lines to the affected engine.

Advanced aircraft will employ a fire extinguishing system where extinguishing chemicals are contained in pressure bottles in the engine compartments. Extinguishing agents can be CO_2 or various forms of HALON. Before activating an extinguishing system, pressurized air from the affected engine should be secured. As with fire extinguishing bottles, the pre-flight inspection should include a check of the proper pressure indications and that the systems will be functional.

A CLASS A FIRE is a fire in an ordinary combustible material. On these water or solutions containing large percentages water are most effective.

A CLASS B FIRE is a fire in a flammable liquid, grease etc. On these fires, a blanketing effect is essential.

A CLASS C FIRE is an electrical fire. On these the use of a non-conducting extinguishing agent is essential.

Carbon Dioxide extinguishers are acceptable when the principal hazard is a Class B or Class C fire.

Water extinguishers are acceptable when the principal hazard is a Class A fire and where a fire might smoulder if attacked solely by other types of extinguishing agents.

Dry chemical extinguishers are best suited for Class B or C fires.

Halon extinguishers suited for all Classes of fire.

2.15 Pneumatic Systems

Pneumatic systems involve the use of compressed air for such requirements as operation of landing gear, flaps, brakes, cargo doors and other forms of mechanical operation. Some aircraft may have a bottle of compressed nitrogen for back-up emergency power to lower landing gear or for brake application on failure of the primary hydraulic system.

As well, pneumatic systems involve positive air pressure that can be used to drive vacuum pumps for powering vacuum operated instruments. The pumps are driven by the engine and pull clear air from the back of the vacuum instruments, thereby creating the suction needed to drive them.

Pneumatics can also be used for ice control systems, such as with de-icer rubberized boots, that inflate and shake off ice when pneumatic pressure is supplied.

On turbine powered aircraft, bleed air from the engines is used to power pneumatics. HIGH PRESSURE BLEED AIR is drawn from the compressor section of the turbine and is used for engine de-ice, cabin pressurization, heating, cooling, powering starters etc. LOW PRESSURE BLEED AIR is obtained as high pressure bleed air is fed through a pressure regulator, and is used for air gyros for flight instruments, de-icing systems, and pressurization outflow valves.

2.16 Oxygen Systems

Supplemental oxygen allows for flight at higher altitudes in non-pressurized aircraft.

With a CONSTANT FLOW oxygen system, a steady metered flow of oxygen is provided to a simple mask having a facepiece and expandible plastic bag. The oxygen comes from a storage bottle via a series of low pressure lines to various mask locations throughout the aircraft. The bag is a REBREATHER, which allows more efficient utilization of oxygenated air escaping from the lungs.

The person using the mask will breathe a combination of fresh oxygen mixed with re-breathed air in the bag. In a constant flow system, the oxygen flows at a fixed rate, or can be manually adjusted by the pilot for different altitudes. Other regulators will provide the appropriate flow of oxygen for the altitude flown at.

The plastic oxygen hoses have a FLOW INDICATOR, which shows a red line if the flow is insufficient. In DEMAND OXYGEN SYSTEMS, oxygen flows into the mask only when the person inhales. This type has a more advanced face mask (there is no re-breather bag).

Before using any oxygen system, the pilot must become fully knowledgeable as to the operation of the system in use. As well, the pilot should know the background medical physiology of oxygen in the human body and the effects of hypoxia.

2.17 Heating Systems

Light reciprocating aircraft often use exhaust muffs for cabin heating, where fresh air is ducted around the exhaust muff and permitted to enter the cabin via mechanical on - off valves in the heating line.

Other aircraft may employ a FUEL HEATER, which is an independent, fuel burning tube. Air flowing around the outside of the tube is heated by it and ducted into the cabin. This greatly reduces the danger of carbon monoxide poisoning. Additional components will include a fuel pump, fuel valve and pressure switch to insure adequate combustion airflow before the fuel flow is turned on, and an ignition system. The system will have an overheat protection switch so the heater will shut down if the thermostat cannot regulate the proper temperature. Fuel heaters can be tricky to operate and the pilot should be fully knowledgable on the proper use of the heater for the aircraft flown.

In turbine powered aircraft, hot bleed air is tapped upstream from the engine combustion chamber, and is mixed with cold outside bleed air to maintain a desired temperature.

2.18 Cabin Pressurization Systems

Engine bleed air is used to drive pressurization systems. OUTFLOW VALVES maintain pressurization. Cruise cabin altitude is normally set on the flight/cabin selector prior to takeoff. After take-off, the cabin altitude will climb to the assigned cruise cabin altitude setting at a specified rate. PRESSURE RELIEF VALVES prevent excessive pressures within the cabin. NEGATIVE PRESSURE RELIEF VALVES open to prevent outside pressure from exceeding cabin pressure, and are activated by pressure. On landing, a relay switch depressurizes the system.

The structure of an aircraft is limited to a MAXIMUM PRESSURE DIFFERENTIAL i.e. the difference between the inside cabin pressure and the outside pressure. The maximum cabin pressure differential is generally from 8.0 - 8.9 PSI, depending on the aircraft type. Differential pressures above this limit will cause the aircraft structure to fail.

As an aircraft climbs, when the maximum pressure differential is reached, the internal cabin altitude must be increased (pressure decreased). Safety valves (outflow valves) will open if the maximum pressure differential is reached.

Cabin pressurization systems are either automatic or manual and vary in sophistication. Generally, a desired cabin pressure and rate of cabin climb or descent can be selected.

To determine the required cabin rate of climb inside the aircraft, first find the change in cabin altitude, and the time to climb to the desired altitude. Then, divide the cabin altitude change by the climb time.

EXAMPLE: An aircraft will climb from an airport at elevation 3,000 feet ASL to FL 290. The expected rate of climb is 1,500 FPM and the desired cabin altitude is 7,000 feet. What is the required cabin rate of climb?

ANSWER: 231 FPM. Change in cabin altitude = 4,000 feet, time taken to reach FL 290 =

$$\frac{26{,}000 \text{ feet}}{1{,}500 \text{ FPM}} = 17.33 \text{ mins}$$

Required cabin rate of climb:

$$\frac{4{,}000 \text{ feet}}{17.33 \text{ mins}} = 231 \text{ FPM}$$

2.19 Air Conditioning Systems

In aircraft air conditioning systems, a special refrigerant fluid is forced to compress by a COMPRESSOR, which heats it up, and is then expanded causing a cooling effect. The compressor is driven by a belt off the engine or by an electric motor. The HEAT EXCHANGER allows for cooling of the fluid. A second exchanger is located in the cabin air ducting, and the expansion of the fluid yields the cooling effect.

Because some air conditioning systems are belted to the engine and extract power, such aircraft will have placards against using the air conditioner during take off and landing. This is not normally an inconvenience because the system will continue to have a cooling effect for a few minutes even after it is turned off.

2.20 Autopilot Systems

An autopilot operates the aeroplane flight controls. Large aeroplane autopilots are generally 3-axis and control aircraft attitude in pitch, roll and yaw. Most autopilots will maintain altitude, hold a desired heading, intercept and hold a VOR radial, intercept a localizer, maintain descent rates and level off at a desired altitude. With the advent of Flight Management Computer (FMC), modern large aeroplane autopilots can store flight parameters including RNAV routes, IAPs, auto-throttle airspeed control and autoland functions.

Even though autopilots can have sophisticated abilities, the aircraft should always be carefully monitored when utilizing an autopilot as they can be complex to operate and not always correctly programmed.

Notes:

Instruments

3.01 Pitot Static System

The pitot static system consists of the pitot tube, the static source and the various tubes connecting the airspeed indicator, the altimeter, and the vertical speed indicator. The openings of both the static source and the pitot tube should be checked during the preflight inspection for obstructions.

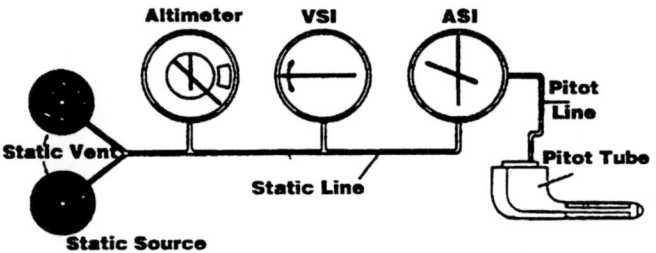

Fig. 3-1: Pitot Static System - representation.

STATIC PRESSURE is atmospheric pressure acting upon an aircraft. It decreases with increased altitude. Static pressure is sensed though holes mounted on the side of the aircraft at a right angle to the relative airflow. Static or ambient air flows into the static source. DYNAMIC PRESSURE is pressure due to the motion of an aircraft. Dynamic or on-rushing air flows into the pitot tube.

The function of the PITOT HEAT is to prevent or remove ice in the pitot tube. The pitot heat control should be activated and tested during the preflight.

IFR aircraft are required to be equipped with an ALTERNATE STATIC SOURCE, which is used if the normal static source becomes blocked. When using the alternate source, since the aircraft is moving, the inside pressure will be slightly less than outside, and this will cause airspeed indications to be slightly high, and the altimeter to over-read.

3.02 Airspeed Indicator

The airspeed indicator (ASI) requires both the pitot and static source. It measures the differential between pitot pressure and static pressure to calibrate airspeed.

Speed is a rate of change of position or distance travelled per unit of time. Airspeed is normally expressed in

knots (nautical miles per hour) or MPH. Note the following definitions:

a) IAS: Indicated Airspeed. Read directly off ASI.
b) CAS: Calibrated Airspeed. IAS corrected for instrument and position error. Refer to owner's handbook for table indicating corrections. This error is greatest at low airspeeds. While in the cruising and higher airspeed ranges, IAS and CAS are about the same.
c) TAS: True Airspeed. CAS corrected for altitude and temperature changes. Use flight computer.
d) EAS: Equivalent Airspeed. EAS compensates CAS for compressibility error.

Refer to S. 5.02 in this text for information and examples of TAS calculations and conversions.

EXAMPLE: Determine TAS from the following.

An ASI measures dynamic pressure and displays the forward speed of an aircraft through the air. It assumes that the surrounding air mass conforms to ICAO Standard Atmosphere Sea Level conditions. With forward motion, the inner capsule expands, and the degree of expansion is translated to pointer movement.

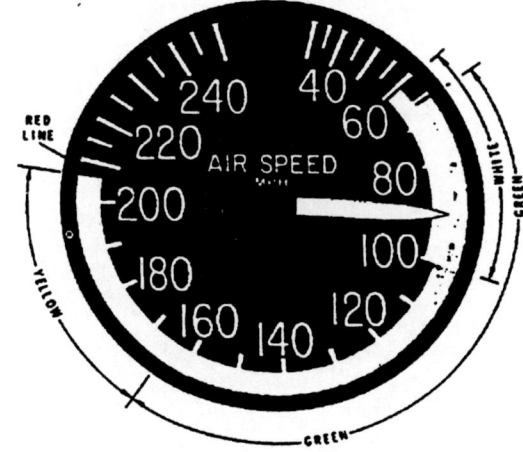

Fig. 3-2: Face of ASI.

The airspeed indicator has various reference markings. The green arc is the range of speeds for normal operations. The yellow arc represents the caution range, and the red line the never exceed speed Vne. The maneuvering speed, Va, is not colour coded.

Errors affecting the ASI are POSITION ERROR (caused by the position of the static ports sensing different readings) and DENSITY ERROR (caused by changes in altitude and temperature). COMPRESSIBILITY ERROR, as stated, is only a factor for high speed aircraft (300+ kts). If the pitot tube becomes blocked, the ASI needle will show a gradual increase in indicated airspeed with height, and it will act "like an altimeter". If the static source becomes blocked, the ASI will under-read in a climb and over-read in a descent.

A useful rule of thumb is that TAS increases by 2% the IAS per 1,000 ft gain in altitude.

EXAMPLE: IAS at 2,000 ft is 100 kts. What will the TAS be at 10,000 ft?

ANSWER: TAS will be 116 kts. [2% X 8 = 16%. 16 kts + 100 kts = 116 kts].

The effect of increased TAS with altitude is very prominent when landing at a high elevation airport, because even though the IAS will remain the same, the TAS and therefore groundspeed at touchdown will be higher.

3.03 Machmeter

The Machmeter is found on high speed, high performance aircraft. It indicates the ratio of aircraft TAS to the speed of sound at flight altitude. The Mach pointer is actuated mechanically by the pressure differential between impact air and static air pressure. Normally, in computing TAS from IAS, air density must be considered. This requires a correction for temperature and altitude. With a Mach number, these corrections are unnecessary because the existing temperature at flight level determines the speed of sound at flight level. The Mach number is determined by the speed of sound, which in turn is determined by air density - therefore, Mach is always a valid index to the speed of the aircraft.

Fig. 3-3: Face of Machmeter.

3.04 Vertical Speed Indicator

The VSI operates solely on the static pressure line, and measures the rate of change of atmospheric pressure when climbing or descending. The VSI is both a TREND instrument (indicates whether aircraft is climbing or descending) and a RATE instrument (gives actual rate of climb or descent in hundreds of feet per minute). A VSI has an aneroid capsule mounted inside a sealed case. The static pressure line is connected directly to the capsule, and the inside of the case is vented to static pressure through a metering device (calibrated leak). A linkage connects the capsule to the pointer.

Fig. 3-4: Face of Vertical Speed Indicator.

Any change of aircraft altitude causes a pressure differential between the static air trapped inside the capsule and the static air that surrounds the capsule inside the case. The resulting expansion or contraction of the capsule produces pointer displacement.

A VSI is subject to LAG ERROR (reliable indications are not presented until pressure differential has stabilized, which may require as long as 6-8 seconds). The VSI is also subject to REVERSAL ERROR where abrupt changes in attitude may cause the VSI to momentarily show a change in the wrong direction.

3.05 Altimeter

HEIGHT is vertical distance measured from a specified datum. ALTITUDE is vertical distance of an object in the air measured from mean sea level. ELEVATION is vertical distance of an object on the Earth's surface measured from mean sea level.

The altimeter is an aneroid barometer which measures changes in atmospheric pressure. Through connections these changes are recorded on the face of the instrument as changes in altitude. The altimeter setting is used to adjust the altimeter to read correctly in non-standard atmospheric conditions. The altimeter face has a long thin pointer representing hundreds of feet and a short, wide pointer indicating thousands of feet. A small triangle indicates 10 thousands of feet. The altimeter setting scale is graduated from 28.0 to 31.0 inches of mercury. A knob is on the left side of the instrument to adjust the altimeter setting.

Fig. 3-5: Face of Altimeter.

On the inside of the altimeter is a partially evacuated capsule which responds to changes in air pressure like a barometer. The altimeter utilizes the static source of the pitot static system. As the aircraft climbs, the capsule expands, causing an indication of increased altitude on the face of the instrument. The altimeter setting mechanism allows selection of standard pressure 29.92 or the local airport altimeter setting.

INDICATED ALTITUDE is the altitude shown by an altimeter with the current altimeter setting applied. CALIBRATED ALTITUDE is indicated altitude corrected for instrument and position error. PRESSURE ALTITUDE is the altitude shown by an altimeter set to standard pressure, 29.92 inches of mercury. DENSITY ALTITUDE is pressure altitude corrected for non-standard temperature and is calculated on the flight computer.

TRUE ALTITUDE is the actual altitude of an aircraft above mean sea level.

ABSOLUTE ALTITUDE is the actual height of an aircraft above the surface. The ALTIMETER SETTING (QNH) is the observed barometric pressure at a selected datum, corrected for temperature, and reduced to mean sea level, assuming ICAO conditions. Once applied to the altimeter subscale, the altimeter displays altitude above mean sea level. One inch on the altimeter subscale represents approximately 1,000 feet change in altitude.

The altimeter has various errors. POSITION ERROR is caused by different indications resulting from changes in the static source angle to the airflow. In some aircraft, position error can be considerable. INSTRUMENT ERROR is caused by the aneroid capsule not assuming the precise size required for a particular pressure difference. MECHANICAL ERROR is caused by misalignment or slippage in the gears and mechanical linkage connecting the altimeter pointer. TEMPERATURE ERROR occurs when extremely cold temperatures prevail, causing the altimeter to over-read (see s. 1.25). ELASTIC ERROR, also called HYSTERESIS, is a lag error in altitude indications caused by the elastic properties of the materials used in the aneroid capsule. Elastic error occurs following large or rapid altitude changes. REVERSAL ERROR occurs during abrupt or rapid attitude changes and is only momentary in duration.

When the local altimeter setting has been set, the altimeter should indicate airport elevation, with a

maximum error of +/- 50 feet.

A RADAR ALTIMETER or RADIO ALTIMETER provides a continuous indication of height above ground. The system accurately measures the distance between the ground and the aircraft in absolute altitude.

Fig. 3-6: Radar Altimeter.

The time interval between a transmitted and received radio signal is automatically converted to an altitude read-out. The read-out can be a logarithmic scale or digital. The main functions of a radar altimeter are to serve as a ground proximity warning, to provide a cross check for the barometric altimeter, and to indicate progress on final approach descent to MDA or DH on an IAP. Radar altimeters usually have a warning marker which can be set to a predetermined altitude AGL to generate a visual or audio warning when the aircraft reaches the preset value. Errors may occur during flight over a surface into which radio waves can penetrate (ice, deep snow).

An ENCODING ALTIMETER is a specialized altimeter that sends out a pressure altitude read-out to ATC and is part of a MODE C transponder system. Refer to s. 7.12 for more detail on encoders.

3.06 Magnetism & Magnetic Compass Systems

The Earth is a magnet, having a north and south pole. Magnetic north is in a different location than geographic north. The difference between magnetic north and true north is termed MAGNETIC VARIATION. The Earth's magnetic lines of force dip towards the poles, which is MAGNETIC DIP.

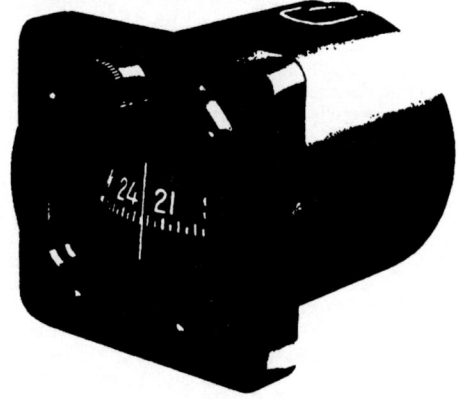

Fig. 3-7: Float based magnetic compass.

The aircraft magnetic compass is based upon two parallel magnetized needles, which are mounted on a float having a compass card.

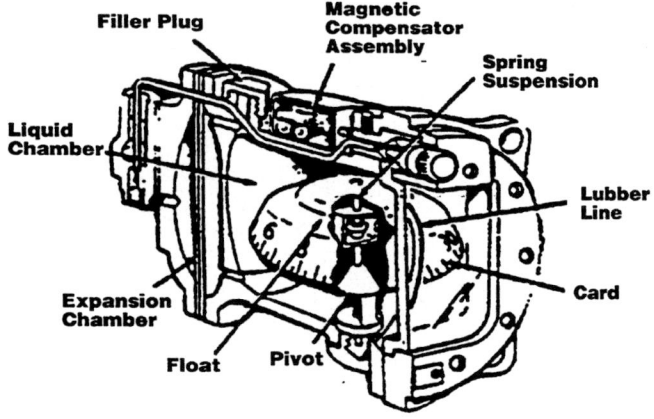

Fig 3-8: Components of Magnetic Compass.

The float assembly is mounted on a pedestal, and is sealed in a case filled with kerosene.

Because the centre of gravity of the compass is below the pivot point, it shows erroneous indications depending on whether the aircraft is accelerating or turning. These errors are caused by magnetic dip. NORTHERLY TURNING ERROR occurs as an aircraft turns to north or south. On turns to or from north, the compass LAGS behind the actual heading. On turns to or from south, the compass LEADS ahead of the actual heading.

ACCELERATION ERROR occurs as the aircraft accelerates or decelerates on east-west headings. As an aircraft accelerates such as during a transition to a climb as power is increased, the compass indicates slightly to the north. As an aircraft decelerates, such as following a power reduction prior to entering a descent, the compass indicates slightly to the south. Acceleration errors can be remembered with the word "ANDS": **A**ccelerate = **N**orth indications, **D**ecelerate = **S**outh indications. To avoid turning error and acceleration error, insure the aircraft is in level, unaccelerated flight prior to relying on indications from the magnetic compass.

EXAMPLE: What compass error results where an aircraft is turning from 330 to 360?

ANSWER: Turning error: compass lags behind actual heading.

EXAMPLE: Which compass error will be greater: a turn from 270 to 300, or from 300 to 330?

ANSWER: The turn from 300 to 330 will indicate greater turning error, because turning error is zero on E - W headings, and increases the closer the aircraft is to N or S.

The compass needle is affected by the magnetic fields generated by the electrical equipment and by metal components in the aircraft. These disturbances deflect the compass needles from alignment, and are termed DEVIATION. To reduce deviation, the compass is SWUNG, which means adjusted such that the compass indicates as correctly as possible. A compass must be swung once per year. The remaining deviation between magnetic heading and compass heading is accounted for on a COMPASS CORRECTION CARD. To fly compass

headings, refer to the compass correction card for corrected headings to steer.

EXAMPLE: If the desired magnetic heading is 090°, and the compass deviation is -1°, what compass heading should be steered?

ANSWER: 089°.

COMPENSATION refers to adjusting the magnetic compass to reduce deviation error. Proper compensation of the compass is best performed by a properly qualified technician. Since the magnetic forces within the aeroplane change, because of landing shocks, vibration, mechanical work, or changes in equipment, the deviation should be checked as required.

To swing a compass, the aircraft is placed on a magnetic compass rose on the ground, the engine started, and electricals normally used (such as radios) are turned on. The aircraft may have to be jacked up into flying position. The aircraft is then aligned with magnetic north indicated on the compass rose and the reading shown on the compass is recorded on a deviation card. The aircraft is then aligned at 30° intervals and each reading is recorded. If the aircraft is to be flown at night, the lights are turned on and any significant changes in the readings are noted. If so, additional entries are made for use at night.

A GYROMAGNETIC REMOTE INDICATING MAGNETIC COMPASS or GYROSIN COMPASS consists of a master unit which is mounted in the rear fuselage to minimize magnetic interference from the engine (s). The master unit consists of an electrically driven gyroscope, the axis of rotation of which is stabilized relative to the Earth and under the influence of a magnetic element. The cockpit display is similar to a conventional compass, with the principle advantage that the system is not affected by magnetic dip errors.

3.07 Outside Air Temperature Gauge

An Outside Air Temperature (OAT) gauge can be either mechanical or electrical. On light aircraft, it is a simple thermometer where airflow enters a tube and the temperature is read off the face of the instrument.

On more advanced aircraft, a mechanical OAT gauge is based on two metal strips within a spring coil which reacts to changes in ambient temperature.

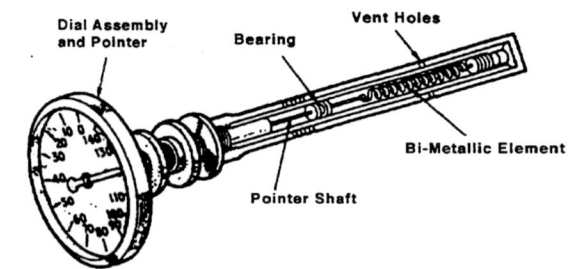

Fig. 3-9: OAT gauge.

3.08 Gyroscopic Instruments

The attitude indicator, heading indicator, and turn coordinator/turn and slip indicator are controlled by gyroscopes. A gyroscope is a spinning mass on an axis. Gyroscopes are classified according to their relationship to the earth's surface, either horizontal or vertical.

All gyroscopes exhibit RIGIDITY IN SPACE and GYRO-SCOPIC PRECESSION. Rigidity in space means that the axis always points to the same position in space, regardless of the position of the earth. Precession means that a force applied to the gyroscope wheel will be felt 90 degrees from the point of impact, in the direction of rotation.

Gyroscopic instruments are typically powered either by suction (heading indicator, attitude indicator) or electricity (turn coordinator/turn and bank indicator). The vacuum system spins the gyro by drawing a stream of air against the rotor vanes to spin the rotor. Either a venturi tube or an engine driven vacuum pump can be used to provide the suction required to spin the rotors. The amount of vacuum required is usually from 3.5" to 4.5" Hg.

The older venturi tube systems were based on a tube mounted on the exterior fuselage. The vacuum would only be effective at speeds of about 90 kts or more, and could not be used for IFR flight.

A typical modern vacuum system consists of an engine-driven vacuum pump, and air/oil separator, a vacuum regulator, a relief valve, an air filter, and tubing and · necessary to complete the connections. A suction gauge on the instrument panel indicates the amount of vacuum in the system.

3.09 Attitude Indicator

The Attitude Indicator duplicates the natural horizon, and indicates pitch attitude (nose up/nose down) and bank attitude (left, right).

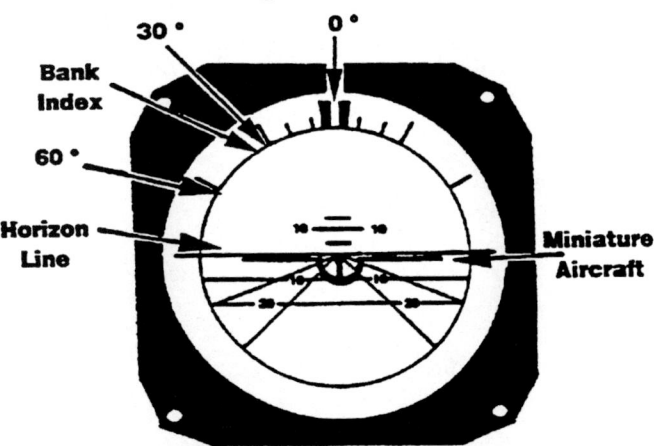

Fig. 3-10: Face of Attitude Indicator.

The attitude indicator utilizes the gyroscopic principles of rigidity in space (to establish a reference plane) and precession (to keep the gyro axis perpendicular to the Earth's surface). The gyroscope of an air-driven attitude indicator is vertically mounted so that its housing is free

to pivot about the vertical axis. The gyro rotor is driven by an air jet which strikes buckets on the rotor.

The horizon bar is linked to the gyroscope by a lever, which is attached to a pivot on the rear of the gyroscope frame and connected to the housing by a pin. The result is that as the aircraft climbs, dives, or banks, the instrument case rotates on the gyroscope gimbals while the bank index and horizon bar remain rigid, reflecting any movement of the aircraft about the pitch and bank axes. The attitude indicator has a correcting mechanism, to maintain the gyroscope's position.

The attitude indicator is subject to errors caused by bearing friction in the instrument. Other errors result from acceleration and deceleration, such that during forward acceleration, for example, the attitude indicator falsely displays a climb. Deceleration has the opposite effect.

3.10 Heading Indicator

The operational principle of the heading indicator is similar to that of the attitude indicator, with the exception that the gyroscope axis of the heading indicator is horizontally mounted.

Fig. 3-11: Face of Heading Indicator.

The gyro rotor is air driven. During turns, the casing turns about the gyro, which is geared to rotate an azimuth compass bearing card, indicating a change in direction.

Bearing friction and the rotation of the earth cause the heading indicator to drift, so the heading indicator should be reset about every 15 minutes. Due to the errors associated with the magnetic compass, the heading indicator should only be set to the compass when in level flight.

3.11 Turn Coordinator

A Turn Coordinator is a combination of two instruments. A miniature aircraft indicates direction and rate of turn, roll rate and turn coordination. The turn coordinator is also a useful secondary bank reference. The ball is a separate instrument, called an INCLINOMETER, which indicates gravity, and whether the aircraft is in coordinated flight, is slipping or is skidding.

A RATE ONE TURN or STANDARD RATE TURN is a turn which provides 360° change in direction in two minutes (3° per second). A HALF STANDARD RATE TURN is 4 minutes for 360° change in direction (1.5° per second) and is the preferred rate turn used by high speed aircraft.

Fig. 3-12: Face of Turn Coordinator.

When there is an in-flight failure of gyro suction, the turn coordinator will continue to function as it is electrically driven, while the attitude indicator and heading indicator will fail. Using the principle that a rate one turn involves a change of direction of 3° for every second, the aircraft can be turned to various headings based on TIMED TURNS. For example, a change in heading of 90° would require a rate one turn for a time of 30 seconds.

The angle of bank for a Rate 1 turn can be calculated based on the TAS: take 10% of the airspeed, then add 7 (if using kts, 5 if using MPH), and this will represent the angle of bank for a Rate 1 turn.

EXAMPLE: What is the angle of bank needed to achieve a standard Rate 1 turn where the TAS is 160 kts?

ANSWER: 23°. Take 10% of 160 = 16, add 7 = 23°.

The turn coordinator's construction is based on a laterally mounted, electric driven gyroscope. The gyro is free to tilt about the aircraft's longitudinal axis and is mechanically joined, through a linkage, to the turn pointer. A restraining spring links the gyro assembly to the instrument case. Mechanical stops prevent the gyro from tilting more than 45° either side of centre.

The TURN AND BANK INDICATOR is an older instrument which provides basically the same information as the turn coordinator (a turn needle is used instead of a miniature aircraft). The turn and bank indicator will display rate of turn and turn coordination, but not roll rate.

The inclinometer consists of a black or white ball inside a sealed, curved glass tube filled with liquid to provide damping action and ensure smooth operation of the ball. The ball is influenced by gravity and centrifugal force. During coordinated flight these two forces are in balance, and the ball rests in the central lower portion of the tube. When the forces become unbalanced, the ball moves away from the centre and indicates uncoordinated flight as a slip or skid.

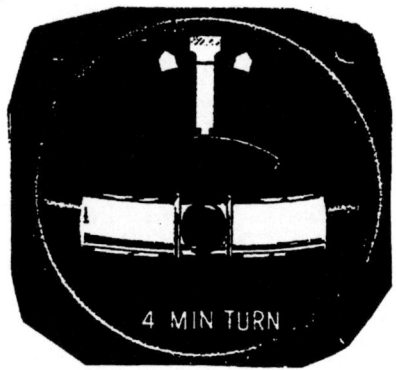

Fig. 3-13: Face of Turn and Bank Indicator. Note this one is calibrated for a 4 minute turn, i.e. 360° will take 4 minutes, or 1.5° per second.

In a SLIP, the rate of turn is too slow for the angle of bank, and the lack of centrifugal force causes the ball to be displaced to the inside of the turn. To correct, decrease the angle of bank, or increase the rate of turn, or both.

In a SKID, the rate of turn is too fast for the angle of bank, and excessive centrifugal force causes the ball to be displaced to the outside of the turn. To correct, increase the bank angle, or decrease the rate of turn, or both.

EXAMPLE: An aircraft turn coordinator shows ball right of centre, needle rate one to the right. What is the aircraft doing, and what is needed to establish a coordinated turn?

ANSWER: The aircraft is slipping to the right. A reduction in bank with left stick pressure coordinated with right rudder pressure will bring the aircraft into a coordinated turn.

3.12 Flight Director

The flight director indicator displays aircraft attitude, computed flight commands, localizer, glideslope, radar altitude for DH, and warning flags for localizer/glideslope failure. It is a multi-purpose display that combines attitude information with navigational information.

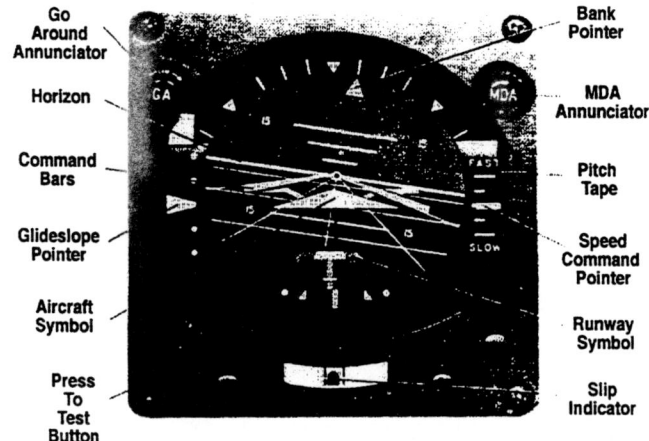

Fig. 3-14: Flight Director Indicator.

By utilizing the command bars, a preprogrammed flight path can be selected. To comply with the command bars, the pilot maneuvers the aircraft to line up the fixed aircraft symbol with the command bars. Various modes can be set up with the flight director control panel, such as for localizer, altitude, vertical speed, heading, automatic approach and go around.

More sophisticated flight directors that receive data from multiple on board computers are referred to as Flight Management Systems (FMS).

3.13 Horizontal Situation Indicator

The HSI in a multi purpose instrument that displays both navigation and heading information at the same time. Displays can be selected for VOR, DME, ILS, localizer or ADF. It is oriented as if looking down on the aircraft's horizontal position in relation to various selected navigational aids.

Fig. 3-15: Horizontal Situation Indicator

The principal advantage to the HSI is that the pilot can maintain more accurate interpretation control for instrument approaches and enroute tracking using radio navigation aids.

3.14 Electronic Flight Instrument System (EFIS)

EFIS displays replace conventional mechanical instrumentation with cathode ray tube (CRT) displays. The most common EFIS systems are based on CRT flight directors and CRT HSI's. The EFIS displays are much larger and various displays can be generated to isolate critical information for a particular flight phase. EFIS displays are also commonly named GLASS COCKPIT displays.

A PRIMARY FLIGHT DISPLAY (PFD) EFIS generates all the flight instrument information that would be provided with conventional instrumentation in a "Standard T" configuration.

Some EFIS displays provide MULTI-FUNCTION DISPLAYS (MFD) such that the pilot can select displays for flight information: Electronic Attitude Director Indicator (EADI), for navigation information: Electronic HSI (EHSI), engine information, weather radar, moving map displays that can be superimposed on HSI indications

along with waypoints, navaids, airports, incorporation of TCAS information, and other data inputs. EFIS displays can also have the ability to enlarge displays i.e. for enroute navigation and approaches.

SYMBOL GENERATORS are the basis of the EFIS display system: the symbol generators receive inputs from various aircraft systems, respond to these inputs, then generate the proper visual display for the EFIS control panel selected.

3.15 Radio Magnetic Indicator (RMI)

The RMI displays the present bearing to or from an NDB or VOR station and the present heading. When an ADF is selected, the pointer head displays the track to the station and the tail displays the track from the station. If a VOR is selected, the tail pointer shows the radial which the aircraft is over. For more information on using an RMI, refer to s. 7.04.

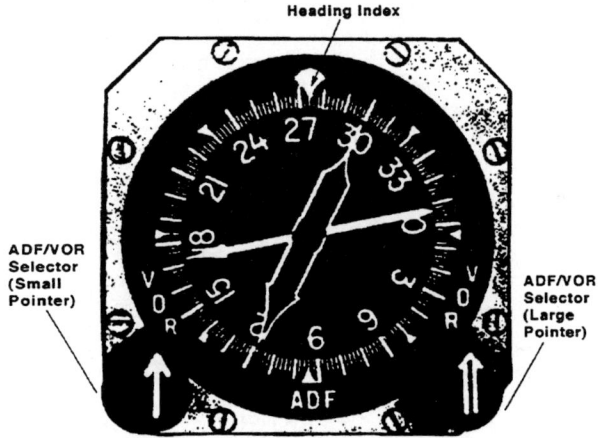

Fig. 3-16: Face of RMI.

3.16 Angle of Attack Indicator

Typically an AOI will consist of a vane or probe that is flush mounted to the side of the aircraft and will give an indication of an aircraft's angle of attack. An AOI system is used to detect critical angles of attack on approach and landing, to provide a basis for stall warning system activation and to assist in verifying in flight attitude and airspeed.

3.17 Engine Instruments

The following engine instruments should be known:

a) TACHOMETER. Indicates speed of engine rotation, calibrated in Revolutions Per Minute (RPM). May be graduated in percent of maximum RPM for the particular engine. Can also be a split needle for dual engines.
b) OIL PRESSURE GAUGE. Indicates oil pressure being supplied to moving engine parts by lubricating system. It warns of impending engine failure which may result from an exhausted oil supply, failure of oil pump, broken oil lines etc.
c) OIL TEMPERATURE GAUGE. Indicates temperature

of oil entering the engine. With a very low oil temperature, the oil may be so thick that it is not able to lubricate the internal moving engine parts. With an excessively high temperature the lubricating oil tends to thin out and decreases the lubricating qualities of the oil.

d) FUEL QUANTITY GAUGES. Allow the pilot to determine the approximate amount of fuel in tanks. Pilots should ALWAYS complete a visual check of fuel prior to flight and never rely on fuel quantity gauges.

e) FUEL PRESSURE GAUGE. Indicates the difference in pressure between the fuel and the air being supplied to the engine, and therefore whether fuel is being supplied and the actual pressure at which it is being forced into the engine. Fuel pressure gauges are very important for low wing aircraft: without fuel pressure from engine driven or electrical fuel pump, fuel will not reach engine.

f) MANIFOLD PRESSURE GAUGE. Indirectly indicates the power output of the engine by measuring the pressure of the air in the fuel/air induction manifold. The higher the manifold pressure, the greater the power being developed by the engine. By means of the throttle, the pilot can control the pressure in the manifold, and thus the power. The manifold pressure gauge is calibrated in inches of mercury, and is found in aeroplanes equipped with a supercharger or a constant speed propeller.

g) CYLINDER HEAD TEMPERATURE GAUGE, Indicates the temperature of the cylinder head of the hottest cylinder (usually one of the rear ones in a horizontally-opposed engine) and gives the pilot a means of determining whether the engine is operating at normal or excessive engine temperatures. It is an important instrument for high compression/high power engines. Operating the engine at a temperature higher than it was designed for will cause loss of power, excessive oil consumption, and damage to the cylinder walls, pistons and valves.

h) EXHAUST GAS TEMPERATURE (EGT) GAUGE. Indicates temperature at exhaust manifold. An important instrument to determine whether proper combustion has taken place. The leaner the mixture, the hotter the EGT. The richer the mixture, the cooler the EGT. Can also be useful in leaning out mixture at altitude, where lean mixture is established just to rich side of peak EGT.

i) TORQUEMETER. Gives an accurate indication of the shaft horsepower input into the reduction gear assembly of a turboprop/turboshaft engine and is used in power control.

j) EPR GAUGE. Gives ratio of tailpipe total pressure divided by compressor inlet total pressure.

k) TURBINE INLET/EXHAUST TEMPERATURE GAUGES. Gives turbine temperatures to insure limits not exceeded.

3.18 Ground Proximity Warning System

The Ground Proximity Warning System (GPWS) is designed to prevent inadvertent flight into terrain (for regulations on GPWS, refer to s. 1.40). The GPWS system continuously monitors the aircraft's flight path with respect to radio altitudes under about 2,500 feet AGL. If the projected flight path would result in impact, the system issues a warning to flight crew. Warnings issued via the GPWS system are both auditory and visual. The visual warning is a red cockpit light, labelled PULL UP, and the audio warnings consist a loud "whooping" signal along with various combinations of spoken words such as "SINK RATE", "PULL UP", "TERRAIN", "DON'T SINK", "TOO LOW GEAR", TOO LOW FLAPS" and "TOO LOW TERRAIN". GPWS units also have warnings when too far below the glideslope that include an amber "BELOW GLIDE SLOPE" cockpit light along with an aural "GLIDE SLOPE" alert. GPWS units can also provide windshear alerts i.e. red warning lights along with a siren and spoken word "WINDSHEAR".

3.19 Central Air Data Computer

High performance aircraft may have a Central Air Data Computer (CADC) that receives inputs from ram air, static air, outside air temperature and converts this data into electrical signals. Instruments that utilize this data include the ASI, VSI and altimeter. Using the CADC system, there are fewer errors associated with mechanical linkages as in conventional instruments. CADC data is also used in aircraft pressurization systems.

Notes:

Notes:

Flight Operations

4.01 General Theory of Flight

Wing Loading: Load Factor

LOAD FACTOR is the weight a wing carries or "G" force. In level flight, lift = weight and the load factor is 1.0. In gusts or turbulence, the load factor increases momentarily (lighter aircraft experience greater load factor in turbulence). Note the following load factors:

Angle of Bank	Load Factor
0°	1.00
15°	1.04
45°	1.41
60°	2.00
75°	4.00

To calculate the stalling speed for a given angle of bank, use this formula:

Vso in Turn = Vso X √LOAD FACTOR

EXAMPLE: For an aircraft with a 60 kt stall speed, what will be the stall speed in a 60° bank?

ANSWER: 84.9 kts.

Range and Endurance

Based on the POWER REQUIRED CURVE, for reciprocating engine, propeller driven aeroplanes, MAXIMUM ENDURANCE is obtained at the point of minimum power required since this would required the lowest fuel flow to keep the aeroplane in steady, level flight. MAXIMUM RANGE would occur where the proportion between velocity and power required is greatest and this point is located by a straight line from the origin tangent to the curve. The maximum range condition is obtained at maximum Lift/Drag ratio. The Lift/Drag ratio is the relationship between lift and drag and is calculated by dividing the lift coefficient by the drag coefficient. It is important to note that max L/D for a particular aeroplane configuration occurs at a particular angle of attack and lift coefficient and is unaffected by weight or altitude. The angle of attack for best L/D varies with wing design, but is substantially smaller than the stalling angle - for many aircraft best L/D is at 4 - 6° angle of attack where stall is at 16° - 18°.

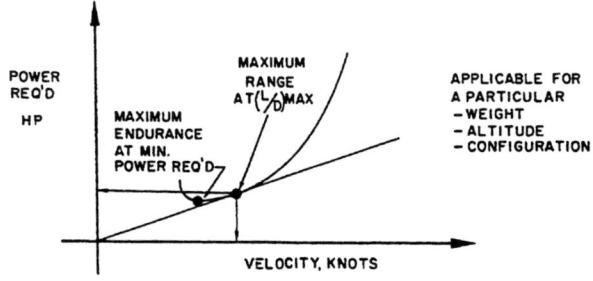

Fig. 4-01: General Range Conditions, Propeller Aeroplane.

In respect of turbojet aeroplanes, maximum endurance is obtained at the maximum L/D since this would incur the lowest fuel flow to keep the aeroplane in steady, level flight. Maximum range is obtained at the aerodynamic condition which produces a maximum proportion between the square root of the Coefficient of Lift and Coefficient of Drag i.e. $(\sqrt{C_L}/C_D)_{max}$. The key consideration with respect to turbo-jet range performance is altitude and no other single item can cause such large variations in range performance i.e. range performance is far better at high altitudes.

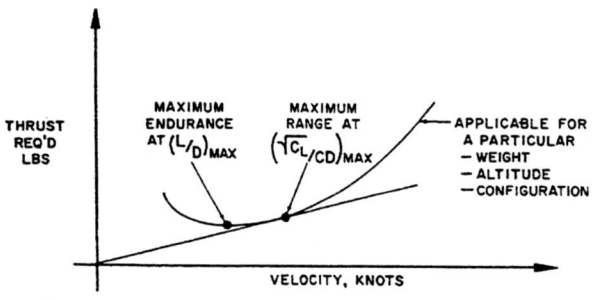

Fig. 4-02: General Range Conditions, Turbojet Aeroplane.

Slow Flight

SLOW FLIGHT is defined as the range of airspeeds from the stall to the speed for maximum endurance. Also known as flight in the region of "reversed command" or "behind the power curve", aircraft handling in slow flight will be sluggish, high power settings will be required and a high nose-up attitude will be required.

Induced Drag

Wing tip vortices cause induced drag. Induced drag is inversely proportional to the square of the velocity: halving the velocity increases induced drag fourfold. Vortex strength depends on wing shape, weight, and speed. Large, heavy, slow aircraft in CLEAN configuration have largest vortices (see s. 4.15 for more information on wake turbulence). Induced drag INCREASES as an aircraft pulls out of ground effect, since the wing has to pull along all the vortices with no relief from ground effect.

Lift vs Drag Formula

The overall lift of an aerofoil can be calculated as follows:

$$\text{LIFT} = CL\,1/2\,pV^2S$$

where
- CL = Coefficient of Lift
- p = Air Density
- V = Velocity
- S = Plan Area of Aerofoil

The overall drag of an aerofoil can be calculated as follows:

$$\text{DRAG} = CD\,1/2\,pV^2S$$

where
- CD = Coefficient of Drag
- p = Air Density
- V = Velocity
- S = Plan Area of Aerofoil

Notice that in both the case of the lift and drag formula that the velocity is squared. This means, for example, if the velocity is doubled, the resulting lift or drag will increase by fourfold. As well, note how the air density and aerofoil area have a direct relationship to the resulting lift or drag.

Dihedral

Positive dihedral enhances roll stability: following a disturbance, aircraft sideslips toward the lower wing, resulting in more lift on the lower wing and bringing aircraft back to level.

4.02 High Speed Flight

Speed of Sound (S)

S is the speed that pressure waves move through the air. S varies only with temperature. S = 660 kts at 15° C, 594 kts at -40° C. Formula:

$$S = 39 \times \sqrt{\text{Temp } °K}$$

where
- S = speed of sound in kts
- °K = degrees Kelvin [°C + 273].

EXAMPLE: What is the speed of sound where the air temperature is 15° C?

ANSWER: 662 kts. 15° + 273° = 288° K. 39 X √288 = 662 kts.

The speed of sound can also be calculated with a flight computer. Set the Mach Index opposite the temperature, then read speed of sound opposite 1.0 on minutes scale.

EXAMPLE: Use a flight computer to the local find speed of sound where the outside air temperature is -50° C.

ANSWER: 580 kts. Using E6B or CR computer, set Mach Index opposite temperature of -50° C in TAS window. Read 580 on outer scale opposite 1.0 on minutes scale.

Mach Number

Mach Number is the ratio of the true air speed of an aircraft to the speed of sound. For example, Mach .5 is half the speed of sound. Use this formula to calculate Mach Number:

$$\text{Mach \#} = \frac{TAS}{S}$$

Example: The local speed of sound is 585 kts, and the TAS is 450 kts. What is the Mach number?

ANSWER: Mach number is 0.77.

To calculate TAS from true or forecast OAT, use the formula TAS = S X Mach No, or, with E6B or CR flight computer, line up Mach Index with True OAT in TAS window, inner scale = Mach number, outer = TAS.

EXAMPLE: Without using a computer and using only the appropriate formulas, given a Mach Number of 0.88 and FD forecast air temperature at altitude of -38° C, what is the resulting TAS?

ANSWER: 526 kts. First determine S. S = 39 X √Temp °K = 39 X √235 °K = 597 kts. Since Mach Number is TAS divided by S, TAS = S X Mach Number or .88 X 597 = 526 kts.

High Speed Aerodynamics

Critical Mach Number or Mcrit is the highest flight speed possible without supersonic flow over any part of the airframe. The TRANSONIC SPEED RANGE is the range of speeds between SONIC and SUPERSONIC where some part of the airflow over the airframe is supersonic. At Mcrit, a small SHOCK WAVE forms on the upper, forward surface of the wing, becoming larger as speed increases.

The shock wave results from the effects of compressibility of air at high speeds. As speed increases beyond Mcrit, the shock wave becomes larger and eventually covers both the upper and lower surface of the aerofoil, creating a block.

The drag associated with the shock wave becomes pronounced at about 5% above Mcrit, called DRAG RISE or DRAG DIVERSION. At about Mach 0.82, the shock wave causes separation of the boundary layer. When the shock wave blocks the boundary layer, the airstream's kinetic energy is converted to a pressure and temperature increase behind the shock wave. The aircraft experiences a sudden increase in drag, movement of centre of pressure and eventual loss of control. Mmo is the "Maximum Mach Operations" speed, normally about M 0.8. Aircraft should not be operated above Mmo due to the effects of drag divergence and increased fuel consumption.

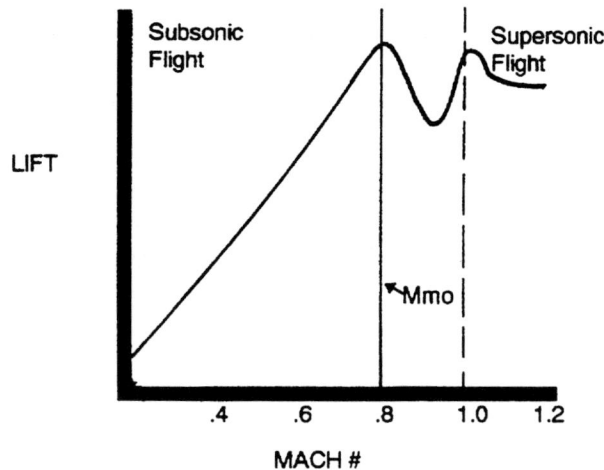

Fig. 4-3: Lift and increasing Mach Number.

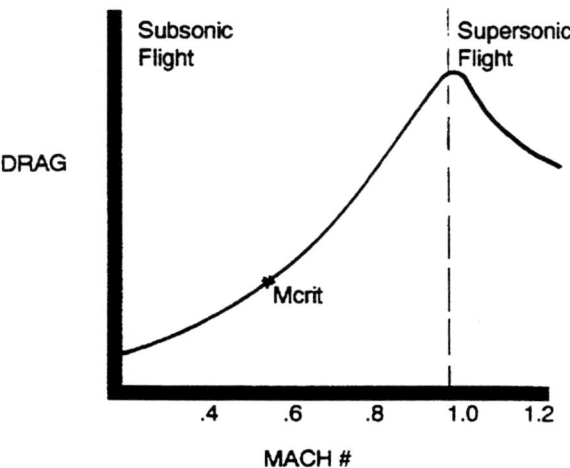

Fig. 4-4: Drag and increasing Mach Number.

Sweepback design

SWEEPBACK is when the wings are angled forward or aft from a right angle to the fuselage (normally aft). 30° of sweep increase Mcrit from about M 0.70 to M .75, delaying shock wave separation and allowing higher cruise speeds.

Lift acts perpendicular to the oncoming airflow. When a wing is swept, the relative oncoming airflow perpendicular to the aerofoil is reduced, meaning a lower amount of lift is obtained for a greater forward speed. VORTEX GENERATORS and BOUNDARY LAYER ENERGIZERS increase Mmo as an alternative to sweepback design (Lear Jet straight wing).

An undesirable effect of sweepback is DUTCH ROLL, which is a yawing and rolling movement caused by the advancing wing meeting greater airflow than the retreating wing. The greater exposure of the advancing wing also yields greater lift due to increased dihedral. The retarding wing is less effective due to reduced spanwise airflow. The vertical fin offsets the excess lift and creates

a restoring force in the opposite direction. The resulting oscillation is dutch roll.

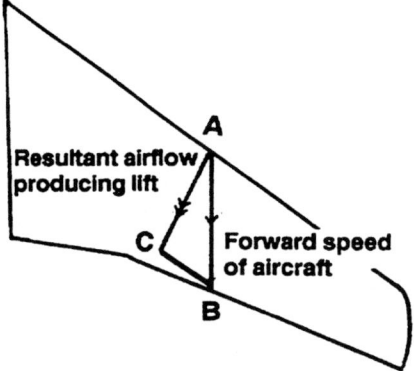

Fig. 4-3: Sweepback increases Mcrit. Lift acts perpendicular to leading edge, therefore, resultant lift producing airflow is from A to C, allowing greater forward speed at A to B.

For subsonic jets, use aileron only to correct dutch roll: do not use rudder. A YAW DAMPER uses a gyro system that senses changes in yaw and causes a corrective force on the rudder to provide the proper corrective rudder deflection, stopping the rolling tendency.

Pitch Trim Compensator

Due to the effects of shock flow induced separation, there is a decreased download on the horizontal stabilizer, and corresponding loss of longitudinal stability. Shock flow also moves the centre of pressure aft. These factors cause a strong diving tendency called MACH TUCK. The PITCH TRIM COMPENSATOR (stick puller) repositions the horizontal stabilizer continuously as a function of Mach number. When flying speed is increased, the pitch trim compensator causes the nose of the aircraft to be trimmed up to maintain level flight.

Mach Buffet

MACH BUFFET occurs when shock waves on the top of the wings interfere with the airflow causing turbulent airflow from the wings to strike the tail surface. The low speed buffet indicates proximity to conventional stall and can be violent. The high speed Mach buffet is not as noticeable initially, but is potentially dangerous, because of Mach buffet there is a complete loss of elevator control. Several accidents have occurred where pilots have disconnected overspeed warning systems and PTC's to obtain higher speeds above certified limits.

AERODYNAMIC CEILING is where the low speed and high speed buffet have merged. The margin between buffet boundaries is reduced by G forces from turbulence and turning. For these reasons, any cruising altitude selected must provide adequate margins between the low and high speed buffet and take into account anticipated G forces for turns and turbulence.

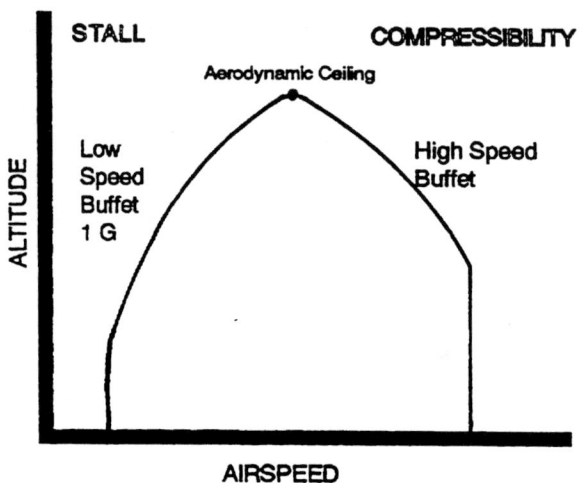

Fig. 4-6: Buffet boundaries chart. Low speed buffet occurs at higher speeds with increasing altitude. High speed buffet occurs at higher speeds with decreasing altitude. Aircraft ceiling is the point at which low speed and high speed buffet intersect.

Review the buffet boundaries chart at Fig. 4-7. Note how as weight and G force increase, buffet margins diminish.

EXAMPLE: Determine the low and high speed Mach buffet speeds for a DC-9 where the gross weight is 105,000 lbs, bank is 52°, and altitude is 25,000 ft.

ANSWER: The aircraft would commence low speed buffet at Mach .61, and high speed buffet at Mach .83.

4.03 Weight and Balance

Aircraft Weights

a) Empty Operating Weight (EOW). Basic aircraft, unusable fluids.
b) Basic Operating Weight (BOW). EOW plus crew, crew baggage, oil, fluids, equipment.
c) Zero Fuel Weight (ZFW). BOW plus PAYLOAD (passengers, passenger baggage, cargo).
d) Landing Gross Weight (LGW). ZFW plus reserve, alternate fuel.
e) Take-Off Gross Weight (TOGW). ZFW plus enroute fuel.
f) Gross Weight for Taxi. TOGW plus ramp fuel (taxi & run-up).

PAYLOAD is what the air carrier gets paid for: passengers, passenger baggage and cargo, therefore, PAYLOAD = ZFW - BOW.

EXAMPLE: Determine payload where total fuel = 2,000 lbs, crew and their baggage = 600 lbs, ZFW = 10,000 lbs, empty operating weight = 6,000 lbs.

ANSWER: 3,400 lbs. EOW + Crew = BOW = 6,600 lbs. ZFW - 6,600 = 3,400 lbs.

Mean Aerodynamic Chord (MAC)

MEAN AERODYNAMIC CHORD is the average distance from the leading edge to the trailing edge of the wing. LEMAC is the leading edge of the mean aerodynamic

chord, TEMAC is the trailing edge of the mean aerodynamic chord.

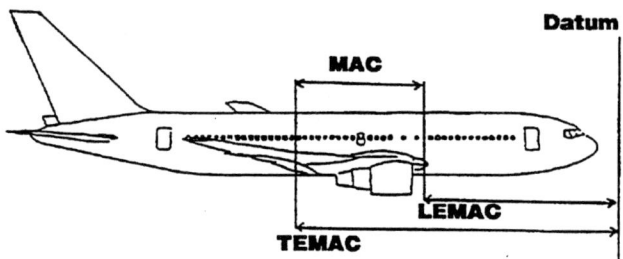

Fig. 4-8: Mean Aerodynamic Chord (MAC).

CG limits are expressed either as inches aft of the datum, or as % MAC. The LEMAC is 0% MAC, and is also identified by inches aft the datum. The TEMAC is 100% of the MAC, and is also identified by inches aft of the datum. The CG operating range lies between the LEMAC and the TEMAC at various points depending on the gross weight of the aeroplane and load configuration. For example, CG limits could be 14% MAC to 38% of MAC aft, for a hypothetical aircraft. Use this formula to determine % MAC:

$$\text{\% MAC} = \frac{\text{ARM (") - LEMAC (")}}{\text{MAC (")}} \times 100$$

EXAMPLE: An aircraft LEMAC is at station 687, MAC is 109", and CG is at station 721.6". Find a) the CG in inches aft of LEMAC and b) the CG expressed as % MAC.

ANSWER:
a) CG = 34.6" aft of LEMAC (721.6 - 687)
b) CG = 31.7% MAC. $\left[\frac{721.6 - 687}{109} \times 100\right]$

EXAMPLE: LEMAC is at 943.0", MAC is 194.0", and the CG is at 19.4% MAC. Find a) CG in inches aft of LEMAC and b) CG in inches aft of DATUM.

ANSWER:
a) 19.4% X 194 = 37.6".
b) 37.6" + 943" = 980.6".

Repositioning of Centre of Gravity

To reposition a CG, weight should be transferred to another station. To determine the minimum weight to be moved a given distance, to bring the centre of gravity within limits, use the following formula:

$$\frac{\text{wt}}{\text{WT}} = \frac{\text{d}}{\text{D}}$$

where wt = weight to be moved
WT = gross weight of aircraft
d = distance CG moves
D = distance item moves

EXAMPLE: An aircraft has a gross weight of 15,000 lbs, with a CG 2" aft of the maximum aft limit. What is the minimum weight to be moved from flight station 320 to flight station 170 to bring the CG just to the aft limit?

ANSWER: 200 lbs. $\frac{\text{wt}}{15,000} = \frac{2}{150}$

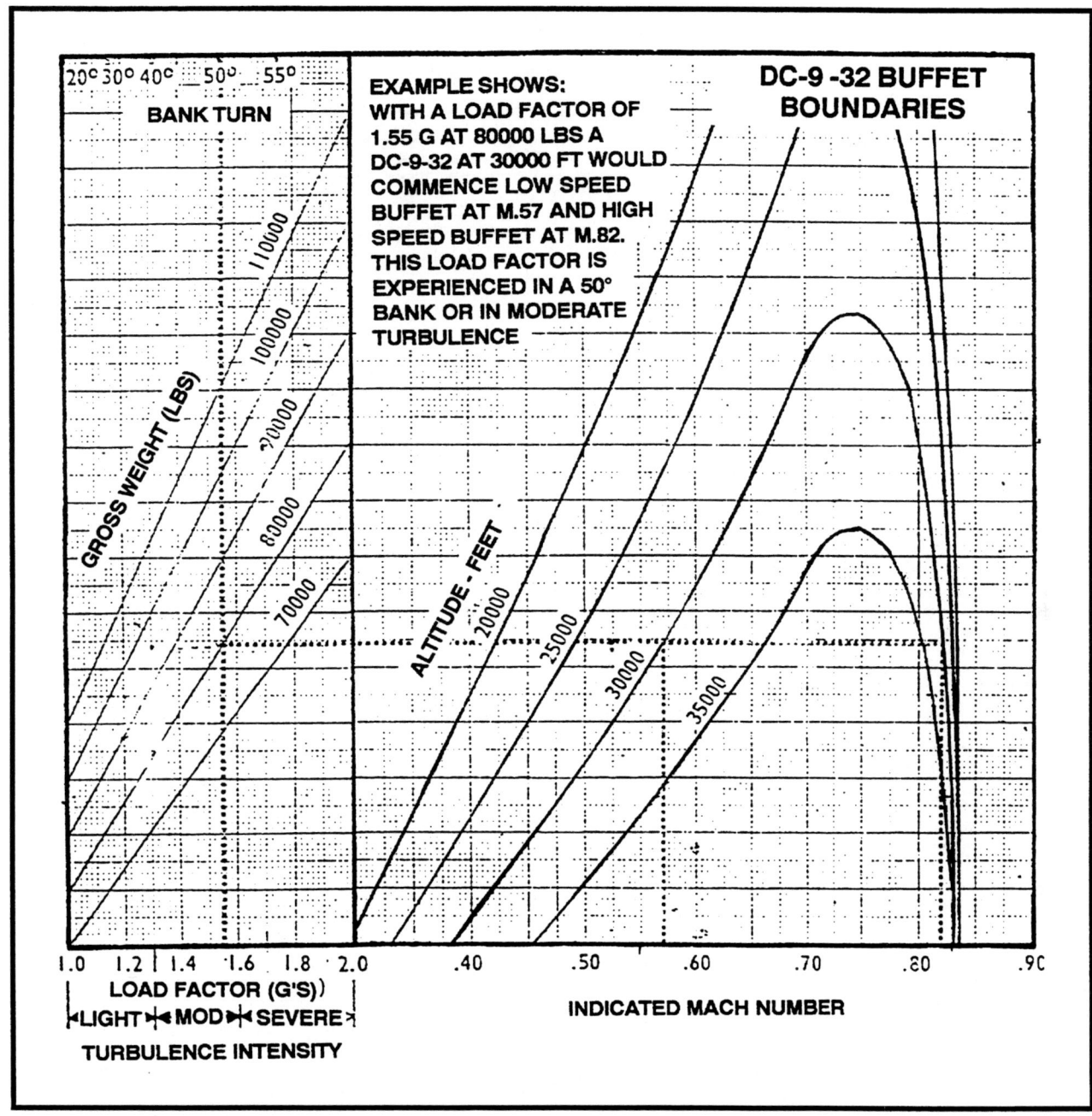

Fig. 4-7: Buffet boundaries chart for McDonnel Douglas DC-9. Note how low speed buffet occurs at higher speeds with increasing altitude, and how high speed buffet occurs at higher speeds with decreasing altitude. Aircraft ceiling is the point at which low speed and high speed buffet intersect.

FUEL LOADING CHART

Fuel Taken as 7.807 lb. per Imp. Gal. / Moments are in in.-lb.

Fuselage and Wing Tanks			Fuselage and Wing Tanks (Cont.)			Fuselage and Wing Tanks (Cont.)			Fuselage and Wing Tanks (Cont.)		
Imp. Gallons	Wt. (lb.)	Mom./1000	Imp. Gallons	Wt. (lb.)	Mom./1000	Imp. Gallons	Wt. (lb.)	Mom./1000	Imp. Gallons	Wt. (lb.)	Mom./1000
10	78	23	300	2,342	682	590	4,606	1,326	880	6,870	1,971
20	156	46	310	2,420	704	600	4,684	1,349	891	6,956	2,000
30	234	68	320	2,498	727	610	4,762	1,370			
40	312	91	330	2,576	748	620	4,840	1,393			
50	390	115	340	2,654	771	630	4,918	1,415			
60	468	137	350	2,732	793	640	4,996	1,438			
70	546	160	360	2,810	815	650	5,075	1,459			
80	625	183	370	2,889	837	660	5,153	1,482			
90	703	205	380	2,967	860	670	5,231	1,504	Wing Tip Tanks		
100	781	229	390	3,045	882	680	5,309	1,526	Imp. Gallons	Wt. (lb.)	Mom./1000
110	859	252	400	3,123	904	690	5,387	1,548	50	390	110
120	937	275	410	3,201	926	700	5,465	1,571	100	781	221
130	1,015	298	420	3,279	949	710	5,543	1,593	150	1,171	331
140	1,093	321	430	3,357	970	720	5,621	1,615	192	1,499	423
150	1,171	343	440	3,435	993	730	5,699	1,637			
160	1,249	366	450	3,513	1,015	740	5,777	1,660			
170	1,327	389	460	3,591	1,038	750	5,855	1,681			
180	1,405	412	470	3,669	1,059	760	5,933	1,704			
190	1,483	435	480	3,747	1,082	770	6,011	1,726			
200	1,561	458	490	3,825	1,104	780	6,089	1,749			
210	1,639	480	500	3,904	1,125	790	6,168	1,770			
220	1,718	503	510	3,982	1,148	800	6,246	1,793			
230	1,796	525	520	4,060	1,171	810	6,324	1,815			
240	1,874	548	530	4,138	1,193	820	6,402	1,838			
250	1,952	570	540	4,216	1,215	830	6,480	1,859			
260	2,030	593	550	4,294	1,237	840	6,558	1,882			
270	2,108	615	560	4,372	1,260	850	6,636	1,904			
280	2,186	638	570	4,450	1,281	860	6,714	1,926			
290	2,264	659	580	4,528	1,304	870	6,792	1,948			

Fig. 4-9: Fuel loading chart.

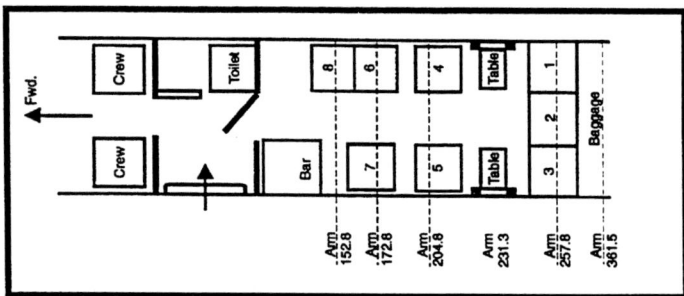

Fig. 4-10: Aircraft stations chart.

Weight and Balance Calculations

To complete a weight and balance, determine the weights at each station, find the moments, and refer to loading chart for CG limits (inches or % MAC). Know the standard jet fuel weights (7.8 lbs/Imp Gal, 6.5 lbs/US Gal) and oil (7.5 lbs per gallon US, 9.0 lbs per gal imp).

Remember these standard formulas:

$$\frac{\text{Moment}}{\text{Weight}} = \text{Arm} \qquad \text{Wt. X Arm} = \text{Moment}$$

Fuel Loading Charts

Loading charts are used to determine moments for weight producing items. Be alert for conditions: below, for example, note separate moments for fuel in tip tanks vs mains.

Aircraft Stations Chart

Use this chart to find the arm for a station such as a passenger seat, baggage compartment or a location where a weight will be placed. For example, arm at passenger seat 8 is 152.8".

Centre of Gravity Envelope

Once the moments and weights have been determined and totalled, refer to CG envelope to insure the aircraft is safely loaded in % MAC. The chart shows if limits are acceptable BEFORE fuel has been loaded: for example, if the ZFW falls in ZONE 1, fuel can be loaded up to the max ramp weight without exceeding CG limits. If the ZFW is in ZONE 2, the fuel quantity to be added must be restricted so as not to be outside loaded CG limits.

EXERCISE: Complete the following weight and balance using the charts at Figs. 4-9 to 4-12.

Conditions:
```
BOW . . . . . . . . . . . . . . . . . . . . . . . . 11,596 lbs
BOW Moment/1,000 . . . . . . . . . . . . . 3,213 " - lbs
Fuel (fuselage and wing tanks) . . . . 590 Imp Gals
Fuel (wing tip tanks) . . . . . . . . . . . 100 Imp Gals
Passengers (6 at seats No. 1 - 6) . . . . 160 lbs each
Baggage . . . . . . . . . . . . . . . 40 lbs per passenger
MAC . . . . . . . . . . . . . . . . . . . . . . . . . . . . . 90.197"
LEMAC . . . . . . . . . . . . . . . . . . . . . . . . . 253.964"
```

01. What is the total moment for the passengers?
02. What is the total moment for the baggage?
03. What is the zero fuel weight?
04. What is the ZFW moment?
05. What is the total fuel weight and moment?
06. What is the ramp weight? What is the moment and arm of the ramp weight?
07. What is the % MAC for the ZFW?
08. What is the % MAC for the ramp weight?
09. What is the loaded CG?
10. Is the aircraft within the CG envelope?
11. How much extra fuel can be loaded? Will it put the CG out of limits?
12. What is the required stabilizer trim setting for take-off?
13. What is the maximum ramp weight?
14. What is the maximum take-off weight?

15. For the given fuel and loading, what is the minimum amount of fuel which must be burned so as not to exceed the maximum landing weight?
16. What are the forward and aft CG limits for the given loaded weight?

ANSWERS:
01. 216,900 "-lbs. Refer to stations chart, multiply weight X arm for each passenger.
02. 86,800 "-lbs.
03. 12,796 lbs. ZFW = BOW + payload.
04. 3,516,700 "-lbs.
05. 4606 lbs (fuselage) + 781 lbs (tips) = 5387 lbs. Moment for fuel: 1,326,000 + 221,000 = 1,547,000 '-lbs.
06. Ramp weight = ZFW + fuel = 12,796 + 5,387 = 18,183 lbs.
07. 23.13%.
$$\% \text{ MAC} = \frac{\text{Arm - LEMAC}}{\text{MAC}} \times 100$$
$$= \frac{274.8 - 253.964}{90.197} = 23.13\%$$
08. MAC = 27.20%. Same calculation as question 7, except use 278.5 for arm.
09. Loaded CG = $\frac{\text{Total Moment}}{\text{Total Weight}}$
$$= \frac{5,063,700}{18,183} = 278.5".$$
10. Yes.
11. 2,817 lbs. ZFW falls in Zone #1, see note on chart: fuel can be loaded up to maximum ramp weight of 21,000 lbs. 21,000 - 18,183 = 2,817 lbs. This won't put the CG out of limits.
12. 3° nose up. Refer to horizontal stabilizer trim setting chart.
13. 21,000 lbs (see CG envelope).
14. 20,700 lbs (see CG envelope).
15. Max landing wt = 19,000 lbs, our ramp weight is 18,183, therefore landing weight not exceeded.
16. 20% - 29% MAC. See left and right edges of envelope for loaded weight.

Horizontal Stabilizer Trim Setting Chart

Use this chart to determine the required stabilizer trim setting for the loaded CG. To use this chart, you must have first converted the CG to % MAC. For example, a CG of 22.5 % MAC would yield a required stabilizer nose up trim setting of 4.5°.

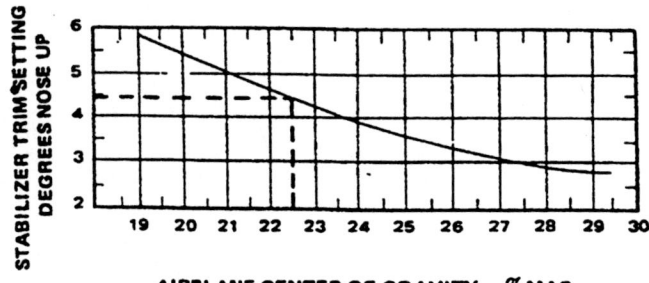

Fig. 4-12: Horizontal Stabilizer Trim Settings for % MAC.

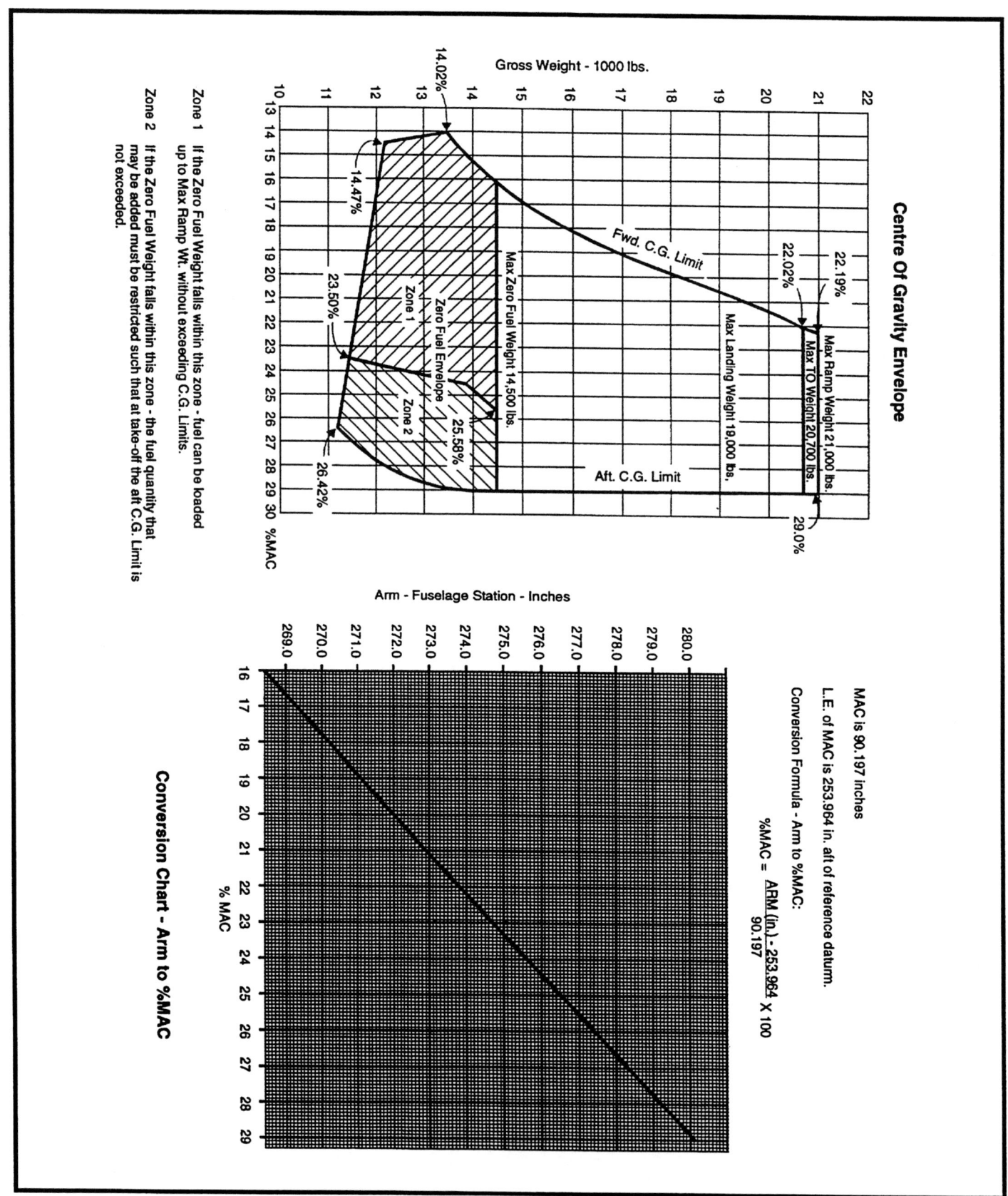

Fig. 4-11. CG Envelope for executive jet. Not how CG envelope is based on % MAC, and conversion table is provided for converting Arm - % MAC.

4.04 V Speeds

Know the following V-speeds:

V1   Critical engine failure recognition speed. Also called Vcef. The speed below which the pilot will reject take-off, and above which take-off will be continued. With failure of critical engine at V1, the distance required to reject the take-off and stop will equal the distance to accelerate and reach V2 at a height of 35' above the end of the runway.

V2   Take-off Safety Speed. The speed reached when 35' following engine failure at V1. Must be at least 20% above the stall speed and 10% above Vmc. Aircraft weight is primary factor determining V2.

V2min   Minimum take-off safety speed.

V3   Flap retraction speed.

Va   Design maneuvering speed. The maximum speed at which rapid full control deflection will not over-stress the aircraft. Also referred to as design speed.

Vb   Speed for maximum gust intensity/turbulence penetration speed. Usually the same or near Va.

Vc   Cruise speed.

Vd   Diving speed.

Vdf   Demonstrated flight diving speed. Also referred to as Mdf.

Vf   Flap speed.

Vfe   Maximum speed with flaps extended.
Vfo   Maximum flap operating speed.

Vh   Maximum level flight speed at maximum continuous power.

Vle   Maximum speed with landing gear extended.

Vlo   Maximum landing gear operation speed (extension or retraction).

Vlof   Lift-off speed.

Vmc   Minimum control speed with critical engine inoperative. Conditions for determination of Vmc: windmilling critical engine prop, other engine(s) producing maximum power, aircraft at maximum gross weight with full aft centre of gravity, landing gear retracted.

Vmo   Maximum operating speed. Also referred to as Vno. Yellow arc is above this speed.

Vmu   Minimum unstick speed. Speed at which aircraft first leaves runway.

Vne   Never exceed speed. Red line.

Vr   Rotation speed.

Vref   Approach reference speed. 1.3 X Vso.

Vs   Stalling speed or minimum steady controllable flight speed.

Vs1   Stalling speed or minimum steady flight speed obtained in a specific configuration.

Vso   Stalling speed in landing configuration (flaps/gear down).

Vsi   Stall speed in clean configuration and power off.

Vx   Best angle of climb.

Vy   Best rate of climb.

Vyse   Best single engine rate of climb speed (blue line).

4.05 Take-Off Performance

ACCELERATE STOP DISTANCE AVAILABLE (ASDA) is the length of the take-off run available plus the length of the stopway, if provided.

A CLEARWAY is a defined rectangular area on the ground or water under the control of the appropriate authority, selected or prepared as a suitable area over which an aeroplane may make a portion of its initial climb to a specified height.

A STOPWAY is a defined rectangular area on the ground at the end of the runway in the direction of take-off prepared as a suitable area in which an aeroplane can be stopped in the case of an abandoned take-off.

TAKE-OFF DISTANCE AVAILABLE (TODA) is the length of the take-off run available plus the length of the clearway.

TAKE-OFF RUN AVAILABLE (TORA) is the length of runway declared available and suitable for the ground run of an aeroplane taking off.

EXAMPLE: Assume a clearway is 1,000 feet, and the stopway is 1,000 feet. If the runway length is 5,000 feet, what is the TODA?

ANSWER: 6,000 feet.

For take-off, the pilot must first determine if there is enough runway to get the aircraft safely airborne, and if there is enough runway available to stop should the take-off be rejected. The CRITICAL ENGINE FAILURE RECOGNITION SPEED, V1 or Vcef, is the speed below which the pilot will reject take-off, and above which take-off will be continued. BALANCED FIELD LENGTH is the shortest distance in which the aircraft can be accelerated to V1, experience a failure of the critical engine and either continue the take-off to reach V2 at a height of 35' over the end of the runway (or clearway), or reject the take-off and stop safely in the remaining runway (including stopway).

V1 is determined once the maximum take-off weight is known. Various factors increase or decrease V1. To determine if a condition increases or decreases V1, ask yourself whether the condition will mean the pilot should commit to flight at a LOWER or HIGHER speed.

Factors Increasing V1

These factors increase V1:

a) Higher gross weight .
b) Headwind (reduces stopping distance).
c) Upslope runway (reduces stopping distance).
d) High airport elevation (reduced performance).
e) High ambient temperature (reduces aircraft perform-ance.
f) Reduced engine output.
g) Excess runway available.

Factors Decreasing V1

These factors reduce V1:

a) Lower gross weight
b) Tailwind (increases stopping distance)
c) Downslope runway (greater stopping distance).
d) Low airport elevation.
e) Low ambient temperature.
f) Anti-skid inoperative (greater stopping distance)
g) Snow or slush on runway.

Take-off Distance Charts

This chart is used to determine take-off distance under given conditions, or to find the limiting weight for take-off with a limited take-off distance. The "chase around" format is typical. First study any examples provided, then mark the intersection of the appropriate values. When moving on to new values, keep parallel to the guidelines (curved lines), move horizontally to the next reference line, then find the next intersection point. Depending on the data, you may have to move in a reverse direction on the chart.

EXAMPLE: Use the chart below to determine the take-off distance under these conditions:

Temperature:	40° F
Surface Winds:	Headwind 30 kts
Field Press Alt.	4,000 ft
Runway Gradient:	Zero
Gross Weight:	16,500 lbs
Anti Skid:	On

ANSWER: 5380 ft. Draw line up from 40° F (1) to 4,000 ft press alt (2). Then, move horizontally to reference line for gross weight. Follow guidelines up to 16,500 lbs (3). Move horizontally to reference line for winds, then follow guidelines to headwind of 30 kts (4). Move horizontally to reference line for runway gradient, which is zero (5), so proceed again to reference line for anti-skid (6). Because anti-skid is operative, proceed horizontally to read balanced field take-off distance 5,380 feet (7).

For take-off calculations, take-off weight is an important consideration. The maximum take-off weight must be computed for each take-off. Some factors which determine maximum take-off weight are runway length, winds, flap position, braking action, pressure altitude and temperature.

EXAMPLE: Determine the limiting weight for take-off using Fig. 4-13 under the following conditions:

Field Length Avail:	3,200 ft
Headwind Component:	30 kts
Anti-skid:	Operative
Pressure Altitude:	MSL
Runway Gradient:	Zero
Temperature:	25° F

ANSWER: 14,200 lbs is maximum weight for take-off. Start at field length of 3,200 ft (A), working horizontally through anti-skid (operational, B), and again horizontally to runway gradient reference line (C). Next, follow guidelines up and left from winds (headwind 30 kts, D), stopping at winds reference line. Then, move horizontally to cross gross weight portion of chart. Go to other side of chart, and move right from temperature 25°F (G), pressure altitude MSL (F), to gross weight reference line horizontally. Follow guidelines up to intersection point of horizontal line from winds for limiting weight (E) at 14,200 lbs.

EXAMPLE: Determine take-off distance using Fig. 4-13 under the following conditions:

Temperature:	75° F
Surface Wind:	Tailwind 10 kts
Pressure Altitude:	8,000 ft
Runway Gradient:	Upslope 3%
Gross Weight:	11,000 lbs
Anti-skid:	Off

ANSWER: 7,500 ft.

Accelerate - Go Chart Format

These charts combine required take-off distances and take-off speeds required.

EXAMPLE: Refer to Fig. 4-14 to complete and determine required field length and V1 for take off under these conditions:

OAT:	20° C
Press. Altitude:	6,000 ft
Weight:	11,000 lbs
Wind:	Headwind 10 kts

ANSWER: V1: 95 kts. Take-off distance: 7,300 ft.

Take-Off Speeds Performance Charts

Take off performance charts are used to determine required take off speeds based on take-off conditions. Read all notes and study any examples provided on the chart.

EXAMPLE: Use the chart at Fig. 4-15 to find V1, Vr and V2 speeds under these conditions:

Airport Temperature:	-20° C
Wind Component:	10 kt tailwind
Pressure Altitude:	8,000 ft
Runway Slope;	2% Upslope
Gross Weight:	17,000 lbs
Anti-Ice:	On

ANSWER: V1 = 116 kts, Vr = 116 kts, V2 = 127 kts. Remember that V1 can never be greater than Vr.

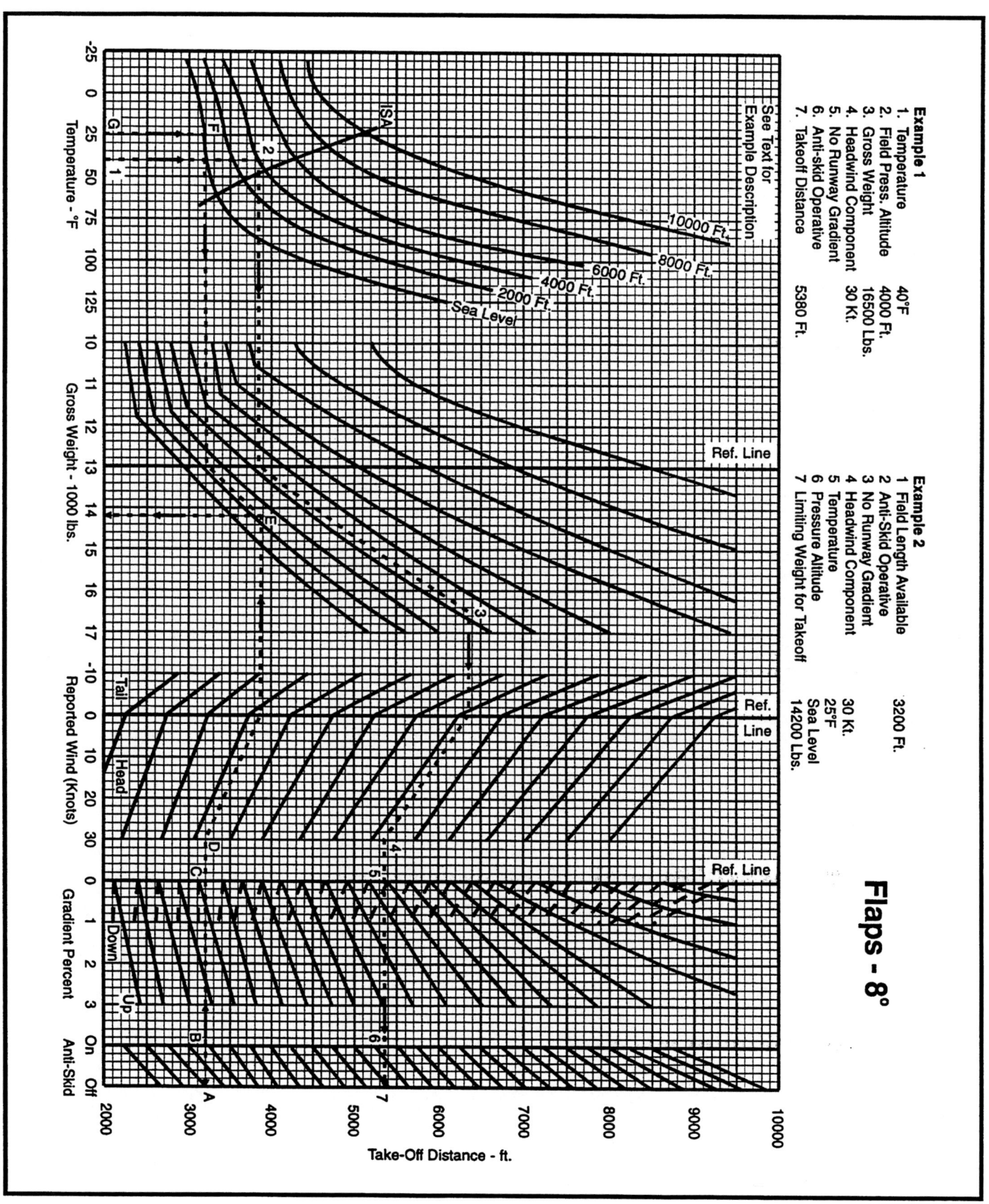

Fig. 4-13: Take-Off Distance chart.

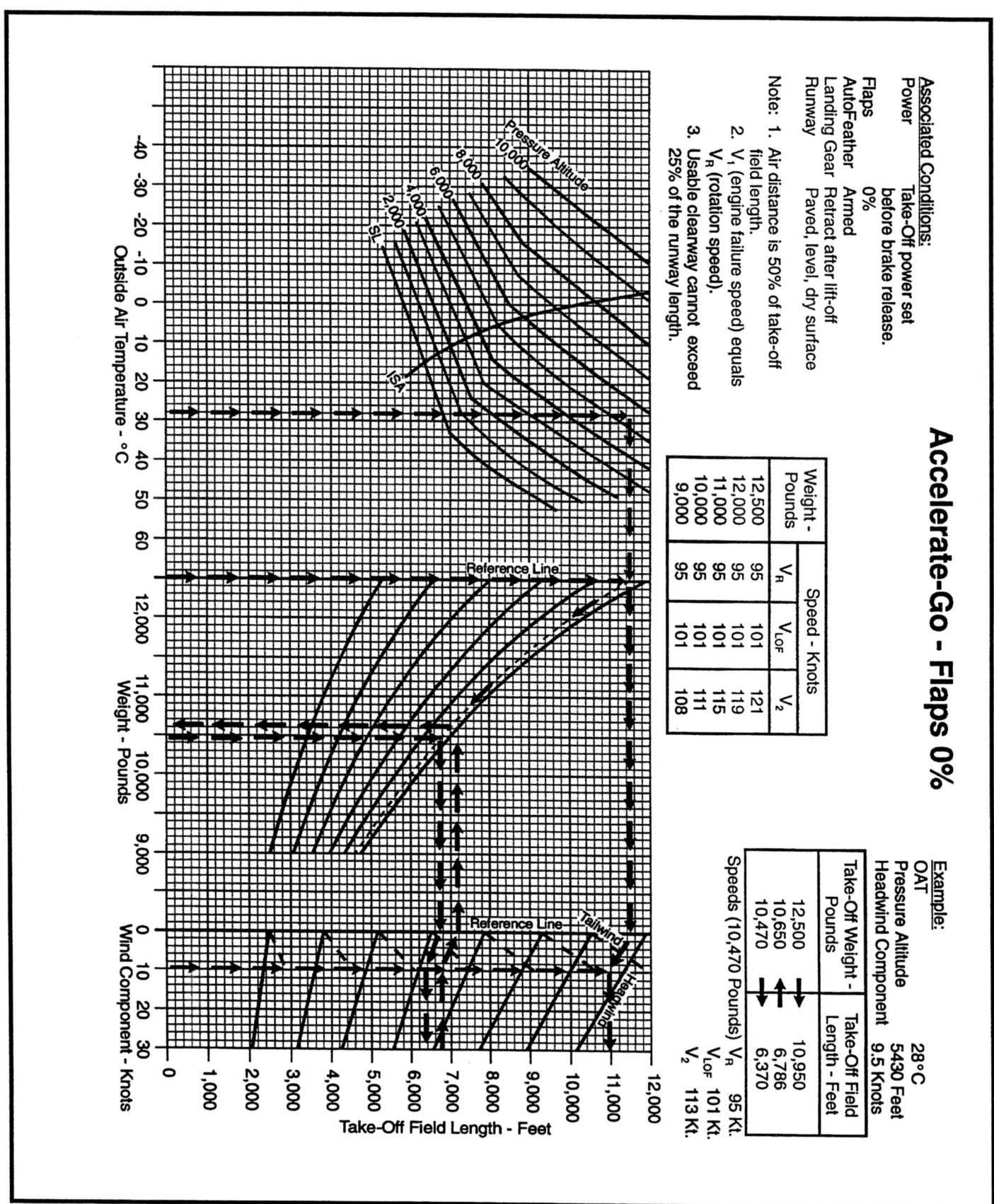

Fig. 4-14: Accelerate-Go Chart.

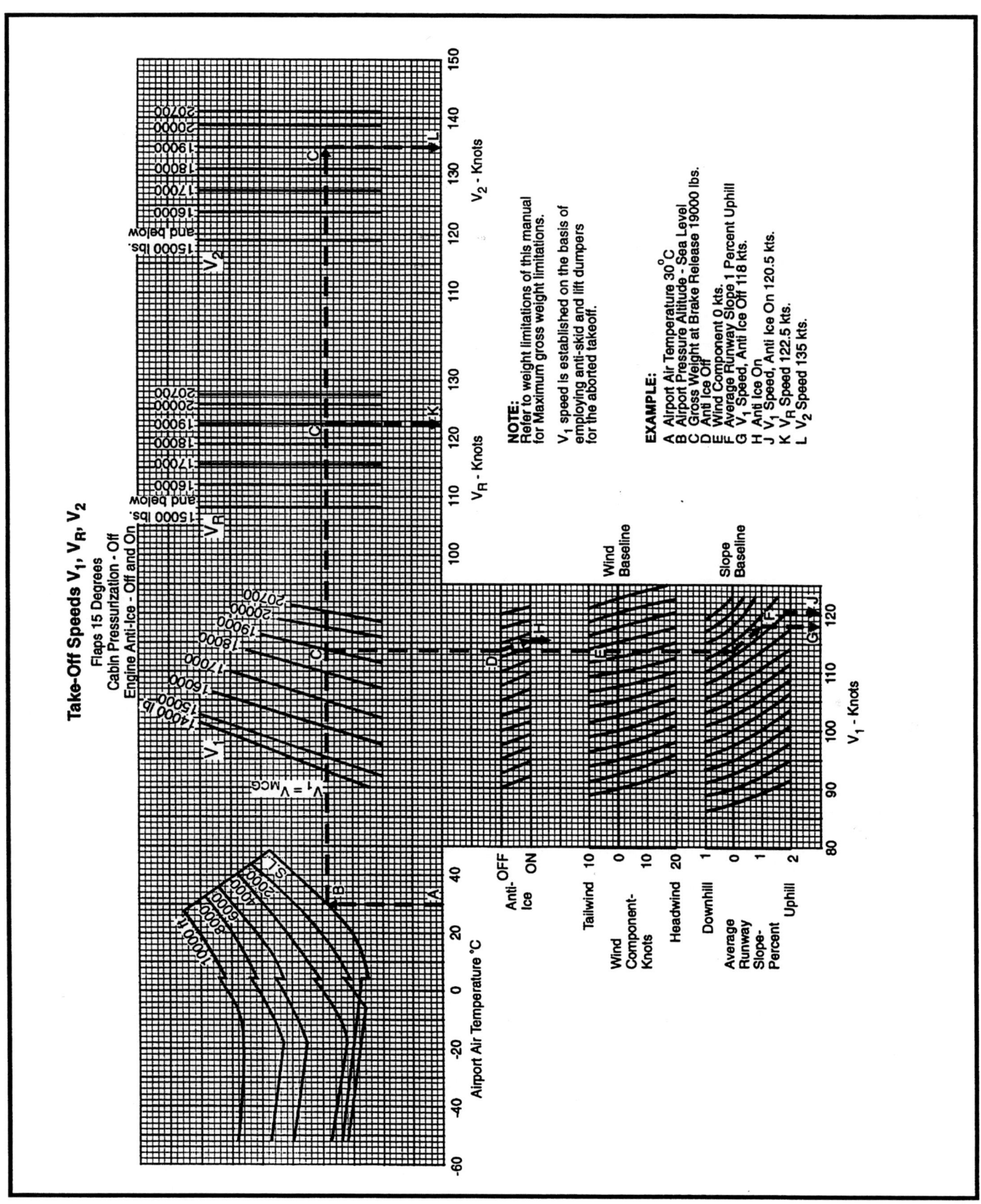

Fig. 4-15: Take-Off Speeds Performance chart.

4.06 Climb Performance

Climb Segments

Following V1 and lift off the following climb segments are defined:

a) REFERENCE ZERO. The point at which take-off distance ends and the climb begins. For Transport or Commuter category aircraft, this is the point at which the aircraft reaches 35 feet at V2.
b) FIRST SEGMENT. Starts at Reference Zero and ends when the landing gear is retracted. Transport Category/Commuter Category aircraft must be capable of positive single-engine climb during this segment.
c) SECOND SEGMENT. Starts at the end of the first segment and continues to the height at which the aircraft is levelled off for cleanup (normally 400 feet AGL). For a Transport or Commuter Category aircraft, speed is V2, flaps and power are at take-off setting, and (where applicable) the propeller of the failed engine is feathered. The required single engine climb performance varies by certification basis and number of engines: for two-engine, transport category aeroplanes, the minimum climb gradient with critical engine inoperative is at least 2.4%, 2.7% for 3-engine aeroplanes, and 3.0% for 4-engine aeroplanes.
d) THIRD SEGMENT. Starts at selected cleanup altitude and ends when the aircraft is ready for enroute climb. Normally conducted in level flight at 400 feet, the aircraft accelerates to Vclimb (Vyse), retracts flaps (normally to zero), and reduces power to Maximum Continuous. (Remember: many engines are rated only for 5 minutes at take-off power).
d) FOURTH (OR FINAL) SEGMENT. Enroute climb from the cleanup altitude to 1,500 feet, higher where obstacles require it. The aircraft is normally in a clean configuration at Vclimb (Vyse) with Max Continuous Power on the good engine and the failed engine shut down/feathered (if applicable).

Climb Performance - Turbo-jet

To obtain maximum climb performance, most jet aircraft are climbed at a constant IAS until a specified Mach number is reached, then Mach Number is held constant until level-off at cruising flight level. At Fig. 4-16 is a typical climb performance chart for an executive transport jet aircraft. The chart is tabulated for temperature variations from standard. Note how the fuel consumption increases with increasing temperature. These charts are normally prepared for zero winds (wind effects on high speed climb are generally ignored).

EXAMPLE: Refer to the chart at Fig. 4-16 to find time, distance and fuel for a climb from MSL to FL 350. Forecast upper temperature for 35,000 ft is -45° C.

ANSWER: ISA temperature at 35,000 ft is 35 X 2° C per 1,000 ft = 70° - 15° = -55° C. -45° C is 10° warmer than ISA, therefore use ISA + 10° on chart = 17.8 mins, 109 NM, 573 lbs.

4.07 Cruise Performance

Cruise flight is the portion of the trip from Top of Climb (TOC) to Top of Descent (TOD). Normally, the objective is to maximize distance for minimum fuel consumed, called flight for RANGE. A secondary factor is to minimize the time required for maximum TAS. LONG RANGE CRUISE (LRC) for turbo-jet aeroplanes involves optimizing range while providing a reasonable TAS and requires a gradual reduction in cruise speed as gross weight decreases with fuel burn-off. The single most important factor for range performance of turbo-jet aeroplanes is ALTITUDE: the higher, the better. MAXIMUM SPEED CRUISE (MSC) is the maximum cruise thrust or Vmo, if limiting, and is used when the flight arrival time is primary. Constant Mach cruise charts are used between LRC and MSC.

SPECIFIC AIR RANGE (SAR) is Nautical Air Miles (NAM) per pound of fuel burned. To determine SAR, use this formula:

$$\text{SAR} = \frac{\text{True Airspeed}}{\text{Fuel Flow}}$$

From the SAR formula, the following can be derived:

$$\text{Fuel Flow} = \frac{\text{True Airspeed}}{\text{SAR}}$$

On the chart at Fig. 4-17, note how LRC is about midway between MSC and minimum cruise speed. Note how SAR increases as gross weight decreases.

EXAMPLE: Using Fig. 4-17, what is the required Mach number for long range cruise at 18,000 lbs gross weight, and the corresponding specific air range?

ANSWER: LRC = M 0.67. SAR = 0.295 NAM/lb.

EXAMPLE: Using the same conditions as the previous example, what is the fuel consumption?

ANSWER: Fuel consumption = $\frac{\text{TAS}}{\text{SAR}}$ = $\frac{385 \text{ kts}}{0.295 \text{ NAM/lb}}$

   = 1305 lbs/hr.

EXAMPLE: Using Fig. 4-17, what is the speed for maximum range at 18,000 lbs weight?

ANSWER: See highest portion of curve at 18,000 lbs = 0.31 NAM/lb at M .59. This speed, while providing max range, would not generally be used as the resulting TAS would be unrealistic.

Factors that affect air range relate only to flight, such as fuel flow, TAS, IAS, time, EPR, temperature, altitude and Mach number. For this reason, range is generally kept in terms of air miles, not ground miles. However, a useful calculation is converting the aircraft's range into nautical ground miles per pound fuel consumed, referred to as SPECIFIC GROUND RANGE (SGR). SGR provides an indication of the most efficient cruising altitude, taking winds into account. The formula used is:

$$\text{SGR} = \frac{\text{Groundspeed}}{\text{Fuel Flow}}$$

Climb Performance Schedule — Two Engines

PRESSURE ALTITUDE - THOUSANDS OF FEET

Alt	RAT °C	ISA − 10°C TIME MIN.	DIST N.M.	FUEL LBS.	ISA TIME MIN.	DIST N.M.	FUEL LBS.	ISA + 10°C TIME MIN.	DIST N.M.	FUEL LBS.	ISA + 15°C TIME MIN.	DIST N.M.	FUEL LBS.	ISA + 20°C TIME MIN.	DIST N.M.	FUEL LBS.
45																
43																
41	−35	18.4	98	579	22.6	141	677									
39	−35	15.4	85	528	18.1	111	589	26.0	165	744	36.3	237	942			
37	−35	13.2	74	481	15.6	94	535	20.9	130	643	25.7	164	740	35.4	231	933
35	−33	11.4	66	447	13.8	82	491	17.8	109	573	21.0	131	640	26.0	165	742
33	−30	10.0	58	412	12.0	71	450	15.0	92	506	17.5	106	546	20.4	126	630
31	−27	8.8	51	381	10.6	61	410	12.8	79	457	15.0	89	491	17.4	103	556
29	−24	8.0	45	351	9.4	53	375	11.3	67	414	13.0	76	444	14.9	87	495
27	−21	7.2	39	323	8.3	47	344	9.9	58	378	11.4	65	403	13.0	74	440
25	−17	6.5	34	298	7.4	40	314	8.9	49	346	10.0	55	369	11.2	62	397
23	−14	5.8	30	270	6.5	35	287	7.8	43	310	8.7	47	333	9.8	54	360
21	−11	5.2	26	245	5.7	30	258	6.8	37	279	7.7	41	299	8.6	46	322
19	−8	4.6	22	220	5.0	26	230	6.0	31	250	6.7	35	268	7.4	39	289
17	−5	4.0	19	198	4.3	22	203	5.2	26	220	5.8	29	231	6.4	33	255
15	−2	3.5	16	178	3.8	19	183	4.5	22	199	5.0	25	211	5.6	28	225
13	+1	3.0	13	151	3.1	15	154	3.8	18	164	4.1	20	178	4.7	23	190
11	+4	2.5	10	130	2.6	12	130	3.1	15	140	3.4	16	151	3.9	19	160
9	+8	2.0	8	108	2.1	9	109	2.5	11	114	2.7	12	123	3.1	15	131
7	+11	1.6	6	84	1.7	7	85	1.9	9	89	2.1	9	98	2.5	11	101
5	+14	1.1	5	63	1.2	5	62	1.4	6	67	1.5	7	71	1.7	8	76

CLIMB LIMITED

→ START CLIMB WEIGHT @ S.L. − 18000 POUNDS
CLIMB SPEED − 250 KIAS, .7 M_I

Fig. 4-16: Turbo-Jet climb performance chart.

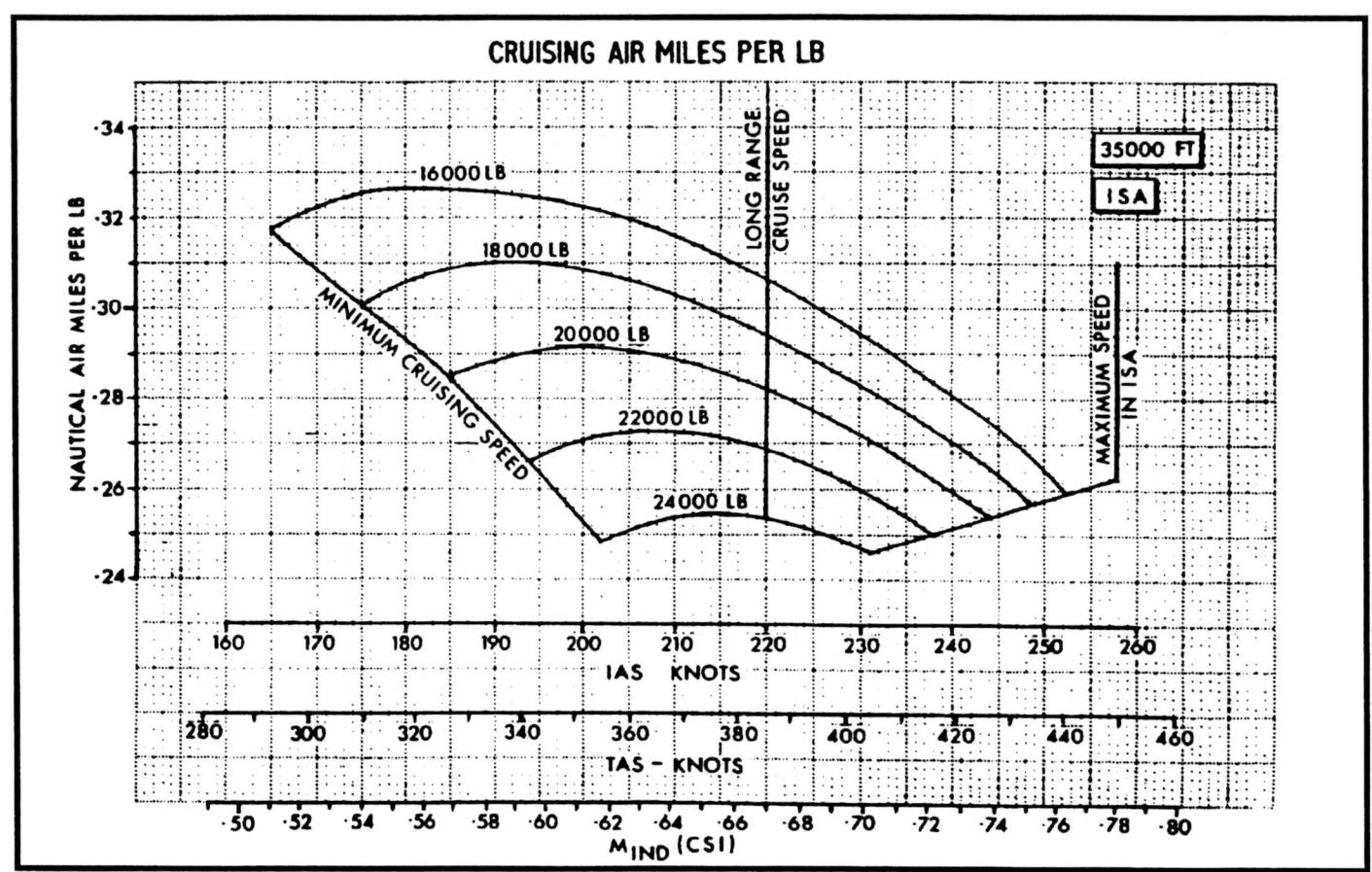

Fig. 4-17: Cruise Performance chart.

EXAMPLE: Find the best westbound altitude for range under these conditions:

	FL 310	FL 350
True Track	270°T	270°T
TAS	330 kts	310 kts
Winds	270° T/30 kts	270° T/60 kts
Fuel Flow	1100 lbs/hr	950 lbs/hr

ANSWER: Best altitude for range is FL 310. Best range will be at maximum SGR. To find SGR for each flight level, first calculate G/S. G/S for FL 310 = 300 kts, G/S for FL 350 = 250 kts. Then find SGR for each flight level:

$$SGR\ FL\ 310 = \frac{300}{1,100} = 0.27\ NGM/lb\ fuel.$$

$$SGR\ FL\ 350 = \frac{250}{950} = 0.26\ NGM/lb\ fuel.$$

There are a wide variety of enroute flight planning charts that give useful trip data based on existing conditions. See for instance Fig. 4-18.

EXAMPLE: Based on Fig. 4-18, determine trip time and fuel required based on the following conditions:

Trip distance	3,000 NGM
Winds	Headwind 50 kts
Cruising altitude	FL 330
Landing weight	475,000 lbs
Cruising Temp.	-41° C

ANSWER: 136,000 lbs fuel, 7.0 hours. From 3,000 NGM, move up to wind reference line, then angle over to 50 kts wind. Then proceed vertically to pressure altitude FL 33 & above, then head horizontally to landing weight reference line. Stop at landing weight reference line, then follow guidelines BACKWARDS (to left and down) to landing weight 475,000 lbs, then proceed from this point horizontally to right to read fuel required 136,000 lbs. For trip time, proceed from pressure altitude FL 33 line to left horizontally to ISA reference line. Since our temperature is 10° warmer than standard (ISA temperature at FL 330 would be -51° C, and since we are -41°C, we are 10° warmer than ISA), slope down to + 10° ISA deviation, and proceed horizontally to left edge of chart to read trip time 7.0 hours.

4.08 Landing Performance, Factoring, Crosswind Limitations and CRFI

Dispatch Limitations

Dispatch limitations apply up to the start of take-off, after that, they are irrelevant.

For PROPELLER DRIVEN LARGE AIRCRAFT, one can only dispatch or conduct a take-off if, based on the weather expected on arrival at the destination and alternate, the AFM landing distance is not more than 70 percent of the landing distance available. You need not provide additional factors for wet runways (unless, of course, the AFM requires it).

For TURBOJET AIRCRAFT, one can only dispatch or conduct a take-off if, based on the weather you expect on arrival at the destination and alternate, the AFM landing distance is not more than 60 percent of the landing distance available. If you expect the runway to be wet when you arrive at your destination (not alternate), you need to increase the flight manual distance by 15 percent, and that new value must not exceed 60 percent of the landing distance available.

Remember that a dispatch limitation applies up to the start of the take-off roll. After that, the 60 percent, 70 percent and 115 percent factors all become irrelevant. Once airborne, you can continue to the destination if you will land within 100 percent of the available landing distance. This will help you cope with a change in ambient weather, a runway change, or an equipment failure that increases the aircraft landing distance.

EXAMPLE: You are operating a turbo-jet aircraft to a destination runway having an LDA of 6,000 feet. Based on a dry runway, what is the maximum landing distance based on the AFM?

ANSWER: Maximum landing distance is 3,600 feet based on 60% dispatch limit for 6,000 ft LDA.

Crosswind Limitations

To use a crosswind chart, refer to the aircraft manual for the maximum 90° crosswind component. A useful approximation for U.S. manufactured aircraft is that the maximum 90° crosswind is 20% Vso.

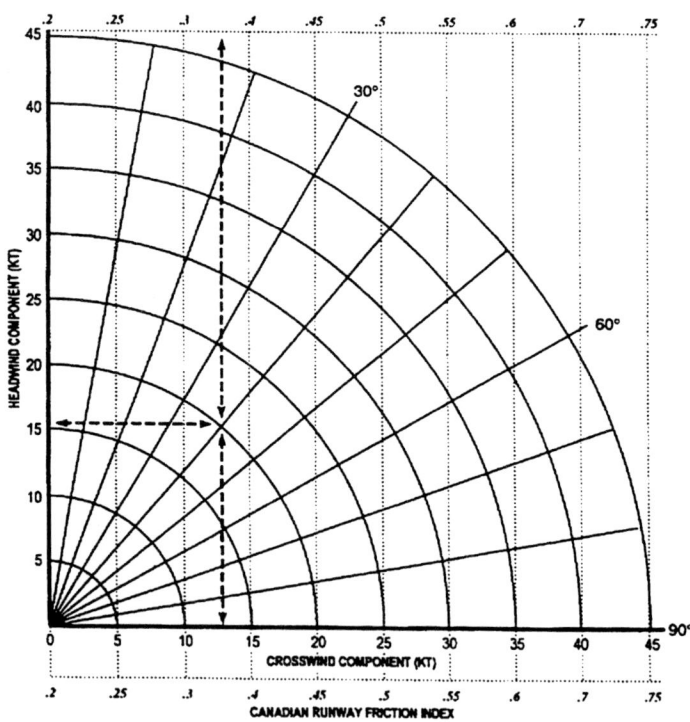

Fig. 4-19: Crosswind limitations and CRFI chart.

To determine whether a given wind exceeds the crosswind limit, use the following steps:

1. Determine the angle between the wind and the runway. Remember: runways are magnetic, winds from FD or TAF are true.
2. Place a small dot where wind angle and wind speed meet. Use curved wind arcs to mark wind speed.

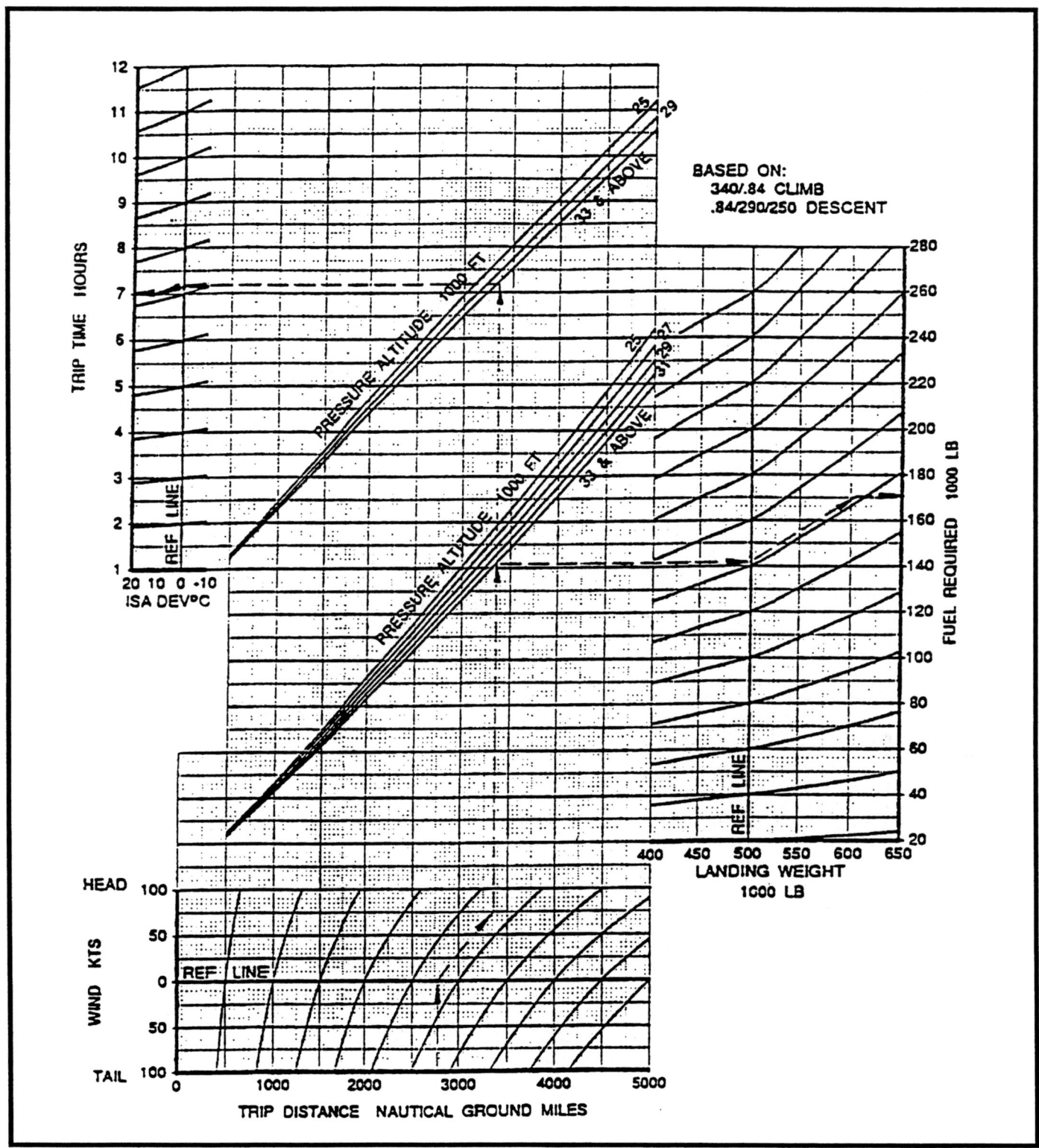

Fig. 4-18: Long range cruise flight planning chart.

3. Draw vertical line down on chart for crosswind component, draw horizontal line for headwind component.EXAMPLE: An aircraft 90° crosswind limit is 20 kts. Using the crosswind chart below, if the surface winds are 263° M at 40 kts, will the crosswind limit be exceeded when using runway 20?

ANSWER: Yes. Crosswind component = 36 kts, headwind = 18 kts.

The crosswind chart is used to determine LIMITING WINDS. The limiting wind is the maximum wind from a given angle off the runway. To determine the limiting winds, draw a vertical line up from the maximum 90° crosswind. The intersection point of the vertical line and wind angle is the limiting wind speed at which the 90° maximum crosswind will be exceeded.

EXAMPLE: What are the limiting winds from 30° for an aircraft with an 18 kt 90° crosswind limit?

ANSWER: 35 kts.

Wet and Contaminated Runways

A runway is generally considered contaminated under the following conditions:

a) standing water or slush more than 0.125" (3.0 mm) anywhere along the proposed take-off run or accelerated-stop surface; or
b) any accumulation of snow or ice along the proposed take-off run or accelerate stop surface.

NOTE: The AFM should be consulted to determine the definition of a contaminated runway for your aircraft.

Apart from dispatch factors, you need to understand how your aircraft handles wet or contaminated runways, and build this understanding into procedures for coping with less than ideal take-off and landing surfaces.

The CANADIAN RUNWAY FRICTION INDEX (CRFI) is a measurement used to indicate how slippery a runway is, due to runway contamination. A CRFI of 1.0 indicates maximum theoretical friction coefficient for a bare and dry runway. In heavy rain or on an ice covered runway, the CRFI may be in the 0.1 to 0.3 range, indicating slippery conditions. With a low CRFI, the aircraft will take a greater distance to stop and will be more prone to drift in a crosswind.

Mechanical and electronic decelerometer measurements used to obtain CRFI values are not accurate for water and slush, thus CRFI values are not provided when these conditions are present.

The greater the crosswind, the greater the required minimum safe CRFI - for a 10 kt crosswind, the minimum CRFI is .3, for a 15 kt crosswind, the minimum CRFI is .4.

EXAMPLE CRFI Report:

```
CYRB CRFI 17T/35T .30, 17A .24, 17B .20,
17C .26 -10  0601191055
```

Meaning: CRFI report Resolute NV for runways 17 and 35, temperature -10° C, CRFI for runway 17/35 is .30, lowest reading on first third of runway 17 is .24, middle third lowest reading is .20, last third lowest reading is .26, date taken January 19, 2006 at 1055 UTC.

A RUNWAY SURFACE CONDITION (RSC) report indicates when conditions yield a slippery runway.
EXAMPLE RSC Report:

```
CYRB RSC ALL RWYS COVERED 1 IN LIGHT SNOW
0601190630
```

Meaning: Runway Surface Condition Report Resolute NV all runways covered in 1 inch light snow January 19, 2006 at 0630 UTC.

EXAMPLE: Using the CRFI crosswind chart, if the active runway is RWY 26, and the surface winds are from 220° M at 20 kts, what is the minimum CRFI required to not exceed CRFI crosswind limits?

ANSWER: Minimum CRFI is 0.35. Complete normal crosswind calculation, read CRFI 0.35 off bottom. A reported CRFI less than 0.35 will be unsafe, and may result in uncontrollable drifting and yawing.

CRFI and Dispatch Limitations Tables

Note the following tables at Fig. 4-21, which show recommended landing distances based on CRFI for no-reverse thrust and for reverse thrust.

To use these tables, determine dry runway landing length requirements from AFM. Then, determine CRFI corrected distances. Finally, apply factors (60%, 70%, or

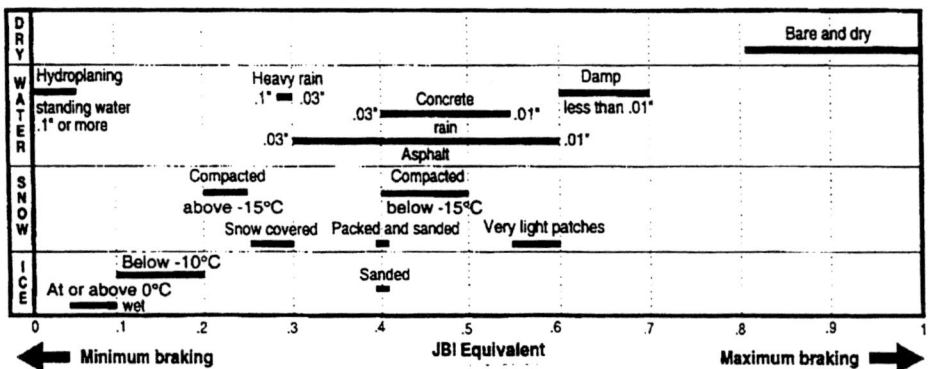

Fig. 4-20: Average equivalent values of CRFI produced by typical runway surface conditions.

Canadian Runway Friction Index (CRFI)
Recommended Landing Distances
(No Reverse Thrust)

Landing Distance Bare and Dry Unfactored	0.60	0.55	0.50	0.45	0.40	0.35	0.30	0.27	0.25	0.22	0.20	0.18	Landing Field Length Bare and Dry 60% Factor	Landing Field Length Bare and Dry 70% Factor
1800	3180	3260	3360	3480	3610	3780	3970	4120	4220	4400	4540	4690	3000	2571
2000	3550	3650	3770	3910	4070	4260	4490	4660	4780	4990	5150	5330	3333	2857
2200	3800	3910	4050	4200	4380	4590	4850	5030	5170	5410	5580	5780	3667	3143
2400	4190	4320	4470	4640	4840	5080	5370	5570	5730	5980	6180	6390	4000	3429
2600	4550	4700	4860	5050	5270	5530	5850	6080	6240	6520	6730	6970	4333	3714
2800	4830	4980	5160	5360	5600	5880	6220	6460	6630	6930	7150	7400	4667	4000
3000	5190	5360	5560	5780	6040	6360	6730	7000	7190	7520	7770	8040	5000	4286
3200	5580	5770	5980	6230	6520	6870	7280	7570	7790	8150	8430	8730	5333	4571
3400	5880	6090	6320	6590	6900	7270	7720	8030	8260	8650	8950	9270	5667	4857
3600	6200	6420	6660	6950	7290	7680	8160	8500	8750	9170	9480	9830	6000	5143
3800	6500	6740	7000	7310	7660	8090	8600	8960	9220	9670	10010	10380	6333	5429
4000	6710	6960	7230	7550	7920	8360	8890	9270	9540	10010	10360	10750	6667	5714

Canadian Runway Frisction Index (CRFI)
Recommended Landing Distances
(Reverse Thrust)

Landing Distance Bare and Dry Unfactored	0.60	0.55	0.50	0.45	0.40	0.35	0.30	0.27	0.25	0.22	0.20	0.18	Landing Field Length Bare and Dry 60% Factor	Landing Field Length Bare and Dry 70% Factor
1800	3130	3200	3270	3350	3450	3560	3690	3790	3860	3970	4060	4150	3000	2571
2000	3500	3580	3660	3760	3870	4000	4160	4270	4350	4480	4580	4700	3333	2857
2200	3740	3830	3930	4040	4160	4310	4480	4600	4690	4840	4950	5080	3667	3143
2400	4130	4220	4330	4460	4590	4760	4950	5080	5180	5340	5460	5600	4000	3429
2600	4480	4590	4710	4840	4990	5170	5380	5520	5630	5810	5940	6080	4333	3714
2800	4740	4860	4990	5130	5300	5490	5710	5860	5970	6160	6300	6450	4667	4000
3000	5100	5230	5370	5530	5710	5920	6170	6340	6460	6670	6820	6990	5000	4286
3200	5480	5620	5780	5960	6160	6390	6660	6840	6980	7210	7380	7560	5333	4571
3400	5780	5930	6100	6290	6510	6750	7040	7250	7390	7640	7820	8020	5667	4857
3600	6080	6250	6430	6630	6860	7130	7440	7660	7820	8080	8270	8490	6000	5143
3800	6380	6560	6750	6970	7210	7500	7830	8060	8230	8510	8720	8940	6333	5429
4000	6590	6770	6970	7200	7450	7750	8100	8330	8510	8800	9010	9250	6667	5714

Fig. 4-21: CRFI Tables.

115%) as appropriate to aircraft dispatch category. Note how for dry runways, 60% and 70% factors are provided, but these are not corrected for CRFI (refer to data in centre of tables for CRFI corrections).

As well, these tables base recommended landing distances on a 95% level of confidence i.e. that in more than 19 landings out of 20, the stated distances will be conservative for properly executed landings on runways surfaces with the reported CRFI. The recommended landing distances for the CRFI are based on standard pilot techniques for minimum distance landings from 50 feet including an approach from a 3° glide slope, a firm touchdown, minimum delay to nose lowering, minimum delay time to deployment of ground lift dump devices and application of brakes, and sustained maximum anti-skid braking until stopped.

EXAMPLE: For a turbo-jet aircraft, based on a dry runway landing distance from the AFM of 3,200 feet, without thrust reversers, what factored runway distance would be required for dispatch assuming dry runway conditions prevail?

ANSWER: 5,333 feet, since no CRFI applies we look at the 60% factor numbers as indicated directly on table.

EXAMPLE: For a propeller driven, large aircraft, given a wet runway with a CRFI of 0.4 and landing distance of 2,000 feet based on dry runway numbers in the AFM, assuming reverse thrust is available, what is the minimum destination LDA for conducting take-off?

ANSWER: Converting for .4 CRFI using table gives 3,870 feet, factoring 70% = 5,529 feet.

Hydroplaning is a function of water depth, tire pressure and speed. DYNAMIC HYDROPLANING is due to standing water on the runway where the tire is completely supported by water. VISCOUS HYDROPLANING involves a damp surface where only a very thin amount of water is involved - it usually occurs on smooth surfaces such as the touch-down zone having rubber smears.

REVERTED RUBBER HYDROPLANING occurs where a hot tire boils water on a wet runway surface after a long skid - the steam prevents the tire from contacting the runway. The minimum speed at which hydroplaning will commence can be calculated by using the following formula:

Hydroplaning Speed (KTS) =

9.0 X √Tire Pressure (PSI)
[7.7 X √Tire Pressure for non-rotating tire]

EXAMPLE: Find the minimum hydroplaning speed for rotating and non-rotating for aircraft tire inflation 49 PSI.

ANSWER: 63 kts rotating, 54 kts non-rotating.

4.09 Visual Approach Systems

Visual Approach Slope Indicator System (VASIS)

An approach slope indicator consists of a series of lights visible from at least 4 NM designed to provide visual indications of the approach slope to the runway threshold (usually 3°). Aircraft following the "on slope" signal have safe obstruction clearance within 6 - 9 ° either side of the extended centreline out to 4 NM from the runway threshold and a safe wheel clearance over the threshold. 3-Bar VASIS systems are based on eye-to-wheel heights of up to 25 ft for normal aircraft (to DC-8), and EWH to 45 ft for wide bodied aircraft.

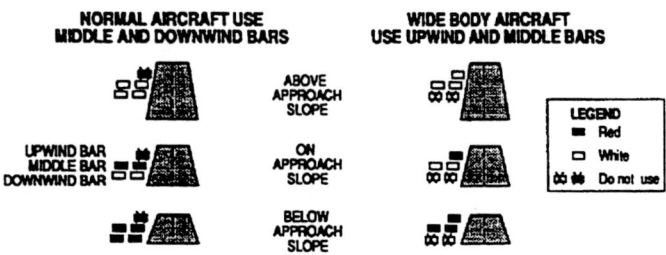

Fig. 4-22: 3-Bar VASIS System.

NOTE: where a VASIS system is provided on a precision approach runway, unless specifically requested by the pilot, the VASIS will be turned off in weather conditions of less than 500 ft ceiling and/or visibility less than one mile. This is to prevent any contradiction between the precision approach equipment and the VASIS.

T-VASIS

A T-VASIS installation consists of 10 light units, each with bisecting longitudinal lines of 6 units. The first three are the FLY UP units, and the last three are the FLY DOWN units. At the middle is a bar of 4 horizontal units that gives the on approach slope indication. All lights are white on a T-VASIS, except for gross undershoot. Following a proper T-VASIS will result in overflying

the threshold at a height of 45 feet.

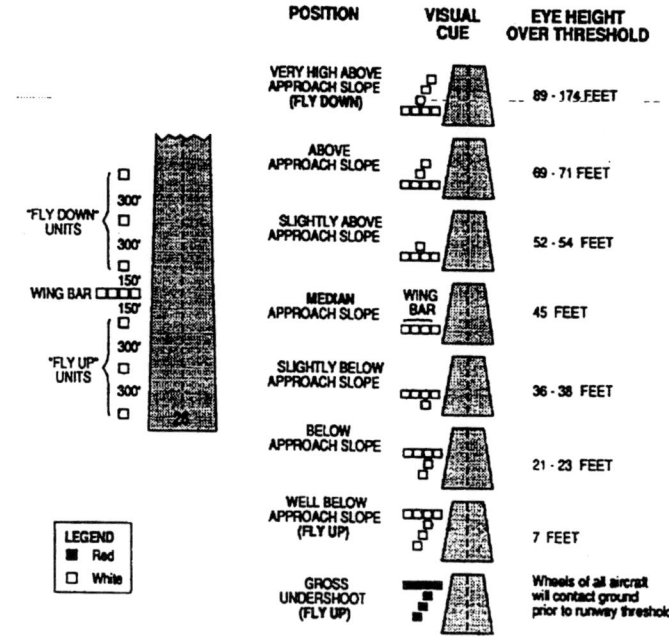

Fig. 4-23: T-VASIS.

Precision Approach Path Indicator (PAPI)

A PAPI has 4 light units on the left side of the runway as a wing bar. When on the proper approach slope, the two units nearest the runway shows red and the two units furthest show white.

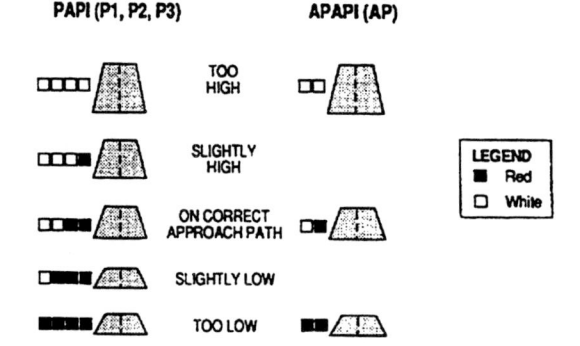

Fig. 4-24: PAPI.

4.10 Runways and Runway Markings

Runway numbering is based on runway direction. The runway number used is the whole number to the nearest one-tenth the magnetic direction of the centerline of the runway, measured from magnetic north. For example, if a runway direction is 098° M, the runway would be numbered 10. In the northern domestic airspace, runways are numbered to the nearest degree in °T (Example: Baker Lake NWT, Rwy 341T).

The take-off or landing area boundaries of aerodromes without runways are indicated by PYRAMIDAL or CONI-

CAL type boundary markers, which are coloured international orange and white (airports) or solid international orange (aerodromes).

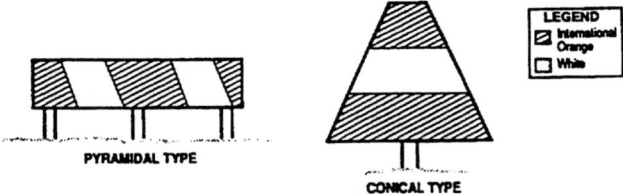

Fig. 4-25: Boundary Markers.

Airplane runway markings for non instrument runways include THRESHOLD MARKINGS, FIXED DISTANCE MARKINGS, and CENTRELINE MARKINGS.

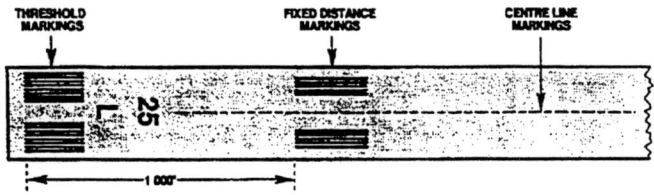

Fig. 4-26: Runway markings, non-instrument runway, over 5,000 ft.

Instrument runways will also have TOUCH DOWN ZONE MARKINGS.

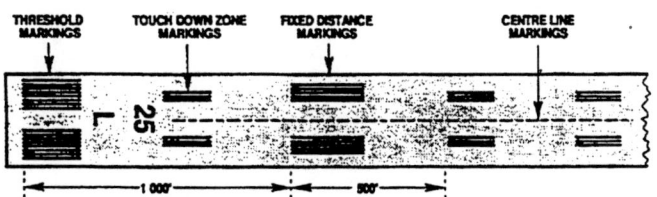

Fig. 4-27: Instrument runway markings.

Taxiways are marked with a centerline painted yellow. The taxiway edge has two yellow, continuous lines. Taxiway HOLDING LINES consist of a single yellow line and a dashed line (non-instrument runway) or double single lines and double dashed lines (instrument runway). Aircraft on the solid line side are required to hold short of the line until further clearance is given. Where no holding position lines are marked, aircraft must remain at least 200 feet from the runway edge.

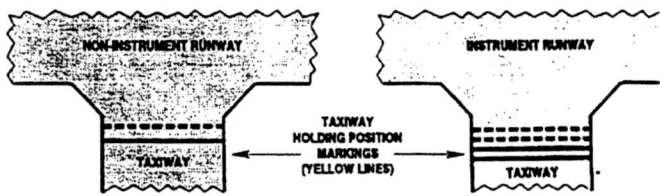

Fig. 4-28: Taxiway holding markings.

4.11 Approach - Runway Lighting Systems

Non-Precision Approach Runways

a) Low Intensity Approach Lighting System. This system is provided on non-precision approach runways and consists of twin aviation yellow fixed intensity light units spaced at 200-foot intervals commencing 200 feet from the threshold and extending back for a distance of 3,000 feet (terrain permitting).

b) Omni-Directional Approach Lighting System (ODALS). This system is a configuration of 7 omnidirectional, variable intensity sequenced flashing lights. ODALS provides circling, offset, and straight-in visual guidance for non-precision approach runways. There are 5 lights on the extended centre line commencing 300 feet from the threshold and spaced 300 feet apart for 1,500 feet. Two lights are positioned 40 feet to the left and the right of the threshold.

c) Medium Intensity Approach Lighting with Sequenced Flashing Lights (MALSE). This system consists of 7 bars of variable intensity lights spaced 200 feet apart. The three bars farthest away from the threshold also contain a sequenced flashing light unit.

Precision Approach Runways

a) High Intensity Approach Lighting System — Cat I. This system consists of rows of 5 white variable intensity light units spaced at 100-foot intervals commencing 300 feet from the threshold and extending back for a distance of 3,000 feet (terrain permitting). Additional light bars have been added to the low intensity system due to the lower landing minima. These are an approach threshold bar (green), contrast bars (red), imminence of threshold bar (red) and 1,000 feet distance bar (white).

b) Medium Intensity Approach Lighting System With Runway Alignment Indicator Lights — Cat I (MALSR). This system consists of a variable intensity approach lighting system extending 2,400 feet from the threshold. This system consists of 7 bars of light spaced at 200 feet over a distance of 1,400 feet, and 5 sequenced flashing lights spaced at 200 feet over a further distance of 1,000 feet.

c) High Intensity Approach Lighting System — CAT II. This system consists of rows of 5 white variable intensity light units commencing 100 feet from the threshold and extending for a distance of 3,000 feet. In view of the very low decision height for CAT II approaches, the following lights are provided in addition: runway threshold (green), 500 feet distance bar (white with red barrettes), side barettes (red).

Runway Identification Lights (RILS)

RILS lights are used where terrain prevents approach light installation, or where unrelated non-aeronautical lights or the lack of daytime contrast reduces the effects of approach lights. RILS are indicated in the CFS with the abbreviation AS. By day, RILS are operated when the visibility is 5 SM or less. By night, RILS are operated in conjunction with the approach and runway lights. RILS can be turned off at the pilot's request.

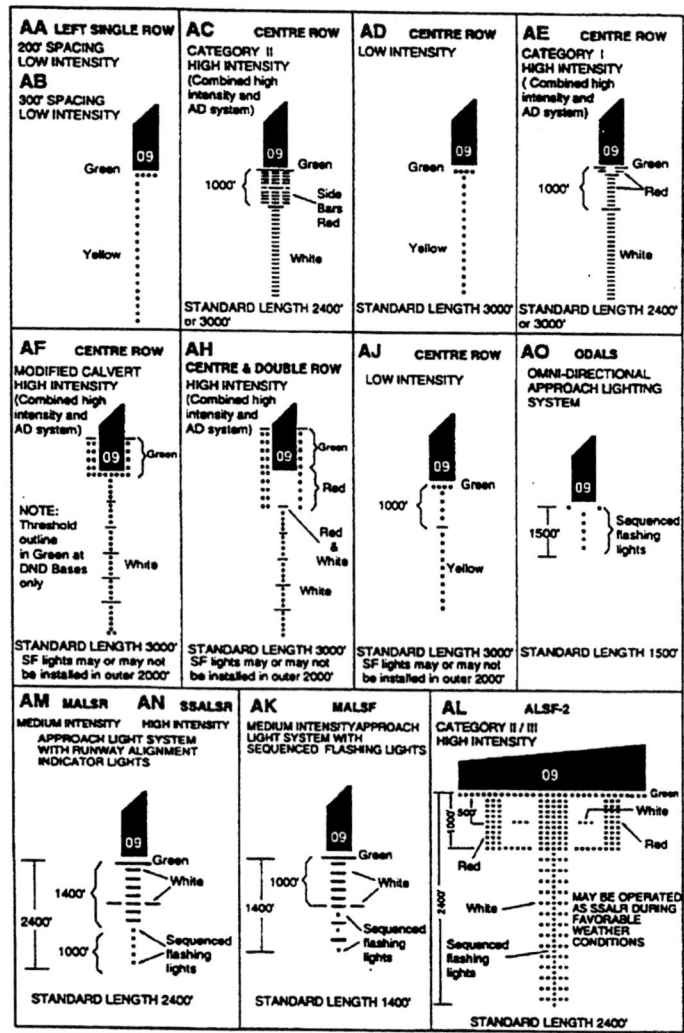

Fig. 4-29: Approach Lighting.

Runway Lighting

A night runway shall display two parallel lines of fixed white lights visible for at least two miles to mark take-off and landing areas. As well:

a) the minimum distance between parallel lines is 75 feet and the maximum is 200 feet;
b) the maximum distance between lights in the parallel lines is 200 feet;
c) the minimum length of parallel lines is 1,400 feet;
d) the minimum number of lights in parallel is 8;
e) each light in the parallel lines is aligned opposite the other and at right angles to the centre line of the take-off and landing area.

Runway Edge Lights

Runway edge lights are variable intensity white lights along the full length of the runway spaced at 200 feet intervals, except at intersections with other runways. They must also be frangible.

Runway Threshold End Lights

Threshold end lights are variable intensity red and green light units in the form of wing bars along the threshold on each side of the runway centre line (for CAT II runways red and green light units extend along the full runway width). Red shows take-off direction, green shows approach direction.

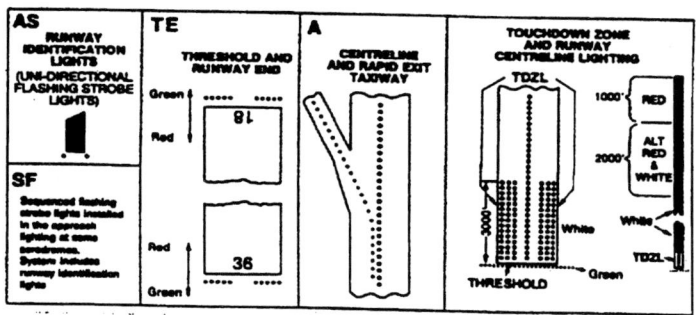

Fig. 4-30: Runway Threshold Lighting.

Displaced Runway Threshold Lighting

Fig. 4-31 below shows the lighting used to show a displaced runway threshold.

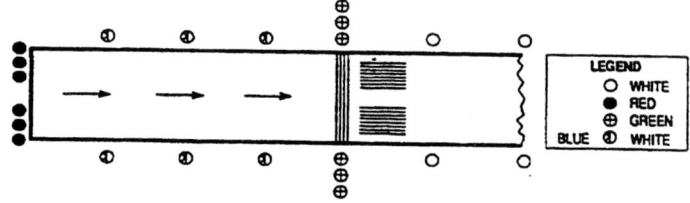

Fig. 4-31: Displaced Runway Threshold Lighting.

Runway Centre Line Lighting

Runway centre line lighting is provided on CAT II runways. It consists of variable intensity lights installed on the runway surface spaced at intervals of 50 feet. The lights leading in the take-off or landing direction are white to a point 3,000 feet from the runway end. They then change to white and red until 1,000 feet from the runway end, at which point they become red.

Runway Touchdown Zone Lighting

Touchdown zone variable intensity white lights are provided on CAT II instrument runways. They consist of bars of three inset lights per bar disposed on either side of the runway centre line, spaced at 100 feet intervals commencing 100 feet from the threshold, extending 3,000 feet down the runway. The lights are uni-directional, showing in the direction of approach to landing.

Rapid Exit Taxiway Lighting

Rapid exit taxiway lights are green in colour and are installed on the taxiway surface commencing approximately 200 feet before the turn and continuing through the rapid exit taxiway to 200 feet beyond the turn.

Taxiway Lighting

Taxiway edge lights are blue and spaced at 200 feet intervals. Where a taxiway intersects another taxiway or a runway, two adjacent blue lights are placed at each side of the taxiway. The intersection of taxiway and parking aprons is indicated by two adjacent yellow lights at taxiway/apron corners. Centre line taxiway lights are green and are installed on the taxiway surface. They are spaced at 200 feet intervals (lesser spacing on curves).

Clearance Bars

Clearance bars may be provided on taxiways, where it is desirable to define a specific aeroplane holding limit. They are located at a point 100 to 200 feet from the near edge of the taxiway and runway intersection. The clearance bars consist of at least three flush-mounted unidirectional yellow lights showing in the direction of the approach to the intersection. They are placed symmetrically about at a 90° angle to the taxiway centre line with individual lights five feet apart.

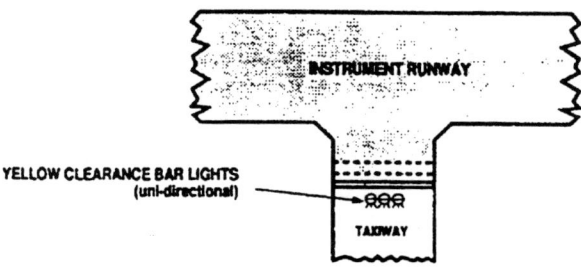

Fig. 4-32: Taxiway Clearance Bars.

Emergency Lighting

Most major airports in Canada are equipped with an emergency power system for lighting visual aids. This system is normally capable of assuming the electrical load within approximately 15 seconds. At airports having non-instrument approach runways, the change-over time may be upwards of two minutes.

Aircraft Radio Control of Aerodrome Lighting (ARCAL)

ARCAL systems are becoming more prevalent as a means of conserving energy, especially at aerodromes and airports not staffed on a continuous basis, or where it is not practicable, to install a landline to a nearby FSS. Aside from obstruction lights, some or all of the aerodrome and airport lighting may be radio controlled. Control of the lights should be possible when aircraft are within 15 NM of the aerodrome or airport. The frequency range is from 118.0 MHz to 136.0 MHz. Activation of the ARCAL system is via the aircraft VHF transmitter and is effected by keying the microphone a given number of times within a specified number of seconds. Each activation will start a timer to illuminate the lights for a period of about 15 minutes. The timing cycle may be restarted at any time during the cycle by repeating the specified keying sequence. The code for the intensity and the lighting period varies for each installation - consult the CFS for the aerodrome involved.

Pilots are advised to key the activating sequence when commencing their approach, even if the aerodrome or airport lighting is already on. This will restart the timing cycle so that the full 15 minute cycle is available for the approach.

A TYPE J ARCAL is activated by keying the mike 5 times within 5 seconds. A TYPE K ARCAL is activated by keying the mike 7 times initially, and the intensity can be adjusted up or down to any one of three settings by keying the mike 7, 5 or 3 times within 5 seconds for high, medium or low intensity settings respectively.

4.12 Winter Operations

Critical Surface Contamination on Ground

CRITICAL SURFACE of an aircraft means the wings, control surfaces, propellers, horizontal stabilizers, vertical stabilizers or any other stabilizing surface on an aircraft and, in the case of an aircraft that has a rear-mounted engine(s), includes the upper surface of its fuselage.

The air regulations prohibit takeoff when frost, ice or snow is adhering to any critical surface of the aircraft. This is referred to as the CLEAN AIRCRAFT CONCEPT. Whenever frost, ice or snow may reasonably be expected to adhere to the aircraft, an inspection of the aircraft is legally required before takeoff. Frost, ice or snow on top of de-icing or anti-icing fluids must be considered as adhering to the aircraft and take-off must not be attempted.

Frozen Contaminants

An aircraft needs a smooth, uninterrupted flow of air over the wings and control surfaces. The slightest accretion of frost can seriously affect performance capabilities, because frost increases the stalling speed. Ice contamination having the thickness and roughness of medium to coarse sandpaper can decrease lift by as much as 30% and increase drag by 40%. Ice or frost can be removed with special de-icing solutions or by putting the aircraft in a heated hangar (ensure the aircraft is completely dry before taking outside). Refer to s. 2.13 for further inormation on de-icing and anti-icing procedures.

The Cold Soaking Phenomenon

Since fuel tanks are normally located in the wings, after a flight, the aircraft and fuel temperature can be much colder than ambient temperatures (fuel temperature drops of 12° to 18° C have been recorded after a flight of 2 hours). An aircraft COLD SOAKED WINGS will conduct heat away from precipitation so that a thin sheet of clear ice can easily form on the wing surface that is hard to detect. The formation of ice on the wing based on the Cold Soaking Phenomenon is dependent on the type, depth and liquid content of precipitation, ambient air temperature and wing surface temperature.

Cold soaking can also cause frost formation at airports much warmer than freezing where there is high relative humidity.

In-Flight Airframe Ice Contamination

Airframe icing can be a serious weather hazard to aircraft in flight. Icing will result in a loss of performance in the following areas:

a) Ice accretions on lifting surfaces will change their aerodynamic properties resulting in a reduction in lift, increase in drag and weight with a resultant IN-CREASE in stalling speed and a reduction in the stalling angle of attack. Therefore, an aerodynamic stall can occur before the stall warning systems activate.
b) Ice adhering to propellers will drastically affect their efficiency and may cause an imbalance with resultant vibration.
c) Ice on the windshield or canopy will reduce or block vision.
d) Carburetor icing can affect engine output.

The three types of ice to contend with are clear ice, rime ice and frost. Review s. 6.13 in this manual for the characteristics of airframe ice. Frost may form on an aircraft in flight when descent is made from freezing conditions to a layer of warm moist air. In these circumstances vision may be restricted as frost forms on the windshield or canopy. Due to these hazards, pilots should insure they keep away from areas of probable icing, or are equipped with suitable anti-ice/de-ice systems.

Landing Wheel Aircraft on Snow Surfaces

Each winter, accidents occur due to pilots attempting to land wheel-equipped aircraft on surfaces covered with deep snow. This usually results in the aircraft nosing-over. Light aircraft should not be landed on surfaces covered with snow unless it has previously been determined that the amount of snow will not constitute a hazard (some aircraft are fitted with skis for winter operations).

Whiteout

Whiteout is defined as:

> An atmospheric optical phenomenon of the polar regions in which the observer appears to be engulfed in a uniformly white glow. Neither shadows, horizon, nor clouds are discernible; sense of depth and orientation is lost; only very dark, nearby objects can be seen. Whiteout occurs over an unbroken snow cover and beneath a uniformly overcast sky, when coupled with the snowblink effect, the light from the sky is about equal to that from the snow surface.

Blowing snow may be an additional cause. Due to the lighting conditions encountered in whiteout, it becomes impossible to see the horizon. The danger is that the pilot will become disoriented. There have been numerous accidents where pilots have flown their aircraft directly into the ground, in whiteout conditions. If whiteout conditions are encountered, or suspected, the pilot should immediately climb if at low level, or level off and turn towards an area where sharp terrain features exist. The following phenomena cause whiteout and should be avoided:

a) WATER FOG WHITEOUT results from thin clouds of super-cooled water droplets contacting cold snow surface. Depending on the size and distribution of the water droplets, visibility may be minimal or zero.
b) BLOWING SNOW WHITEOUT results from fine snow being plucked from the surface by winds of 20 knots or more. Sunlight is reflected and diffused resulting in zero visibility.
c) PRECIPITATION WHITEOUT results from small wind-driven snow crystals falling from low clouds above which the sun is shining. Light reflection complicated by spectral reflection from the snow flakes and obscuration of landmarks by falling snow may reduce visibility and depth perception to zero.

4.13 Flight Operations in Mountainous Areas

The importance of proper training, procedures and preflight planning when flying in mountainous regions can never be over emphasized.

In the Pacific area, the combined effect of the great mountain system and the adjacent Pacific Ocean lead to extremely changeable weather conditions and a variety of weather patterns. Some of the factors to be taken into consideration regarding the effect on aircraft performance when operating under these conditions are:

a) elevation of the airport;
b) temperature and pressure;
c) turbulence and wind effect;
d) determination of safe take-off procedures to insure clearance over obstacles and intervening high ground.

IFR route selection in mountainous areas should be made with extreme care, paying particular attention to minimum altitudes as they relate to aircraft performance and anti-ice/de-ice systems. Review the procedures for flight within designated mountainous regions at s. 1.24 in this manual.

Remember also that greater care should be taken in determine appropriate departure procedures and corresponding obstacle clearance requirements.

4.14 Flight in Rain

An error in vision can occur when flying in rain. The presence of rain on the windscreen, in addition to causing poor visibility, introduces a refraction error. This error is because of the reduced transparency of the rain covered windscreen which causes the eye to see a horizon below the true one, because of the eye response to the relative brightness of the upper bright part and the lower dark part. As well, the shape and pattern of the ripples formed on the windscreen, particularly on sloping ones, causes objects to appear lower. The error may be present as a result of one or other of the above causes, or both, which can amount to a cumulative error of a 5° angle. Therefore, a hilltop or peak 1/2 NM ahead of an aircraft could appear to be approximately 260 feet lower than it actually is.

Pilots should remember this additional hazard when flying in conditions of low visibility in rain and should maintain sufficient altitude and take other precautions, as necessary, to allow for this error. As well, pilots should insure proper terrain clearance during en route flight and on final approach to landing.

4.15 Wake Turbulence

Wake turbulence is caused by wing tip vortices. The most powerful vortices are caused by heavy, slow moving aircraft in CLEAN configuration. Vortices are most likely encountered behind descending or climbing aircraft.

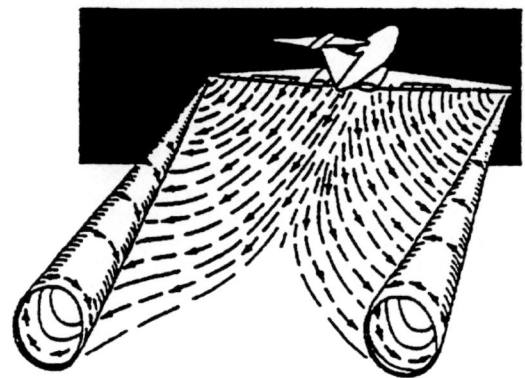

Fig. 4-33 Vortex generation.

Vortices persist for a minimum of two minutes. To avoid wake turbulence upon arrival, plan an approach path that is higher than the path of the generating aircraft, and plan to touchdown beyond the touchdown point of the generating aircraft. To avoid wake turbulence on departure, plan to lift-off well in advance of the lift-off point of the generating aircraft, and do not fly through its flightpath.

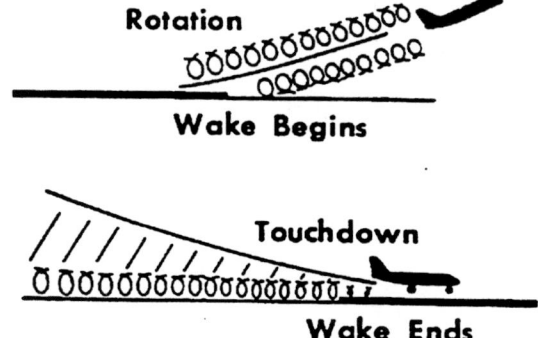

Fig. 4-34: Where vortices begin and end. Visualize and avoid these areas in take-off and landing.

Fig. 4-35: Vortex Roll.

Flight into a vortex core can produce roll rates of 80 degrees per second, and downdrafts of 1,500 feet per minute. It is extremely important to visualize the location of vortices and avoid flying into these areas.

When the vortices of a large aircraft sink to within about 200 feet of the ground, they tend to move laterally at about 5 kts. In a 5 - 10 kt crosswind, the upwind vortex remains stationary with the runway.

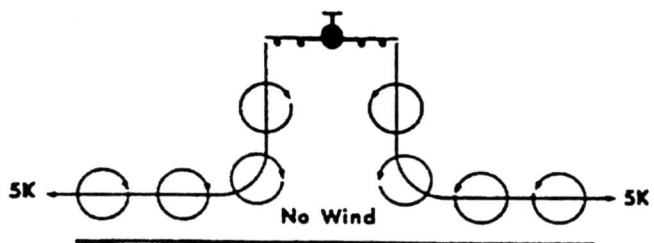

Fig. 4-36: Vortex movement in ground effect, no wind.

A tailwind can push the vortices back on the runway following landing of a heavy aircraft. The worst ground conditions would be a light, quartering tailwind, because the vortices will creep forward and remain over the runway.

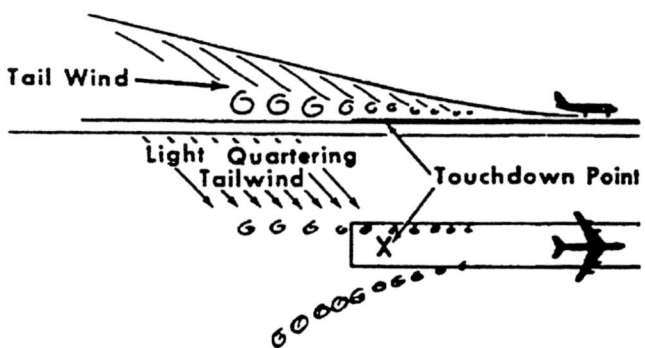

Fig. 4-37: Vortex movement in wind.

In forward flight, HELICOPTERS generate a pair of trailing vortices similar to wing-tip vortices of fixed wing aircraft. Helicopter vortices are highly concentrated and the size of the helicopter is not a significant factor in vortex strength. Pilots of small aircraft should avoid the vortices as well as the downwash.

Pilots should also know that in operations in an ATC environment, the receipt of any traffic information, instructions to follow another aircraft, or the acceptance of approach clearance does not make ATC responsible for wake turbulence.

THE PILOT IS ALWAYS RESPONSIBLE TO INSURE COMPLIANCE WITH ATC WILL PERMIT A SAFE AP-PROACH AND LANDING.

If in doubt, advise ATC and request alternate instructions.

4.16 Flight Procedures for Low Level Wind Shear

WIND SHEAR occurs when the wind through which an aircraft is flying changes so rapidly that it causes a change in the airspeed. When the wind changes at a greater rate than the aircraft can accelerate or decelerate because of its inertia, the airspeed changes. Wind shears are either INCREASED PERFORMANCE SHEAR (airspeed increases due to entering increased headwind or decreased tailwind) or DECREASED PERFORMANCE SHEAR (airspeed decreases due to entering decreased headwind or increased tailwind). With an increased performance shear, lift increases. With a decreased performance shear, lift decreases.

Wind shears are caused by thunderstorms, lee side mountain waves, fronts, valley winds, convective thermals and low level inversions. Thunderstorms cause downbursts. The effect of a downdraft can be added to a decreased performance shear to increase the hazard.

To recover from decreased performance shear, increase power to maintain glidepath and IAS, when shear is cleared, reduce power to below original setting. Airspeed can be exchanged for height. Don't pitch down: if low, the aircraft may strike the ground (it is better to fly at reduced airspeed until shear is cleared).

To recover from increased performance shear, reduce power to maintain glidepath and IAS, when aircraft stabilizes, increase power to higher then before shear.

Microburst Wind Shear

Microburst wind shear may create a severe hazard for aircraft within 1,000 feet of the ground, particularly during the approach to landing and landing and take-off phases. The impact of a microburst on aircraft which have the unfortunate experience of penetrating one is characterized at Fig. 4-40. The aircraft may encounter a headwind (increased performance wind shear) followed by a downdraft and tailwind (decreased performance wind shear).

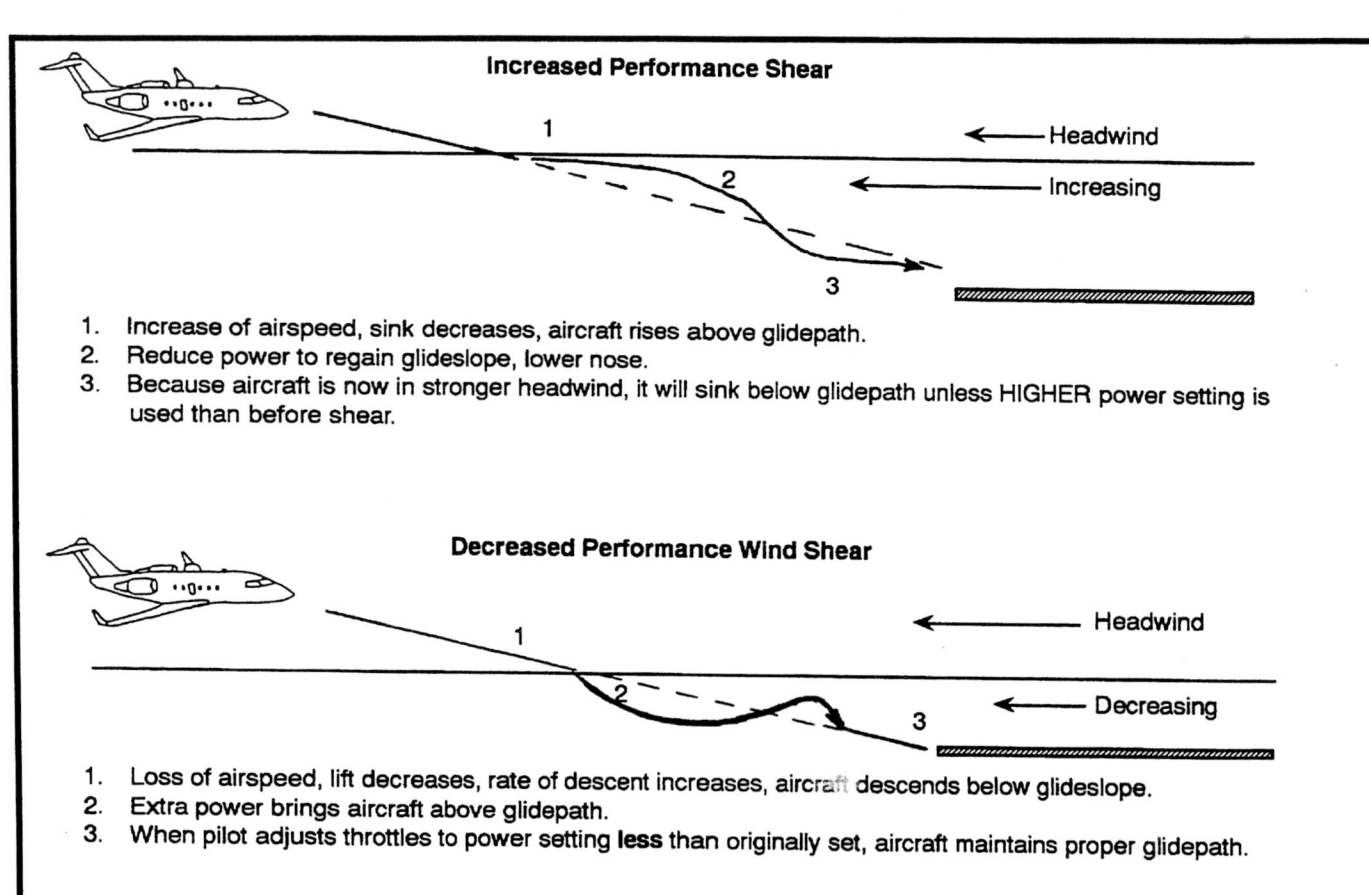

Increased Performance Shear

1. Increase of airspeed, sink decreases, aircraft rises above glidepath.
2. Reduce power to regain glideslope, lower nose.
3. Because aircraft is now in stronger headwind, it will sink below glidepath unless HIGHER power setting is used than before shear.

Decreased Performance Wind Shear

1. Loss of airspeed, lift decreases, rate of descent increases, aircraft descends below glideslope.
2. Extra power brings aircraft above glidepath.
3. When pilot adjusts throttles to power setting **less** than originally set, aircraft maintains proper glidepath.

Fig. 4-38: Increased performance and decreased performance wind shear.

Angle of Attack and Pitch Attitude in Downbursts

Entering a downburst reduces the aircraft angle of attack, due to the change in relative airflow. Remember that angle of attack is the angle between chord line and relative air.

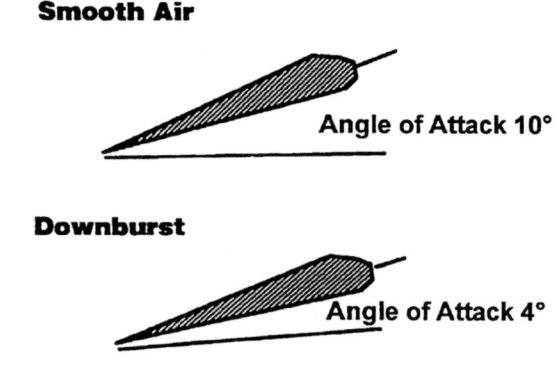

Smooth Air

Angle of Attack 10°

Downburst

Angle of Attack 4°

Fig. 4-39: Effect of downburst where pitch attitude is maintained: reduced angle of attack.

To correct for the decrease in the angle of attack on entry into a downburst, the aircraft must be placed in an abnormally HIGH pitch attitude. When leaving the downburst, be prepared to quickly reduce the pitch attitude to that required for the new situation.

4.17 Landing Techniques

A successful landing of a heavy transport aeroplane, to a large degree, depends on a successful descent and approach. Avoid high descent rates - a useful rule of thumb is to take 50% of the ground speed and convert to FPM ie for GS of 120 kts, use descent rate on GP of 600 FPM.

On large transport aircraft, due to the high weights and momentum forces, do not put yourself in the position of having to make any large or abrupt maneuvering on final.

Do not "duck under" the glideslope and a proper landing will involve holding the glideslope all the way to flare (plan touchdown around ILS reference point at first 1,000 feet). Do not take off power too high in the flare as transport aircraft have very high sink rates. Keep enough stabilizer trim for full elevator effectiveness for flare.

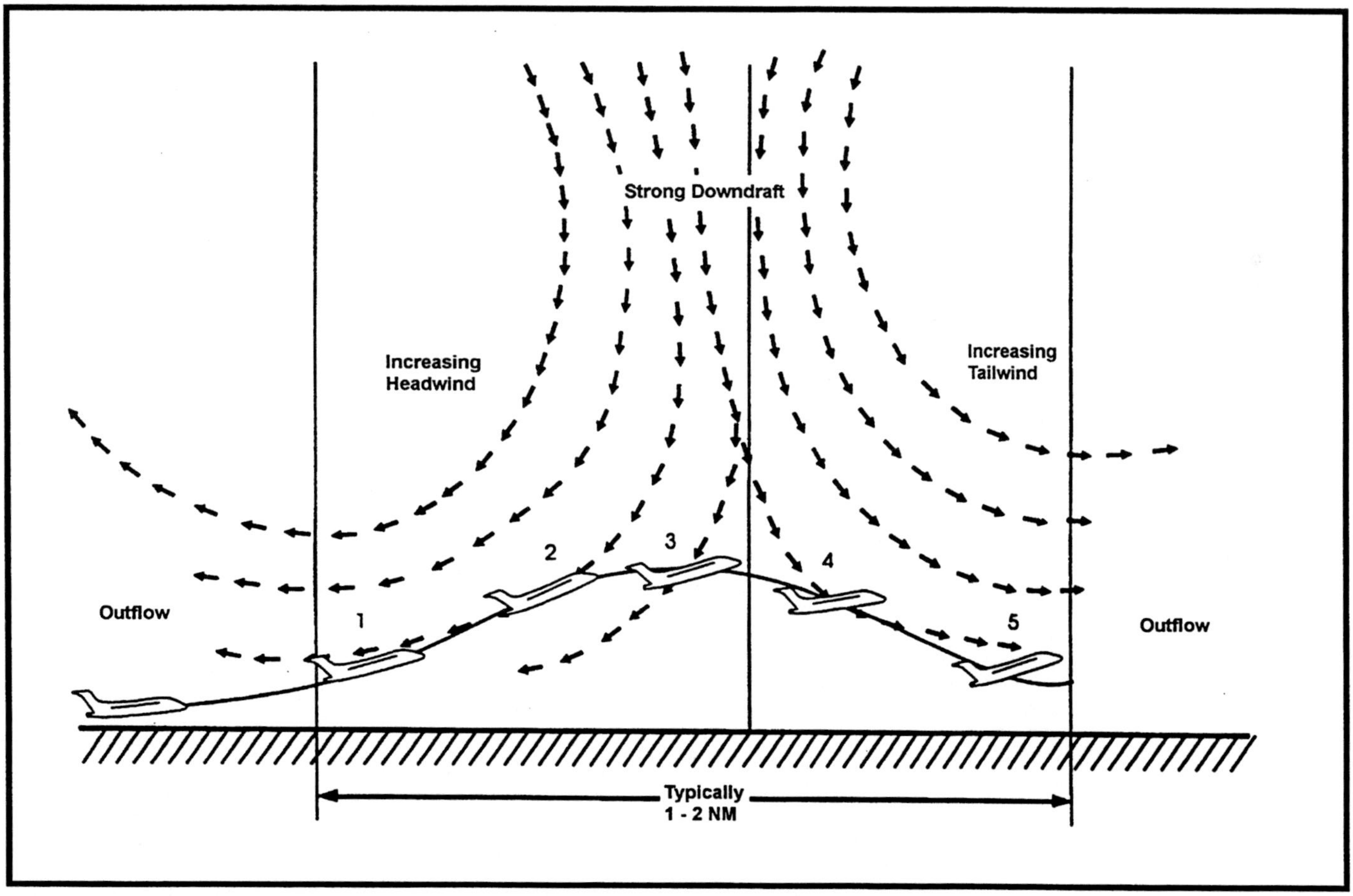

Strong Downdraft

Increasing Headwind

Increasing Tailwind

Outflow

1 2 3 4 5

Outflow

Typically 1 - 2 NM

Fig. 4-40: A microburst encounter during take-off. The aeroplane first encounters a headwind and experiences increasing performance (1), this is followed shortly by a decreasing headwind component (2), a downdraft (3), and finally a strong tailwind (4), where 2 through 5 all result in decreasing performance of the aeroplane. Position 5 represents and extreme situation just prior to impact.

Do not have too high a speed on short final.

From the threshold, flare (if necessary), reduce to idle thrust, push off drift (if necessary), and land firmly if the runway is wet. On contact, push controls forward to hold the nose down, keeping controls laterally level on ailerons, raise spoilers, apply reverse thrust and start braking. Keep the aircraft straight and get the maximum benefit from reverse thrust, cancelling only when aircraft is at taxi speed.

For a crosswind landing, remember that extending the spoilers may result in some loss of lateral control.

For maximum braking effectiveness, with anti-skid functional, apply brake pressure just into cycling range as will be confirmed by foot thumpers. If landing without anti skid, use only light braking pressure at high ground speeds and slowly increase pressure as speed diminishes (tires may blow with maximum brake pressure).

4.18 Volcanic Ash

Flight operations in areas of volcanic ash is very dangerous since powerplant failures (both recip and turbine) are likely. Areas of ash can reach heights of up to FL 600. As well, the risk of entering ash in night or in IMC conditions is particularly dangerous since there is no clear visual warning. Pilots encountering such conditions where nothing was anticipated should file an urgent PIREP.

4.19 Aviation Physiology

Effect of Various Gases

Gases diffuse from high pressure to low pressure. ALVEOLI are tiny air sacs in the lungs that allow the air in the lungs to contact blood capillaries. The blood picks up oxygen from the air and returns CO_2 from body tissues.

HYPOXIA is a shortage of oxygen in the blood. The higher the altitude, the lower the oxygen available for respiration. It is a dangerous condition, because it severely impairs one's ability to operate an aircraft. HYPOXIC HYPOXIA is caused by a shortage of oxygen with altitude. ANEMIC HYPOXIA is caused by an inadequate supply of blood haemoglobin to carry oxygen, in such cases as those having recently donated blood, or those suffering from stomach ulcers, menstrual periods or heavy smokers. A heavy smoker's ground altitude is already at an equivalent atmospheric altitude of 5,000 to 7,000 feet. An early symptom of hypoxia is a sensation of well-being referred to as EUPHORIA. Increasing hypoxia interferes with muscular coordination, mental calculation and reasoning ability. Unconsciousness may occur before the pilot has realized the problem. TIME OF USEFUL CONSCIOUSNESS (TOC) refers to the time in which we can recognize a hypoxia problem and prevent it. TOC varies with altitude as follows:

Altitude	TOC
10,000 ft	hours
20,000 ft	5 - 12 minutes
30,000 ft	45 - 75 seconds
40,000 ft	13 - 30 seconds
45,000 ft	12 - 15 seconds or less

To avoid hypoxia, do not fly above 10,000 ft without supplemental oxygen. There are three basic types of oxygen systems: CONTINUOUS FLOW, DILUTER DEMAND SYSTEMS and PRESSURE DEMAND systems. If available, oxygen should be used from the ground up at night if cabin altitude is going to be excessive. Above 10,000 feet, cabin altitude oxygen should always be available and flight crew should watch each other carefully for symptoms of breathlessness, increasing instrument errors, personality changes or evidence of poor judgment.

CABIN PRESSURIZATION is the maintenance of a cabin altitude lower than the actual flight altitude. This is accomplished by compressing air in the aircraft cabin. Heated air from the engine or bleed air from turbine engines is typically used to power aircraft pressurization systems. DECOMPRESSION is the inability of the aircraft's pressurization system to maintain its designed pressure. EXPLOSIVE DECOMPRESSION is a change in cabin pressure faster than the lungs decompress, with possible lung damage. Decompression in less than 0.5 seconds is considered explosive. RAPID DECOMPRESSION is a change in cabin pressure where the lungs can decompress faster than the cabin and will not cause lung damage. During a decompression there may be noise, fog, dust or flying debris. TOC is considerably shortened following rapid decompression.

The ear consists of the outer ear, middle ear and inner ear. The eardrum is a thin membrane about .004 inches thick. The EUSTACHIAN TUBE connects the middle ear cavity to the throat. Inner ear discomfort is caused when one is unable to equalize the pressure differential through the eustachian tube during a climb or descent. Normally, there is little difficulty equalizing pressure during descent by swallowing or yawning. During sleep, the rate of swallowing slows down, so sleeping passengers should be awakened prior to descent to permit ear ventilation. If inner ear pain persists, the VALSALVA PROCEDURE should be used: close the mouth, hold the nose and blow. This forces air up the eustachian tube and into the middle ear. Painful ear block generally occurs when descent is made too rapidly: to relieve this pain, ascent to a higher altitude is recommended, followed by a slower descent.

Nitrogen gas diffused in the body's tissues can bubble and cause DECOMPRESSION SICKNESS. When the barometric pressure falls, the alveoli nitrogen pressure falls. The pressure differential causes nitrogen bubble formation in the blood and body tissues. Pain associated with nitrogen bubbles forming in the joints is called BENDS. Older or obese persons are more likely to get bends. After bends develop, the only cure is to descend below the altitude at which the problem occurred. The pain becomes worse if ascent is made to higher altitude.

CARBON MONOXIDE is an odourless, colourless gas that results from combustion. If there are leaks in the exhaust system, there is a danger of carbon monoxide poisoning. The symptoms of carbon monoxide poisoning are blurred thinking or headache, and cherry-red lips. If carbon monoxide poisoning is suspected, immediately shut off the heater, open the air ventilators, descend to lower altitudes, and land at the nearest airfield. Consult an Aviation Medical Examiner. It may take several days to fully recover and clear the blood supply of the carbon monoxide.

Vision

The RETINA of the eye is extremely sensitive to shortages in oxygen. Pilots flying at night will experience reduced vision at lowered altitudes due to hypoxia: if oxygen is available, for night flying, use it from the ground up. Sunglasses should be used during daytime to avoid eye fatigue and improve dark adaptation. Conscientious scanning for traffic, especially at night should be routine. Drugs, alcohol, smoking and fatigue all adversely affect day and night vision.

Hypothermia

HYPOTHERMIA is severe accidental cooling of the body. Prolonged exposure to water or cold air temperatures following an aircraft accident typically lead to this condition. Air temperatures need not be freezing to lead to hypothermia. Basic first aid treatment is to provide blankets and lukewarm, sweet drinks to raise the lowered blood sugar level. Do not use hot water bottles or other direct heating as this will cause sudden opening of superficial blood vessels, taking away blood from deep tissues and essential organs.

Alcohol

Do not fly while under the influence of alcohol. Although the air regulations require that a minimum of 8 hours lapse before flying, following alcohol consumption, a good rule is to allow 24 hours, and 48 hours after heavy drinking. Alcohol is selectively concentrated in certain areas of the body and can remain in the fluid of the inner ear even after all traces in the blood have disappeared. This is why it is difficult to balance during a hangover.

Medication

Do not fly while under medication, except on the advice of an Aviation Medical Examiner. Drugs associated with aviation accidents are: antihistamines, tranquillizers, appetite suppressing drugs, barbiturates, nerve tonics and digestive pills. Following adminstration of spinal or general anesthetics, or following serious medical operations/surgeries, do not fly until your doctor says it is safe to do so. Following administration of local anesthetics such as for dental work, common sense suggests that you wait at least 24 hours before flight.

SCUBA Diving

It is dangerous to fly an aircraft immediately following scuba diving: due to changes in the gas pressure of nitrogen, bubbles can form in the blood stream causing difficulty in breathing or death. Allow at least 24 hours prior to flying after scuba diving.

Blood Donation

Do not donate blood while actively flying: it takes several weeks for circulation to return to normal, and these effects are amplified when flying. Consult an Aviation Medical Examiner prior to resuming flight, and do not fly for at least 48 hours.

Hyperventilation

HYPERVENTILATION is overbreathing caused by anxiety or panic resulting in a lower percentage of blood carbon dioxide. Sensations of hyperventilation are a tingly feeling and shortness of breath, followed by dizziness and eventual unconsciousness.

The essential precaution against hyperventilation is training. Hyperventilation is likely to occur when you are excited, tense, or scared, because a natural urge exists to overbreathe when you face psychological or physical stress. Stress and fatigue are built into flying, especially instrument flying, in inverse proportion to your knowledge, skill, and level of proficiency. An instrument pilot could face difficulties with hyperventilation by attempting a stress-inducing flight with insufficient pilot proficiency, inadequate equipment, or inadequate planning to prevent fatigue and tension.

The basic remedy for hyperventilation is to slow down your rate of breathing and intermittently hold your breath to allow the carbon dioxide to build up to its normal level. Other techniques include breathing into paper bag, reducing breathing rate to 10 - 12 breaths per minute, or talking aloud. The difficulty with symptom recognition is that the victim of hyperventilation is not likely to be aware of the seriousness of the condition, nor able to take the required corrective action.

Several single pilot, high performance jet aircraft accidents have been attributed to hyperventilation.

Fatigue

Fatigue slows reaction times, reduces concentration and leads to attention errors. Fatigue may be caused or aggravated by lack of sleep, insufficient rest, business pressures or other stresses. Boredom and fatigue worsen each other. To overcome boredom, make frequent groundspeed checks, weather updates, fuel consumption calculations, flight planning for diversions etc. as needed to keep alert.

Orientation and Disorientation

ORIENTATION means using our sensory apparatus to tell us which way is up. Our sensory system for this purpose consists of our eyes, vestibular organs of our inner ears and kinesthetic muscle senses.

When flying, many conditions encountered can cause conflicts or illusions to our sensory system. DISORIENTATION refers to a loss of proper bearings, state of mental confusion as to position, location or movement.

SPATIAL DISORIENTATION can occur in conditions of reduced visibility, such as in whiteout conditions, in night flying, or when flying under IMC.

Kinesthetic Illusions

Relying on normal muscle sensations to determine orientation becomes unreliable when flying under IMC. This is because we are used to a downward pulling force telling us this is down, but when flying, this downward force acts outward from the turn and tricks our sense of orientation. Accordingly, where there is a conflict between attitude information from our flight instruments and our muscle sensations, the flight instruments must be relied on as being correct.

Visual Illusions

Visual information comes from two sources: central or focused vision and peripheral vision. It is peripheral vision that assists in our sense of orientation. We use central vision for object recognition. CENTRAL VISUAL ILLUSIONS are misunderstandings of what we see, PERIPHERAL ILLUSIONS are false impressions of movement or rotation.

Central visual illusions are affected by what we are expecting. A pilot used to flying from a wide runway who encounters a narrow runway can lead to a late flare and early touchdown, since the narrow runway will appear to be longer and farther away.

Landing at an UPSLOPE runway will tend to be short, since there is an illusion of overshooting caused by the taller runway image. Landing at a DOWNSLOPE runway will tend to be long, since there is an illusion of under-shooting caused by the low profile of the runway image.

The FALSE HORIZON effect results from sloping cloud formations, dark scenes spread with ground lights and stars, and certain geometric patterns of ground lights can create illusions of not being aligned correctly with the actual horizon. The disoriented pilot may place the aircraft in a dangerous attitude. Spatial disorientation from these illusions can be prevented only by visual reference to reliable, fixed points on the ground or to flight instruments.

When flying in high wind conditions at low altitude, turning UPWIND yields the illusion of a skidding turn and decreased airspeed, and turning DOWNWIND yields the illusion of a slipping turn and increased airspeed.

Vectional Illusions

VECTIONAL ILLUSIONS are illusions of movement caused by outside movements. For example, raindrops striking an aircraft windshield during flight in cloud can give a false sense of vertical movement.

The AUTOKINETIC EFFECT refers to a fixed light in the distance at night which appears to suddenly move as the eye loses fixation and jumps back to the target: it is the eye that moves, not the target. The effect can result in the feeling that the distant object is moving. The feeling can be overcome by looking away from the object for a moment, then looking back again.

Vestibular Illusions

Vestibular illusions are the most complex and danger-ous, and result from errors of orientation associated with our inner ear organs. The inner ear consists of OTOLITHS that are sensitive to linear acceleration and the SEMICIRCULAR CANALS that are sensitive to angular acceleration.

Linear Accelerations

With respect to the otolith organs, they can yield PITCH UP and PITCH DOWN illusions. The PITCH UP ILLUSION is caused during aircraft acceleration, such as a take-off roll. The brain mistakes the acceleration for a sense of aircraft pitch up, resulting in a dangerous tendency to pitch the aircraft nose down. The pitch up illusion is more pronounced during a night take-off from a well-lit field into a black hole. The PITCH DOWN ILLUSION is caused by deceleration giving a sensation that the nose of the aircraft is dropping and can be experienced by dropping speed brakes or lowering flaps. The pitch down illusion is most likely on final approach at slow speed and the reaction of pulling back on the controls may cause a stall.

Angular Accelerations

The semicircular canals respond to angular accelerations. There are three planes in the inner ear, each corresponding to yaw, pitch and roll.

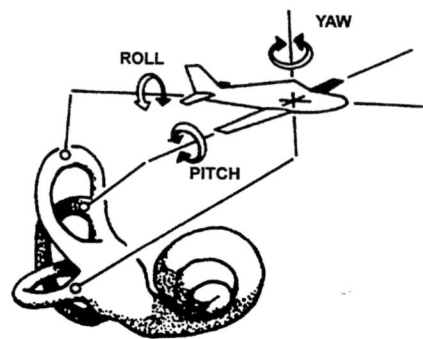

Fig 4-41: The semi circular canals of the inner ear are filled with fluid and are located at right angles.

Once fluid in any part of the semicircular canals begins to rotate, the result is a sensation of movement in the particular plane. The OPPOSITE TURNING ILLUSION occurs in an aircraft after levelling out from a turn as inertia cause the inner fluid to continue to move. This sensation can last for 10 - 20 seconds.

The opposite turning effect can also cause a GRAVE-YARD SPIN when an inexperienced pilot enters a spin where after the pilot recovers from a 2 - 3 turn spin, the pilot has the sensation of turning in the opposite direc-tion, resulting in miscorrecting and re-entering another spin. Eventually the aircraft nose pitches to an extreme nose up attitude, then falls to an extreme nose low attitude and tight spiral as the aircraft enters a severe GRAVEYARD SPIRAL and eventual ground contact - many accidents have occurred this way attributable to a non-IFR pilot encountering IMC.

The COREOLIS ILLUSION is one of the most dangerous of the vestibular illusions and occurs following an inappropriate head movement. The semi circular canals involved sense new movements in a new plane in differ-ent canals and the result is a violent sensation of tum-bling. Accordingly, pilots should remember that turning the head sharply while in IMC, particularly if the move-ment is against the direction of turn, is extremely hazardous.

THE LEANS can result from an abrupt correction of a banked attitude, which has been entered too slowly to stimulate the motion sensing of the semi circular canals, creating an illusion of banking in the opposite direction. The disoriented pilot will roll the aircraft back into its original dangerous attitude, or if level flight is maintained, will feel compelled to lean in the perceived vertical plane until this illusion subsides.

ALTERNOBARIC VERTIGO is a short lasting but acute sense of spinning which is caused by a sudden difference in pressure in the two inner ears when flying with a cold as one ear clears before the other on an ascent.

Prevention of Sensory Illusions

Most sensory illusions are problems that usually show up under conditions of poor visibility. Whenever the visibility is poor enough to prevent you from checking your equilibrium senses with your eyes, your equilibrium system is undependable. That is why your aircraft is provided with an artificial equilibrium system for indicating bank angle, aircraft altitude, rate of climb etc. This system is much more reliable than your own sensory system. Be aware of the various sensory difficulties you have, and insure that you are as current and proficient as possible by virtue of being competent at IFR skills in IMC.

Overcoming Disorientation

If you become aware of a sense of disorientation, remember the following points:

a) focus on you instrument scanning and ignore visual cues or bodily sensations;
b) minimize head movements;
c) if possible, bring the aircraft to basic straight and level flight;
d) transfer aircraft control to another pilot if operating with two crew;
e) use the autopilot if equipped;
f) if necessary, declare an emergency.

4.20 Aviation Psychology

Pilots make better decisions when they understand the factors and process of decision making. An awareness of decision making will help the pilot make safer choices both for in flight situations and for go/no-go decisions before flight.

The Decision Making Process

To make safe decisions, the pilot must have SITUATIONAL AWARENESS, by finding out all they can about important information in a flying situation. Next, the pilot should identify as many OPTIONS as possible to choose from. Once the options are worked out, the pilot must CHOOSE the best option for the situation at hand. Finally, once an option is chosen, the pilot must ACT and EVALUATE the results to confirm the best option has been chosen. If the best option has not been chosen, the decision making process will have to be repeated.

Many factors affect the decision making process, such as job demands, stress, monetary gain, skill and level of training.

Factors that Influence Decision-Making

Anything that gets in the way of clear thinking (such as fatigue, a bad cold etc.) can harm good decision-making. The more the background KNOWLEDGE available, such as pre-flight planning and preparation, the better. SKILLS should be practised until they become automatic reactions and must be maintained long after the flight test. EXPERIENCE usually aids decision making, although getting away with a sloppy practise in the past can lead to unsafe flying. REASONING is the process of thinking carefully to recognize early the situations that may lead to trouble. RISK ASSESSMENT is important to evaluate how serious a problem is and how much time is available to deal with it. STRESS can place the pilot into problem situations and negatively affect the pilot's ability to deal with them. A good ATTITUDE goes a long way to avoiding dangerous situations.

4.21 Pilot—Equipment Materials Relationship

The human mind is often unreliable: pilots should be in the habit of double-checking information received from aircraft displays. Such errors can be found in instrument flying, when an instrument is incorrectly read, resulting in a control error.

Often, a pilot may completely forget a crucial check - such as fuel tank selection. It is for this reason that checklists have been developed. A checklist should be accurate and followed to the letter - it is a good operating practise to read each item out loud as the checks are being carried out. As well, all checklists should be up to date and properly organized. A poor or outdated checklist, or a checklist for a different aircraft can lead to dangerous situations.

Errors can also result from improper map reading. Know what the various symbols indicate on IFR charts.

To insure that all pilots follow the same procedure to operate a given aircraft, air carriers develop STANDARD OPERATING PROCEDURES (SOP's). Provided an air carrier's SOP's are respected and followed by all flight crew, this will avoid cockpit disagreements and insure that a standardized and consistent procedure is applied for all aircraft operations.

Old IFR navigation equipment can be poorly designed. For example, some VOR's display the OBS selected radial at the top of the display, while some display it at the bottom. Some fixed card ADF's have a rotatable card, while some do not. Some RNAV units do not have readout to show what location the equipment is set up on. The basic rule is to learn any idiosyncracies of your equipment and not fly an aircraft with unfamiliar equipment until you have learned the applicable differences. The pilot seat position should be adequate to provide optimum view of flight and navigation instruments, but should also provide the best possible outside view for traffic, runway lighting etc. The cockpit windshield always should be clean.

4.22 Interpersonal Relations

CRM or "Crew Resource Management" is the effective use of all resources including the aircraft and its systems, printed materials and computer software, and people, to achieve the highest level of safety possible. The term "people" includes pilots, flight attendants, air traffic controllers, flight dispatchers and anyone else who may interact with the cockpit crew.

Key concepts to interpersonal relations and CRM are firstly, good COMMUNICATION. That means listening carefully and attentively to others, thinking about what they say, asking questions when unclear, and being able to effectively communicate to others. Good communications includes being able to give, and to receive, useful criticism.

Another key concept to interpersonal relations and CRM is TEAMWORK, that is, the ability of the group to work together well so as to complete a flight safely and efficiently, without letting any personal differences or conflicts develop.

Other important concepts involving CRM and interpersonal relations include good Pilot Decision Making (PDM) skills and, effective workload management and SITUATIONAL AWARENESS.

Notes:

Notes:

Notes:

Notes:

Notes:

Notes:

Navigation General

5.01 Basic Navigation Concepts

A MERIDIAN is a line joining the north pole to the south pole. The PRIME MERIDIAN runs through Greenwich, England. LONGITUDE is formally defined as "the smaller arc of the equator intercepted between the Greenwich meridian and a position". Longitude is the angular measurement of the number of meridians east or west of the prime meridian. The earth is divided into 360 meridians: 180° west and 180° east of the prime meridian. 180° opposite the prime meridian is the INTERNATIONAL DATE LINE. LATITUDE is measured north and south of the equator, 0 through 90°. The north pole is 90°N latitude. Every degree of latitude is divided into MINUTES: there are 60 minutes to each degree. One minute of latitude is 1 NM.

Fig. 5-1: Latitude, Longitude.

A GREAT CIRCLE is a circle whose plane passes through the centre of the earth. The equator is a great circle. A great circle is the shortest distance between two points on the globe. A RHUMB LINE is a curved line that crosses all meridians at a constant angle. The advantage to flying a rhumb line is that a constant heading can be maintained.

VARIATION is the angular displacement between true north and magnetic north. Variation is either east or west. An ISOGONAL LINE joins all points having equal variation. The AGONIC LINE joins all points having zero magnetic variation. To convert from degrees true to degrees magnetic, add variation west or subtract variation east.

HEADING is the angular measurement of the longitudinal axis of the aircraft while in flight. TRUE HEADING is measured from true north, MAGNETIC HEADING is measured from magnetic north. Track is the angular measurement of a line or path over the ground (the United States term for track is COURSE). TRUE TRACK is measured from true north, MAGNETIC TRACK is measured from magnetic north.

EXERCISE: Complete the following conversions:.

	True	Variation	Magnetic
1.	090	2° W	____
2.	180	10° E	____
3.	____	22° W	055
4.	____	12° E	356
5.	185	____	180
6.	290	____	298

ANSWERS:
1. 092
2. 170
3. 033
4. 008
5. 5° E
6. 8° W

AIRSPEED is the speed of the aircraft moving through the air, either indicated or true. GROUNDSPEED is the speed of the aircraft over the ground. A TRIANGLE OF VELOCITIES consists of the WIND VECTOR (wind direction, wind speed), the AIR VECTOR (true airspeed, true heading), and the GROUND VECTOR (groundspeed and true track). DRIFT is the lateral displacement of the aircraft from its heading to its track (downwind) caused by wind. WIND CORRECTION ANGLE is the correction in aircraft heading to counteract drift.

TRACK MADE GOOD is the actual path flown over the ground. If winds are from behind, the groundspeed is greater than the true airspeed, and vice versa if flying into a headwind. A flight computer is used to calculate true heading (where to steer to eliminate drift) and groundspeed (actual speed over the ground). To use the flight computer, enter TAS (obtain TAS from operator's handbook), true track (measure off map), wind speed and wind direction (weather information). In the same way, if flight planning for airways, use magnetic track (airway data) and magnetic winds (convert winds to magnetic using variation).

DEAD RECKONING NAVIGATION consists of calculating the heading, track, groundspeed and position of an aircraft in flight from a knowledge of the probable wind speed and wind direction, and true airspeed.

5.02   Navigation Calculations

Wind Drift Calculations

Be able to use a fligtht computer of your choice e.g. circular, E6B or electronic to complete basic wind drift calculations.

EXERCISE: Complete the following using a flight computer.

W/V	TT	TAS	WCA	TH	GS
1. 204/38	092° T	250 kts	___	___	___
2. 247/15	046°T	295 kts	___	___	___
3. 053/35	355°T	130 kts	___	___	___

ANSWERS:
1. 8° R, 100° T, 262 kts.
2. 1° L, 045° T, 309 kts.
3. 13° R, 008° T, 108 kts.

Be able to calculate in-flight winds from a known TAS, true heading true track and ground speed.

EXERCISE:

W/V	TT	TAS	WCA	TH	GS
1. _____	360	160 kts	___	006	140 kts
2. _____	290	130 kts	___	280	109 kts
3. _____	050	150 kts	___	055	165 kts

ANSWERS:
1. 041/25 kts, 6° R.
2. 240/30 kts, 10° L.
3. 191/20 kts, 5° R.

Time - Speed - Distance Calculations

Be able to complete standard time-speed-distance calculations using a flight computer of your choice.

Complete the exercise below.

EXERCISE:

	G/S	Distance	Time
1.	125 kts	63 NM	_____
2.	110 kts	307 NM	_____
3.	106 kts	29 NM	_____
4.	____ kts	50 NM	36 mins
5.	____ kts	91 NM	1 hr 5 mins
6.	____ kts	47 NM	19 mins
7.	74 kts	____ NM	34 mins
8.	95 kts	____ NM	39 mins
9.	67 kts	____ NM	1 hr 17 mins
10.	69 kts	____ NM	13 mins

ANSWERS:
1. 30 mins.
2. 2 hrs 47 mins.
3. 16 mins.
4. 83 kts.
5. 84 kts.
6. 148 kts.
7. 42 NM.
8. 62 NM.
9. 86 NM.
10. 15 NM.

Actual True Airspeed

A standard aviation calculation is to solve for Actual TAS (sometimes termed "Corrected TAS") on the basis of a known Calibrated Airspeed (CAS) and Pressure Altitude (PA). Although this calculation is familiar to pilots of low speed aircraft, for high speed, high altitude aircraft, a number of factors need to be considered before attempting to solve for TAS.

For high speed aircraft, compressibility causes erroneously higher indications on the ASI as well as falsely high indications for the indicated temperature as read from the temperature probe. These errors need to be accounted for when making TAS calculations.

As well, different methods need to be applied to solve for TAS depending upon what type of flight computer is used. The CR type circular computers are normally set up to automatically take into account compressibility error. Standard electronic flight computers are also normally set up to automatically compensate for compressibility error. However, standard E6B type flight computers do not automatically compensate for compressibility.

EXAMPLE: Determine the actual TAS on the basis of the following:

Indicated Airspeed:	230 kts
Airspeed Calibration Error:	+2 kts
Compressibility Error:	-10 kts
Indicated OAT:	-28° C
Temperature Rise:	+20° C
Cruising Altitude:	FL 350

ANSWER: 404 kts. To solve this problem using a CR type circular flight computer, the first step is to determine the CAS, so 230 kts IAS + 2 kts = 232 kts CAS. Next, since the CR circular computer already compensates for compresibility error and temperature rise, we ignore the compressibility and temperature rise data given in the problem. Thus, we simply input the given CAS of 232 kts and place this opposite FL 350 in the CAS/PA window (don't use the smaller TAS/PA/TEMP window closer the centre of the computer!). Next we take the cursor and move it over the indicated temperature of -28° C at the point it lines up with the inner spiral line. Reading the cursor at the point it crosses the TAS scale yields a TAS of 404 kts.

To solve this problem using an electronic computer, select the function "ACT TAS", then enter pressure altitude 35,000 feet and OAT -28° C. The computer yields an actual TAS of 403.4 kts. Using this method, there was no need to apply the compressibility data or temperature rise data provided in the problem.

To solve this problem using a standard E6B type flight computer, since standard E6B's do not compensate for compressibility or for temperature rise, we need to use all of the data provided in the problem. Use the sequence "ICE-T" = Indicated, Calibrated, Equivalent, True i.e. first calculate CAS = 232 kts, then subtract the comressibility error of -10 kts for an EAS of 222 kts. Then determine the true outside air temperature = -28° C, minus a further 20° C for compressibility heating of the temperature probe = -48° C. Finally, as per the standard TAS calculation method using an E6B, line up the pressure altitude 35,000 feet with OAT -48° C in the TAS window, find EAS 222 on inner scale of outer edge and read actual TAS of 404 kts on outer scale.

True Air Temperature Using CR Flight Computer

NOTE: It is not a mandatory requirement of the ATPL or IATRA writtens that you be able to use a circular flight computer to calculate true air temperature. The following discussion on the use of the CR flight computer is optional.

A CR flight computer can be used to calculate the temperature rise at the temperature probe. The rate of heating of the temperature probe is expressed as a TEMPERATURE RECOVERY COEFFICIENT, which varies with the installation and design of the temperature probe on the particular aircraft. Once a recovery coefficient is known for an aircraft, it will not vary significantly with speed or altitude. The temperature probes on most jets have recovery coefficients of either .8 or 1.0. These coefficients are on the hairline of the cursor on the computer. Some CR computer cursors have two lines for the Ct 1.0 reference line, one solid, and the other dashed: the dashed line is for the standard sea level temperature of +15° C, and the solid line is for standard stratosphere temperature of -55° C, and you should interpolate in such cases. For our purposes, assume Ct = 1.0.

To find true air temperature, set the cursor above the indicated air temperature, using the proper recovery coefficient. Rotate the inner scale until the cursor lines up over TAS above the spiral line inside the air temperature window. Read temperature rise through temperature rise window.

EXAMPLE: Find true air temperature given indicated air temperature 5° C and TAS 292 kts.

ANSWER: -6° C. Line up cursor so Ct 1.0 is over intersection of 5° C and spiral line (A). Rotate inner scale to select TAS 292 kts (B). Temperature rise (C) is +11° C. True air temp = 5° - 11° = -6° C.

EXERCISE: Using a CR flight computer, calculate temperature rise and true air temperature.

	True Airspeed	Indicated Air Temp	Temp Rise	True Air Temp
1.	172 kts	10° C	_____	_____
2.	360 kts	14° C	_____	_____
3.	292 kts	-20°C	_____	_____

ANSWERS

1. 4°C, 6°C.
2. 17°C, -3°C.
3. 11°C, -31°C.

True Air Temperature Using Ram Air Temperature Tables

An alternate method to calculate the temperature rise is to utilize a RAM Air Temperature (RAT) Chart. The chart provides the expected indicated temperature on the aircraft thermometer for a given Mach Number and altitude. To use the chart, first determine the expected temperature for the altitude, using the upper winds and temperatures forecast or ISA. Then, take the difference from the value obtained on the chart: the result will be the temperature rise. To determine ISA temperature, use standard lapse rate of 2° C per 1,000 feet to 35,000 feet ASL, then use standard stratosphere temperature of -56.5° C to 65,000 feet.

EXAMPLE: Using the above RAT table at Fig. 5-1 (b), determine the temperature rise at FL 350, given Mach 0.82, and expected upper air temperature of -48° C.

ANSWER: 23° C. Expected temp -48° C, RAT for FL 350 and Mach Number .82 = -25° C, difference = 23° C.

EXAMPLE: Using the RAT table, determine the temperature rise at FL 250, given Mach .76 and expected upper air temperature is unknown.

ANSWER: 27° C. Expected air temp = -35° C using standard lapse rate, RAT for FL 250 and Mach Number .76 = -8° C. Difference = 27° C.

The RAT table can also be used to find the EXPECTED indicated Ram Air Temperature. Note the instructions within the box below the table that state "The differenced between indicated RAM air temperature and RAM air temperature for ISA is approximately the differenced in OAT above or below ISA". Thus, find difference between ISA and forecast temperature, then add or subtract this difference to the Ram Air Temperature for ISA from the table.

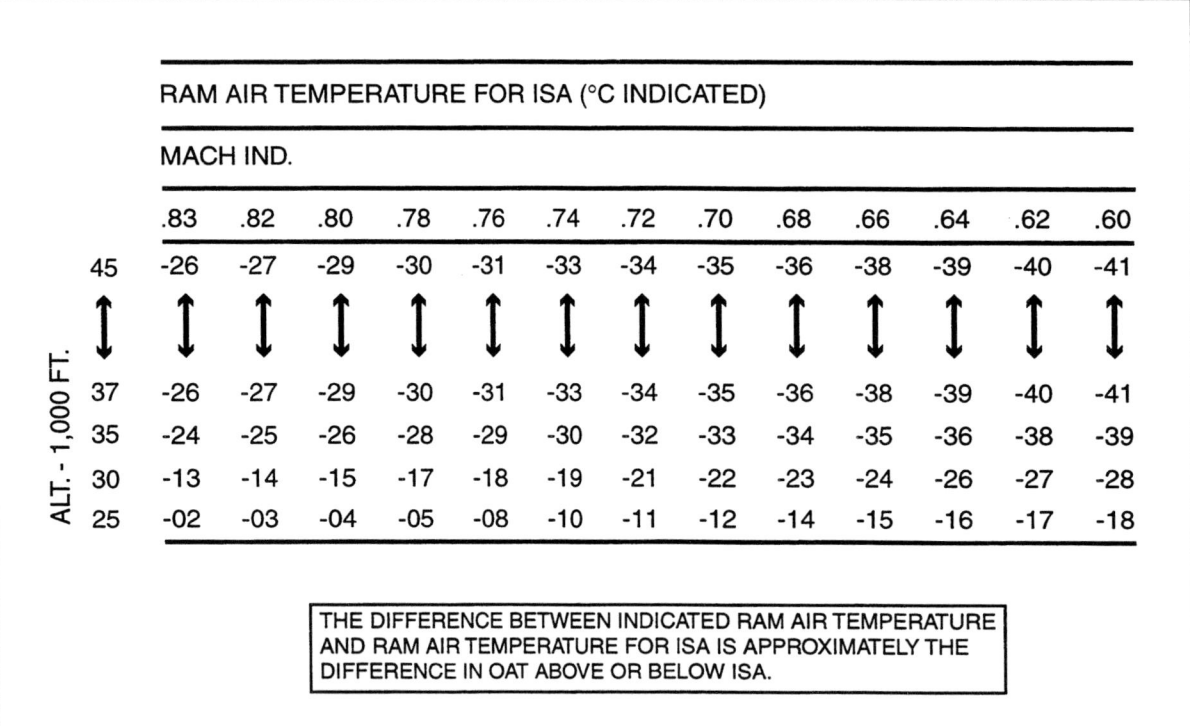

RAM AIR TEMPERATURE FOR ISA (°C INDICATED)

MACH IND.

ALT. - 1,000 FT.	.83	.82	.80	.78	.76	.74	.72	.70	.68	.66	.64	.62	.60
45	-26	-27	-29	-30	-31	-33	-34	-35	-36	-38	-39	-40	-41
37	-26	-27	-29	-30	-31	-33	-34	-35	-36	-38	-39	-40	-41
35	-24	-25	-26	-28	-29	-30	-32	-33	-34	-35	-36	-38	-39
30	-13	-14	-15	-17	-18	-19	-21	-22	-23	-24	-26	-27	-28
25	-02	-03	-04	-05	-08	-10	-11	-12	-14	-15	-16	-17	-18

THE DIFFERENCE BETWEEN INDICATED RAM AIR TEMPERATURE AND RAM AIR TEMPERATURE FOR ISA IS APPROXIMATELY THE DIFFERENCE IN OAT ABOVE OR BELOW ISA.

Fig. 5-1(b): Ram Air Temperature for ISA.

EXAMPLE: Find expected indicated RAM air temperature using the above table. Conditions:

Mach Ind.   0.70
Cruising Altitude: FL 300
Forecast OAT: -40° C

ANSWER: Expected Ram air temperature under these conditions will be -7° C. ISA at FL 300 = 30 X 2°/1,000 = 60° - 15 ° = -45° C. Difference between indicated RAM air temperature and RAM air temperature for ISA = 5° warmer, therefore, take RAM air temperature for ISA from table above = -22°, expected indicated RAM air temperature will be 5° C warmer = 5° + (-22°) = -17° C.

True Airspeed Calculations with CR Computer

NOTE: The following information should be considered optional and will not be a formal requirement of the ATPL/IATRA writtens.

With a CR model computer, TAS can be calculated directly from the CAS, Pressure Altitude and indicated OAT. CR model flight computers automatically compensate for the temperature rise and compressibility error. To do the calculation, line up the pressure altitude with the CAS in the PA/CAS window, then line up the cursor with the recovery coefficient over the temperature and spiral line. Read TAS through cursor. (Assume Ct = 1.0).

EXAMPLE: Use a CR computer to determine TAS for FL 280, CAS 385 kts and -30° C indicated OAT.

ANSWER: 528 kts. Line up CAS 385 kts with pressure altitude 28,000 in CAS/PA window (A). Then set cursor with Ct 1.0 line over spiral line and indicated temperature -30° C (B). Read TAS 528 kts in TAS window (C).

EXERCISE: Find TAS using CR model computer.

	CAS	PA	Ind. OAT	TAS
1.	650 kts	24,000 ft	-20° C	_____
2.	450 kts	18,000 ft	20° C	_____
3.	390 kts	30,000 ft	15° C	_____

ANSWERS:
1. 755 kts.
2. 570 kts.
3. 605 kts.

Calculation of True Altitude

TRUE ALTITUDE is the actual altitude above sea level. An altimeter only indicates true altitude when ICAO/ISA standard conditions exist. Since the atmosphere is rarely standard, the indicated altitude is normally in error.

True altitude can be calculated either by formula or computer.

True Altitude: Formula Method

The most accurate method to calculate true altitude is to use the following formula:

Ind. Altitude - Gnd Elev. 1,000 feet	X	Diff. between ICAO Temp. & actual temp	X 4 ft =	Amount to be subtracted from Ind. Alt.

EXAMPLE: Find the true altitude of an aircraft above a ridge under the following conditions:

Ground elevation of airport:	3,000 feet
Altimeter Setting:	30.43"
Indicated OAT at 6,500 feet:	-40° C
Indicated Altitude:	6,500 feet
Height of Ridge:	5,850 feet

ANSWER: 62 feet. First, find difference between indicated altitude and the elevation of the airport from which the altimeter was set, = 3,500 feet. Divide 3,500 by 1,000 feet = 3.5. Then, find the difference between indicated OAT (-40° C) and ISA temp (+2° C) = 42° C. Multiply 3.5 X 42 X 4 feet = 588 feet. Subtract 588 feet from the indicated altitude of 6,500 = 5,912 feet true altitude. Clearance above ridge is 5,912 - 5,850 = 62 feet.

True Altitude: Flight Computer Method

To determine true altitude using the flight computer, follow these steps:

1. Calculate pressure altitude. To calculate pressure altitude, take difference between altimeter setting and 29.92, and convert to feet. 1 inch Hg = 1,000 ft. Subtract from elevation if altimeter is HIGHER than 29.92, ADD to elevation if altimeter is LOWER than 29.92.
2. Set pressure altitude opposite OAT (°C) in true altitude window of flight computer.
3. Read true altitude on outer scale above indicated altitude on inner scale. If ground level is known, subtract elevation of ground station from indicated altitude. Calculate true altitude for above ground elevation, then add this to ground elevation to obtain true altitude.

EXAMPLE: Using a flight computer, find true altitude under the following conditions:

Pressure altitude:	10,000 feet
OAT° C:	-13° C
Indicated altitude:	12,000 feet
Elev of ground station:	5,400 feet

ANSWER: 11,800 feet. Set 10,000 feet pressure altitude opposite -13° C, in true altitude window of flight computer. Subtract ground station elevation from indicated altitude = 6,600 feet. Read 6,400 feet true altitude on outer scale opposite 6,600 on inner scale. Add 5,400 to 6,400 = 11,800 feet true altitude.

5.03 Time and Longitude

Because there are 360 meridians of longitude, and the earth takes 24 hours to complete 1 rotation, each hour or time zone represents a change in longitude of 15°. Therefore, if you know the longitude of a place, you should be able to determine the approximate number of hours away from GMT.

EXAMPLE: A place has longitude 110° W. How many hours is this place from UTC (assume standard time in effect)?

ANSWER: 110° divided by 15° = 7.3 hours.

Know the various time zones in Canada and how to convert from one time to another.

Time Zone	Conversion
Newfoundland Std	UTC - 3.5 hours
Atlantic Std	UTC - 4 hours
Eastern Std	UTC - 5 hours
Central Std	UTC - 6 hours
Mountain Std	UTC - 7 hours
Pacific Std	UTC - 8 hours

When daylight saving time is in effect (abbreviated DT), use 1 less hour in the conversion.

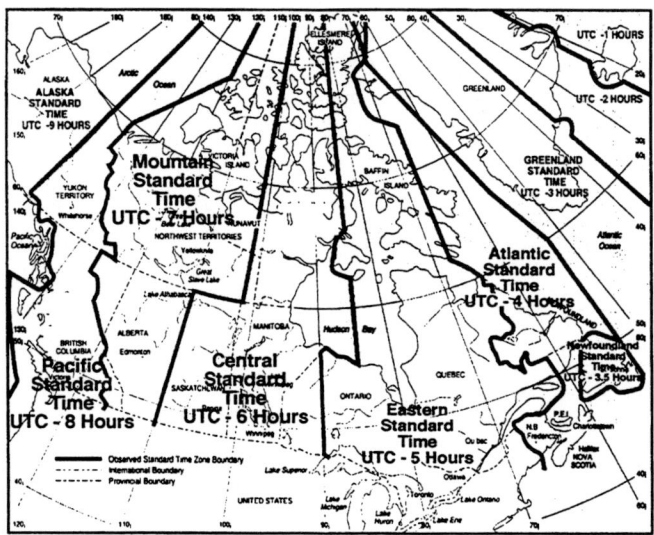

Fig. 5-2: Time zones in Canada.

EXERCISE: Complete the following time zone conversions.

1. 1600 PST = _____ UTC.

2. 1000 AST = _____ UTC.

3. 0000 MDT = _____ UTC.

4. _____ CST = 0900 UTC.

5. _____ EDT = 2300 UTC.

6. _____ PST = 1430 UTC.

ANSWERS:
1. 2400 UTC.
2. 1400 UTC.
3. 0600 UTC.
4. 0300 CST.
5. 1900 EDT.
6. 0630 PST.

5.04 Sun's True Bearing

When operating in the northern regions of Canada, the magnetic compass is unreliable, and a means must be utilized to set the aircraft heading indicator. Several methods are available, such as using a GPS fix, LORAN, astro navigation or using the Sun's true bearing.

The Sun will be in the same position relative to a given latitude, longitude, time of day and date from year to year. The Sun's true bearing tables are simply a summary of average times based on the Sun's position relative to true north. Since all directions in the northern areas of Canada are based on the True North reference, setting the aircraft heading indicator to true north is accomplished by referencing the Sun's true bearing tables. Once the bearing has been pulled from the tables, simple line up the longitudinal axis of the aircraft with the Sun, then set the bearing obtained on the DG.

A few calculations are required to use the tables. Since time zones are actually very wide, you need a more accurate time. The time must be referenced to your precise longitude to find the LOCAL CIVIL TIME (LCT). Since every 15° of longitude is one hour of time, divide your present longitude by 15, and convert this to hours and minutes. Then, subtract this time from GMT. You always subtract because in Canada, all longitude will be west, and the time difference is therefore subtracted from GMT to obtain local time (in Russia, east of the prime meridian, you would add). With the LCT, look up the Sun's bearing based on the LCT, latitude and date.

An obvious limitation with this method is that the Sun must be visible: cloudy weather or night will make the method useless.

EXAMPLE: Refer to the Sun's true bearing table below. Based on a GMT of 22:16, longitude of 114° W, latitude of 55° N and date of September 7, what is the Sun's true bearing?

55 Degrees North Latitude

DATE	HOURS	1400	1408	1416	1424	1432	1440	1448	1456	1504	1512	1520	1528	1536	1544	1552
DEC 31		207.0	208.7	210.5	212.3	214.0	215.7	217.4	219.1	220.8	222.5	224.2	225.8	227.5	229.1	230.7
JAN 10		206.3	208.1	209.9	211.7	213.4	215.2	216.9	218.6	220.3	222.0	223.7	225.4	227.1	228.7	230.3
JAN 20		206.0	207.8	209.7	211.5	213.3	215.1	216.8	218.6	220.3	222.1	223.8	225.5	227.2	228.8	230.5
JAN 30		206.1	208.0	209.9	211.8	213.6	215.4	217.2	215.0	220.8	222.6	224.3	226.1	227.8	229.5	231.2
FEB 9		206.7	208.7	210.6	212.5	214.4	216.3	218.1	220.0	221.8	223.6	225.4	227.2	228.9	230.7	232.4
FEB 19		207.8	209.8	211.8	213.7	215.7	217.6	219.5	221.4	223.3	225.1	227.0	228.8	230.6	232.3	234.1
MAR 1		209.3	211.3	213.4	215.4	217.4	219.4	221.3	223.3	225.2	227.1	228.9	230.8	232.6	234.4	236.2
MAR 11		211.1	213.3	215.4	217.4	219.5	221.5	223.5	225.5	227.5	229.4	231.3	233.2	235.0	236.9	238.7
MAR 21		213.3	215.5	217.6	219.8	221.9	224.0	226.0	228.0	230.0	232.0	233.9	235.8	237.7	239.6	241.4
MAR 31		215.6	217.9	220.1	222.3	224.4	226.6	228.6	230.7	232.7	234.7	236.7	238.6	240.5	242.4	244.2
APR 10		218.1	220.4	222.7	224.9	227.1	229.2	231.4	233.4	235.5	237.5	239.5	241.4	243.3	245.2	247.0
APR 20		220.5	222.9	225.2	227.4	229.7	231.8	234.0	236.1	238.1	240.2	242.1	244.1	246.0	247.9	249.8
APR 30		222.7	225.1	227.5	229.8	232.0	234.2	236.4	238.5	240.6	242.6	244.6	246.6	248.5	250.4	252.2
MAY 10		224.6	227.1	229.5	231.8	234.1	236.3	238.5	240.6	242.7	244.7	246.7	248.7	250.6	252.5	254.3
MAY 20		226.1	228.6	231.0	233.4	235.7	237.9	240.1	242.2	244.3	246.4	248.4	250.3	252.2	254.1	255.9
MAY 30		227.0	229.5	232.0	234.4	236.7	238.9	241.1	243.3	245.4	247.4	249.4	251.4	253.3	255.2	257.0
JUN 9		227.3	229.8	232.3	234.7	237.1	239.3	241.6	243.7	245.8	247.9	249.9	251.9	253.8	255.7	257.5
JUN 19		226.9	229.5	232.0	234.5	236.8	239.1	241.4	243.6	245.7	247.7	249.8	251.7	253.7	255.5	257.4
JUN 29		226.1	228.7	231.2	233.7	236.1	238.4	240.6	242.8	245.0	247.0	249.1	251.1	253.0	254.9	256.7
JUL 9		224.9	227.5	230.0	232.5	234.9	237.2	239.5	241.7	243.8	245.9	247.9	249.9	251.9	253.8	255.7
JUL 19		223.5	226.1	228.6	231.1	233.4	235.8	238.0	240.2	242.4	244.5	246.5	248.5	250.5	252.4	254.3
JUL 29		222.1	224.7	227.2	229.6	231.9	234.2	236.5	238.7	240.8	242.9	245.0	247.0	248.9	250.9	252.7
AUG 8		220.9	223.4	225.8	228.2	230.5	232.8	235.0	237.1	239.3	241.4	243.4	245.4	247.3	249.3	251.1
AUG 18		219.8	222.2	224.6	226.9	229.2	231.4	233.6	235.7	237.8	239.9	241.9	243.8	245.8	247.7	249.6
AUG 28		218.9	221.3	223.6	225.8	228.0	230.2	232.3	234.4	236.5	238.5	240.5	242.4	244.3	246.2	248.1
SEP 7		218.2	220.4	222.7	224.9	227.0	**229.1**	231.2	233.2	235.2	237.2	239.2	241.1	243.0	244.9	246.7
SEP 17		217.6	219.7	221.9	224.0	226.1	228.1	230.2	232.1	234.1	236.0	238.0	239.8	241.7	243.5	245.3
SEP 27		216.9	219.0	221.1	223.2	225.2	227.2	229.1	231.1	233.0	234.9	236.8	238.6	240.4	242.2	244.0
OCT 7		216.3	218.3	220.3	222.3	224.3	226.2	228.1	230.0	231.9	233.7	235.5	237.4	240.9	242.7
OCT 17		215.5	217.5	219.5	221.4	223.3	225.2	227.0	228.9	230.7	232.5	234.3	236.0	237.8	239.5	241.3
OCT 27		214.7	216.6	218.5	220.3	222.2	224.0	225.8	227.6	229.4	231.2	232.9	234.6	236.4	238.1	239.8
NOV 6		213.5	215.5	217.3	219.1	220.9	222.7	224.5	226.3	228.0	229.7	231.5	233.2	234.9	236.5	238.2
NOV 16		212.4	214.2	216.0	217.8	219.6	221.4	223.2	224.8	226.6	228.3	229.9	231.5	233.3	234.9	236.6
NOV 26		211.1	212.9	214.7	216.5	218.2	219.9	221.7	223.4	225.1	226.7	228.4	230.1	231.7	233.3	234.9
DEC 6		209.8	211.6	213.3	215.1	216.8	218.5	220.2	221.9	223.6	225.3	226.9	228.6	230.2	231.8	233.4
DEC 16		208.5	210.3	212.1	213.8	215.5	217.2	218.9	220.6	222.3	224.0	225.6	227.2	228.9	230.5	232.1
DEC 26		207.5	209.2	211.0	212.7	214.4	216.2	217.9	219.6	221.2	222.9	224.6	226.2	227.8	229.5	231.1

Fig. 5-3: Extract, Sun's True Bearing Table

ANSWER: 229°. For LCT, take 114° and divide by 15 = 7.6 hours, convert to 7 hours and 36 minutes. Subtract 7 hrs 36 mins from GMT to convert to LCT = 14:40. Go to table for date and LCT. Sun's true bearing = 229.1 (round off the .1). If the aircraft is pointed to the Sun and DG set to 229, the DG is set to True North.

5.05 Astro Navigation

NOTE: TC no longer requires a detailed understanding of Astro Nav for the purposes of the ATPL or IATRA writtens. However, since GHA is still a required subject, more detailed information on Astro Nav has been provided below.

One major benefit of Astro Navigation is that it is completely independent of electronics: with so many advanced navigational systems currently available, it is hard to imagine a system that was never dependent on ground based stations, satellite stations or advanced, expensive electronics. Even though the system seems hardly relevant today, it offered an inexpensive, reliable method of long distance navigation when used by a properly trained pilot or navigator.

An ASTRO COMPASS is a sighting device that, when lined up with a heavenly body such as the Sun, various stars and constellations or the Moon, will indicate a bearing which can be set on the HI when the aircraft is lined up with the astrological body. It is useful in northern areas because no magnetic reference is needed. Using the Sun as the astrological body, the following factors are needed to line up the sights on the astro compass: present latitude, declination, and Local Hour Angle (LHA).

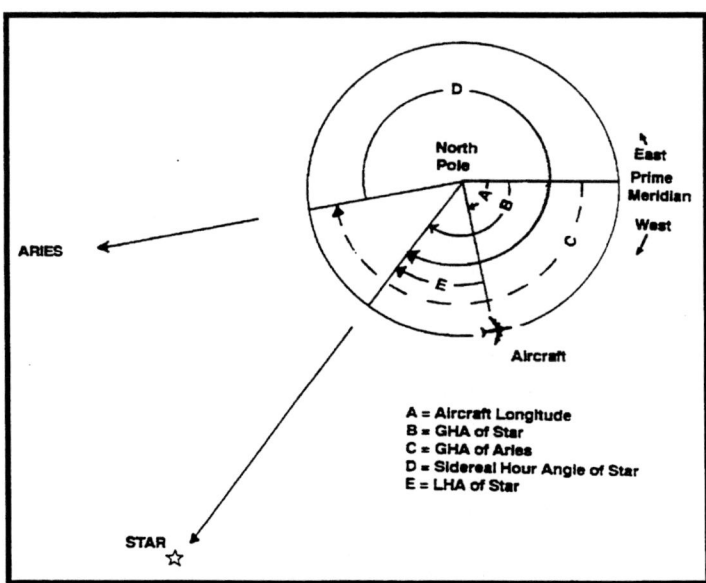

A = Aircraft Longitude
B = GHA of Star
C = GHA of Aries
D = Sidereal Hour Angle of Star
E = LHA of Star

Fig. 5-4: Concepts of Astro Navigation.

The prime meridian is a standard point of reference. The angle between the Sun and the prime meridian is referred to as the GREENWICH HOUR ANGLE (GHA). The angle between your present longitude and the Sun is the Local Hour Angle (LHA). The difference between the GHA and LHA is your present longitude. Know this formula:

LHA = GHA +E -W Longitude

To complete the calculations, refer to the AIR ALMANAC for the date, which provides the GHA and declination for the time of day in GMT.

EXAMPLE: Given 10:00 AST and longitude 70° W, using the Sun as a reference body, what is required to use an astro compass to set HI to true north?

ANSWER: First, convert local time to GMT, which is 14:00 GMT. Then calculate LHA. LHA is GHA - 70° W longitude. From table on next page, GHA at 14:00 is 29° 44', so LHA = 29° 44' - 70° 00' = 319° 44' (add 360° to the 29° 44' to make the subtraction easier). Also from table, declination is N 8° 22'. Then set the latitude, LHA and declination on the astro compass, and when the Sun is sighted, read the true north bearing on the bearing plate and set HI to this number.

For stars, as opposed to the Sun, moon or planets, another calculation must be made. To find the GHA for a star, look up the SIDEREAL HOUR ANGLE (SHA) of the star. Then use this formula:

GHA (Star) = GHA (Aries) + SHA (Star)

5.06 Maps and Charts: Projections

Maps and charts have various inaccuracies, because maps are flat and the earth is a sphere. The basis upon which maps are constructed is termed a PROJECTION. The most common aeronautical chart projection is a LAMBERT CONFORMAL CONIC projection, where a straight line between two points on the map is a GREAT CIRCLE, and meridians converge towards north. WAC charts (1:1,000,000) and VNC charts (1:500,000) are based on conic projections. Another type of projection is a MERCATOR projection, where longitude and meridians are straight and parallel lines. A straight line on a mercator projection represents a RHUMB LINE. Canadian Pilotage Charts (CPC) have a scale of 1:500,000 and are VFR charts used in northern operations. CPC charts are based on a TRANSVERSE MERCATOR projection, having the same characteristics as a conic projection (straight line = great circle). VTA terminal area charts are based on the transverse mercator projection. For low level IFR navigation, the ENROUTE LOW ALTITUDE CHART is utilized (LO). HIGH LEVEL ENROUTE CHARTS (HE) are used for operations above 18,000 ft.

When navigating in the polar regions, conventional charts are not practical due to excessive convergence of meridians and errors in scale. As a result, POLAR STEREOGRAPHIC charts are used, where a great circle is a straight line. GRID NAVIGATION involves imposing an arbitrary grid over charts used in northern navigation. GRID NORTH is the selected datum, normally the Greenwich meridian.

Using the grid system, although a great circle track cuts true meridians at different angles, the grid meridians are crossed at a constant angle.

202   GREENWICH P. M.   APRIL 11 (SUNDAY)

GMT	SUN GHA	SUN Dec.	ARIES GHA T	MARS −0.4 GHA	MARS −0.4 Dec.	JUPITER −1.6 GHA	JUPITER −1.6 Dec.	SATURN 1.3 GHA	SATURN 1.3 Dec.	MOON GHA	MOON Dec.	Lat.	Moon. set	Diff.
h m	° ′	° ′	° ′	° ′	° ′	° ′	° ′	° ′	° ′	° ′	° ′	°	h m	m
12 00	359 43.8 N 8	20.5	19 29.4	217 38 N10 48		324 01 N18 59		34 48 S 8 19		231 33 N18 00		N		
10	2 13.9	20.6	21 59.8	220 08		326 31		37 18		233 58 17 58		72	□	
20	4 43.9	20.8	24 30.3	222 39		329 01		39 48		236 22 56		70	06 58	.
30	7 13.9 .	20.9	27 00.7	225 09 .		331 31 .		42 19 .		238 47 . 55		68	06 12	.
40	9 44.0	21.1	29 31.1	227 40		334 02		44 49		241 11 53		66	05 42	−06
50	12 14.0	21.2	32 01.5	230 10		336 32		47 19		243 36 51		64	05 19	+01
13 00	14 44.0 N 8	21.4	34 31.9	232 41 N10 48		339 02 N18 59		49 50 S 8 19		246 00 N17 49		62	05 01	05
10	17 14.0	21.5	37 02.3	235 11		341 33		52 20		248 25 47		60	04 45	08
20	19 44.1	21.7	39 32.7	237 42		344 03		54 50		250 49 45		58	04 32	10
30	22 14.1	21.8	42 03.1	240 12 .		346 33 .		57 21 .		253 14 . 44		56	04 21	12
40	24 44.1	22.0	44 33.5	242 43		349 04		59 51		255 39 42		54	04 11	13
50	27 14.1	22.1	47 04.0	245 13		351 34		62 22		258 03 40		52	04 02	15
14 00	29 44.2 N 8	22.3	49 34.4	247 44 N10 48		354 04 N18 59		64 52 S 8 19		260 28 N17 38		50	03 54	16
10	32 14.3	22.5	52 04.8	250 14		356 35		67 22		262 52 36		45	03 37	18
20	34 44.2	22.6	54 35.2	252 44		359 05		69 53		265 17 34		40	03 23	20
30	37 14.3 .	22.8	57 05.6	255 15 .		1 35 .		72 23 .		267 41 . 33		35	03 11	21
40	39 44.3	22.9	59 36.0	257 45		4 06		74 53		270 06 31		30	03 01	23
50	42 14.3 .	23.1	62 06.4	260 16 .		6 36		77 24		272 30 29		20	02 43	25
15 00	44 44.3 N 8	23.2	64 36.8	262 46 N10 48		9 06 N18 59		79 54 S 8 19		274 55 N17 27		10	02 27	26
10	47 14.4	23.4	67 07.2	265 17		11 37		82 24		277 20 25		0	02 11	28
20	49 44.4	23.5	69 37.6	267 47		14 07		84 55		279 44 23		10	01 56	30
30	52 14.4 .	23.7	72 08.1	270 08 .		16 37 .		87 25 .		282 09 . 21		20	01 40	32
40	54 44.4	23.8	74 38.5	272 48		19 08		89 56		284 33 20		30	01 21	34
50	57 14.5	24.0	77 08.9	275 19		21 38		92 26		286 58 18		35	01 10	35
16 00	59 44.5 N 8	24.1	79 39.3	277 49 N10 48		24 08 N19 80		94 56 S 8 18		289 23 N17 16		40	00 58	36
10	62 14.5	24.3	82 09.7	280 19		26 39		97 27		291 47 . 14		45	00 43	38
20	64 44.6	24.4	84 40.1	282 50		29 09		99 57		294 12 12		50	00 24	40
30	67 14.6 .	24.6	87 10.5	285 20 .		31 39 .		102 28 .		296 36 . 10		52	00 15	41
40	69 44.6	24.7	89 40.9	287 51		34 09		104 58		299 01 08		54	00 05	43
50	72 14.6	24.9	92 11.3	290 21		36 40		107 28		301 25 06		56	25 25	45
17 00	74 44.7 N 8	25.1	94 41.8	292 52 N10 48		39 10 N19 00		109 59 S 8 18		305 50 N17 04		58	25 16	47
10	77 14.7	25.2	97 12.2	295 22		41 40		112 29		306 15 03		60	25 06	49
20	79 44.7	25.4	99 42.6	297 53		44 11		114 59		308 39 17 01		S		
30	82 14.8 .	25.5	102 13.0	300 23 .		46 41 .		117 30 .		311 04 16 59				
40	84 44.8	25.7	104 43.4	302 54		49 11		120 00		313 28 57				
50	87 14.8	25.8	107 13.8	305 24		51 42		122 30		315 53 55				
18 00	89 44.8 N 8	26.0	109 44.2	307 55 N10 48		54 12 N19 00		125 01 S 8 18		318 18 N16 53			Moon's P. in A.	
10	92 14.9	26.1	112 14.6	310 25		56 42		127 31		320 42 51				
20	94 44.9	26.3	114 45.0	312 55		59 13		130 02		323 07 49				
30	97 14.9 .	26.4	117 15.4	315 26 .		61 43 .		132 32 .		325 31 . 47				
40	99 44.9	26.6	119 45.9	317 56		64 13		135 02		327 56 45			° ′ ° ′	° ′ ° ′
50	102 15.0	26.7	122 16.3	320 27		66 44		137 33		330 21 43			0 + +	+ +
19 00	104 45.0 N 8	26.9	124 46.7	322 57 N10 48		69 14 N19 00		140 03 S 8 18		332 45 N16 42			7 59	54 34
10	107 15.0	27.0	127 17.1	325 28		71 44		142 33		335 10 40			13 58	55 33
20	109 45.1	27.2	129 47.5	327 58		74 15		145 04		337 35 38			17 57	56 33
30	112 15.1 .	27.3	132 17.9	330 29 .		76 45 .		147 34 .		339 59 . 36			20 56	57 32
40	114 45.1	27.5	134 48.3	332 59		79 15		150 04		342 24 34			22 55	58 31
50	117 15.1	27.6	137 18.7	335 30		81 45		152 35		344 48 32			25 54	60 30
20 00	119 45.2 N 8	27.8	139 49.1	338 00 N10 48		84 16 N19 00		155 05 S 8 18		347 13 N16 30			27 53	61 29
10	122 15.2	28.0	142 19.6	340 30		86 46		157 36		349 38 28			29 52	62 28
20	124 45.2	28.1	144 50.0	343 01		89 16		160 06		352 02 26			31 51	63 27
30	127 15.3 .	28.3	147 20.4	345 31 .		91 47 .		162 36 .		354 27 . 24			33 50	64 26
40	129 45.3	28.4	149 50.8	348 02		94 17		165 07		356 52 22			34 49	65 25
50	132 15.3	28.6	152 21.2	350 32		96 47		167 37		359 16 20			36 48	66 24
21 00	134 45.3 N 8	28.7	154 51.6	353 03 N10 48		99 18 N19 00		170 08 S 8 18		1 41 N16 18			38 47	67 23
10	137 15.4	28.9	157 22.0	355 33		101 48		172 38		4 06 16			39 46	68 22
20	139 45.4	29.0	159 52.4	358 04		104 18		175 08		6 30 14			41 45	69 21
30	142 15.4 .	29.2	162 22.8	0 34 .		106 49 .		177 39 .		8 55 . 12			42 44	71 19
40	144 45.4	29.3	164 53.3	3 05		109 19		180 09		11 20 10			43 43	72 18
50	147 15.5	29.5	167 23.7	5 35		111 49		182 39		13 44 08			45 42	73 17
22 00	149 45.5 N 8	29.6	169 54.1	8 06 N10 48		114 20 N19 00		185 10 S 8 18		16 09 N16 06			46 41	74 16
10	152 15.5	29.8	172 24.5	10 36		116 50 00		187 40		18 34 04			48 40	75 15
20	154 45.6	29.9	174 54.9	13 06		119 20		190 10		20 58 02			49 39	76 14
30	157 15.6 .	30.1	177 25.3	15 37 .		121 51 .		192 41 .		23 23 16 01			50 38	77 13
40	159 45.6	30.2	179 55.7	18 07		124 21		195 11		25 48 15 59			51 37	78 12
50	162 15.6	30.4	182 26.1	20 38		126 51		197 42		28 12 57			53 36	79 11
23 00	164 45.7 N 8	30.5	184 56.5	23 08 N10 48		129 22 N19 00		200 12 S 8 18		30 37 N15 55			54 35	80 10
10	167 15.7	30.7	187 26.9	25 39		131 52		202 42		33 02 53			55 34	
20	169 45.7	30.9	189 57.4	28 09		134 22		205 13		35 26 51				
30	172 15.7 .	31.0	192 27.8	30 40 .		136 53 .		207 43 .		37 51 . 49				
40	174 45.8	31.2	194 58.2	33 10		139 23		210 13		40 16 47			. Sun SD 16′.0	
50	177 15.8	31.3	197 28.6	35 41		141 53		212 44		42 40 45			Moon SD 16′	
													Age 10′	

Fig. 5-5: Air Almanac Extract.

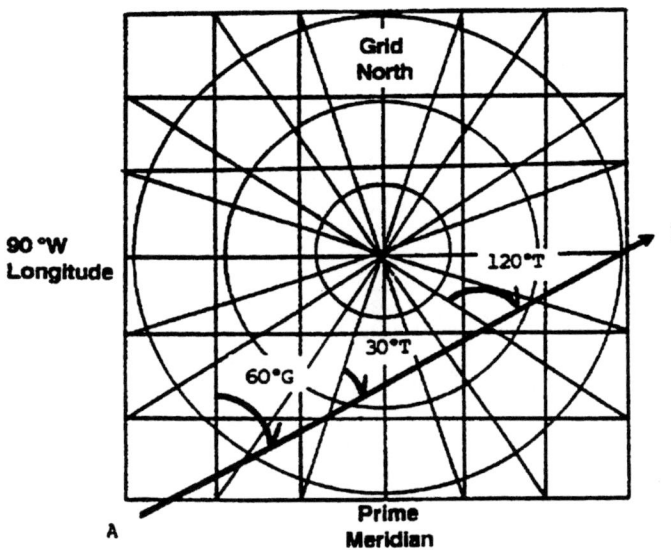

Fig. 5-6: Grid Navigation.

The angular difference between grid north and true north is GRIVATION. Dotted lines indicating grivation are ISOGRIVS.

To determine grid heading, use one of these formulas:

MH + Griv E/- Griv W = GH

TH -E long/+long W = GH

EXAMPLE: Longitude is 90° W, grivation is 105° E, magnetic heading is 165° M. Find grid heading, true heading and magnetic variation.

ANSWER: Grid heading = 270 (MH + Griv E = GH, therefore GH = 165 + 105 = 270). True heading = 180 (TH + long W = GH, therefore GH - Long W = TH. 270 - 090 = 180). Magnetic variation = 15° E (TH - Var E = MH, therefore for TH 180, MH 165, variation is -15 = 15° E).

5.07 Point of No Return (PNR)

PNR is defined as "the farthest point that an aircraft can proceed to, under a given wind condition, from the departure point and still have enough fuel to return to the point of departure." The PNR is also referred to as the RADIUS OF ACTION, because based on flying a given track under given winds and fuel consumption, the edge of the radius is the limiting distance that can be travelled where the same fuel would be used to return to base.

PNR is a fuel problem, because it depends on fuel endurance in hours. PNR is not related to distance. Situations could exist where the PNR is beyond the destination.

Use this formula to calculate distance to the PNR:

Dist to PNR = $\dfrac{\text{End X G/S Out X G/S Home}}{\text{G/S Out + G/S Home}}$

In zero winds, the time to the PNR will be 1/2 of endurance. In a headwind, the time to PNR is longer. In a tailwind, it takes less time to reach the PNR. Therefore, the PNR "moves into the wind" from the midpoint. To calculate the time to the PNR, remember to use the groundspeed out.

To calculate time to PNR, use this formula:

Time to PNR = $\dfrac{\text{Dist to PNR}}{\text{GS Out}}$

EXAMPLE: Calculate time and distance to PNR under the following conditions:

Rhumb line track:	220° T
Winds:	170° T at 30 kts
TAS:	180 kts
Distance:	1,240 NM
Endurance:	8 hrs

ANSWER: 705.5 NM. Find G/S out using flight computer: enter winds, TAS, true track; result = G/S out 159 kts. Next, find G/S home using flight computer: enter winds, true track of reciprocal track 040°, TAS; G/S home 198 kts. Apply formulas:

$$\frac{\text{End X Out X Home}}{\text{Out + Home}}$$

$$= \frac{8 \text{ X } 159 \text{ X } 198}{159 + 198} = 705.5 \text{ NM.}$$

Time to PNR = $\dfrac{\text{Dist to PNR}}{\text{G/S Out}}$ = 4 hrs 26 mins.

5.08 Critical Point (CP)

CRITICAL POINT is the point along the planned track from which it takes the same time to return to the departure point as to continue on to destination. The principal used is that an aircraft has some form of power loss, on an extended over-water flight (or other remote surface) and the pilot must know whether to carry on to destination or return to base. If the pilot knows the critical point in advance, from which it takes the same time to continue to destination as to return to base, the pilot can quickly act should the power difficulties occur.

Remember that critical point is entirely a distance calculation, and has nothing to do with fuel. In zero winds, the CP is halfway to destination. In a headwind, the CP moves closer to destination. Distance to CP is less in a tailwind. Therefore, the CP "moves into the wind" from the midpoint. The formula used is as follows:

CP Dist = $\dfrac{\text{D X Hr}}{\text{Or + Hr}}$

where D = total distance
       Hr = reduced G/S home
       Or = reduced G/S out

EXAMPLE: Find distance to critical point, given an engine-out eventuality, under these conditions:

Track:	016° T
Winds:	220° T at 40 kts
4 Engine TAS:	230 kts
Distance:	1,128 NM
3 Engine TAS:	190 kts

ANSWER: 455 NM. Use flight computer to find 4 engine G/S out and 4 engine G/S home by entering winds, track (home = recip of out), and TAS. 4 engine G/S out = 266 kts, 4 engine G/S home = 193 kts. Then calculate 3 engine G/S out and G/S home, using reduced TAS. 3 engine G/S out = 226 kts, 3 engine G/S home = 153 kts. Apply formula:

$$\frac{D \times Hr}{Or + Hr}$$

$$= \frac{1128 \times 153}{226 + 153} = 455.4 \text{ NM.}$$

This formula is used to find time to CP:

Flight time to CP = $\dfrac{\textbf{Distance to CP}}{\textbf{Normal G/S out}}$

EXAMPLE: What is the flight time to the CP, using the figures from the above example?

ANSWER: $\dfrac{455 \text{ NM}}{266 \text{ kts}}$ = 1 hr 43 mins.

Notes:

Notes:

6
Meteorology

6.01 The Earth's Atmosphere

The Earth's atmosphere consists of 78% nitrogen, 21% oxygen, and 1% argon. Important physical properties are pressure, temperature, humidity, and radiation. The lowest layer of the Earth's atmosphere is the TROPO-SPHERE, from the ground to 60,000 feet at the equator, and to 35,000 feet at the poles. Above the troposphere lies the STRATOSPHERE. The TROPOPAUSE separates the troposphere from the stratosphere.

AIR DENSITY is the weight of air per cubic foot. AT-MOSPHERIC PRESSURE is the weight of an imaginary column of air. Sea level pressure is 29.92 inches of mercury, and decreases 1.00 inch of mercury per 1,000 ft gain in height. MSL pressure is calculated by reducing station pressure to MSL using the average temperature for the previous 1 - 2 hours. The STANDARD ATMOS-PHERE (ISA) is MSL pressure 29.92" Hg, MSL tempera-ture +15° C, normal lapse rate of 1.98° C per 1,000 ft, tropopause height 36,000 ft, temperature of tropopause -56.5° C, with temperature constant from tropopause to top of standard atmosphere.

EXAMPLE: What is the ISA upper air temperature a) at 20,000 ft ASL and b) at 40,000 ft ASL?

ANSWERS: a) -25° C (20 X 2.0° C/1,000 = 40° C; 15° C - 40° C = -25° C.) b) -56.5° C (don't apply the 2° C/1,000 ft rule if the resulting temperature will be less than -56.5° C.)

STATION PRESSURE is the atmospheric pressure at a station. Pressure varies horizontally. ISOBARS are lines of equal MSL pressure at 4 millibar intervals. CON-TOURS indicate a particular pressure level in upper charts and are generally presented at 60 metre intervals.

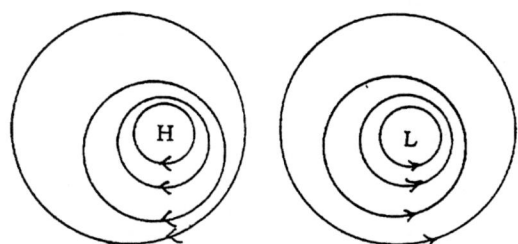

Fig. 6-1: High pressure and low pressure systems.

A HIGH or ANTI-CYCLONE is when atmospheric pres-sure increases towards the centre. A LOW or CYCLONE is when atmospheric pressure decreases towards the centre. Highs circulate clockwise, lows counter-clock-wise. Highs have descending air or SUBSIDENCE, while lows have ascending air caused by CONVERGENCE. A TROUGH is an elongated area of low pressure, similar to a valley. A RIDGE is an elongated area of high pressure, similar to a ridge. A COL is an neutral area between highs and lows.

PRESSURE TENDENCY refers to if pressure is rising or falling at a given location. PRESSURE GRADIENT refers to the horizontal change of pressure with distance. If the isobars or contour lines are spaced closely on a weather chart, it indicates high winds because of the steep pressure gradient.

The pressure pattern, both vertically and horizontally, determines the resulting drift.

EXAMPLE: An aircraft is on a flight from CYQL (Lethbridge, AB) to CYWG (Winnipeg, MB) at FL 250 and is experiencing sustained south drift. From this fact, what conclusion can be reached about the pressure pattern along the flight route?

ANSWER: The aircraft is entering lower pressures and is either leaving a high or entering a low or both.

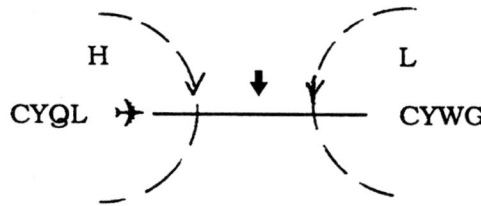

EXAMPLE: An aircraft on a westbound flight experiences a strong, sustained tailwind while maintaining a con-stant pressure level. Where is the area of lowest pres-sure?

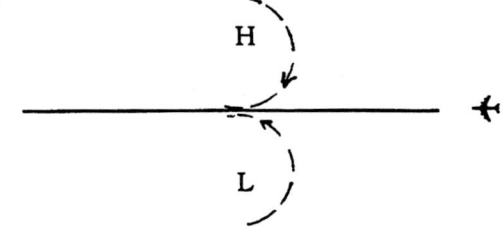

ANSWER: To the south.

Temperature influences the height of pressure levels. If the temperatures are cold, a given pressure level will be found at a lower altitude than in warmer temperatures.

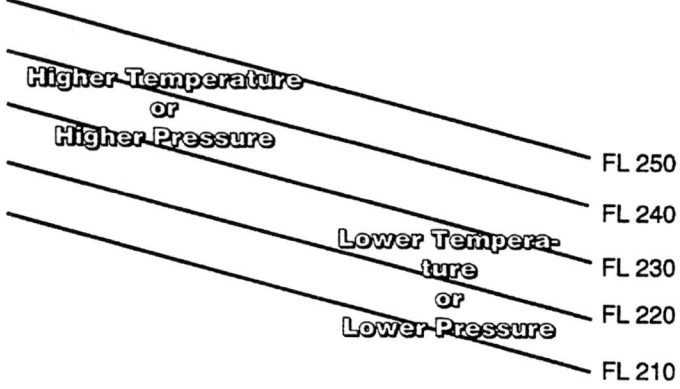

Fig. 6-2: Pressure levels are lower in colder air and in areas of lower pressure.

Remember that this theory assumes that you are within the same air mass (i.e. a Pacific low pressure system based on mP air will be WARMER than a high based on mA air).

EXAMPLE: If the winds were 090° at 45 kts, the temperatures below the aircraft flight level would be higher to the north or south?

ANSWER: To the north.

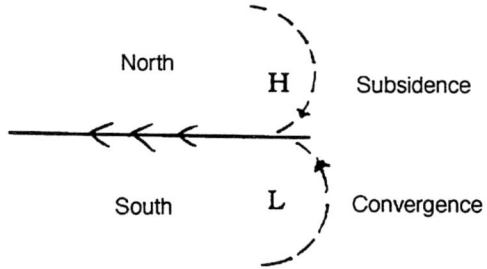

EXAMPLE: A large area to the southeast of a station has higher temperatures than to the NW. Where is the wind from?

ANSWER: SW.

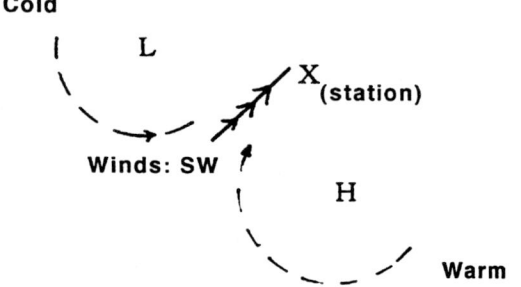

6.02 Meteorological Aspects of Altimeters

PRESSURE ALTITUDE is the equivalent altitude in the standard atmosphere for a given pressure. To calculate pressure altitude, take the local altimeter setting and calculate the difference from 29.92 inches of mercury, then convert to feet, and add to airport elevation where the altimeter setting is lower than 29.92. [NOTE: pressure altitude can also be determined by setting an aircraft altimeter to 29.92 and reading off the resulting indication.]

EXAMPLE: Calculate pressure altitude.

1. Altimeter setting 29.62, airport elev. 3,000 ft.
2. Altimeter setting 30.12, airport elev. 1,500 ft.

ANSWERS:
1. 3,300 ft pressure altitude. [29.92 - 29.62 = +.30 = +300 ft. 300 + 3000 = 3,300]
2. 1,300 ft pressure altitude. [29.92 - 30.12 = -.20 = -200 ft. 1,500 - 200 = 1,300]

The ALTIMETER SETTING is station pressure reduced to MSL under standard conditions. The altimeter must be set to the current local altimeter setting so that the it indicates the correct altitude.

When flying from high pressure areas to low pressure areas, without resetting the altimeter, the altimeter indicates high. A useful rule to remember is from HIGH TO LOW, LOOK OUT BELOW.

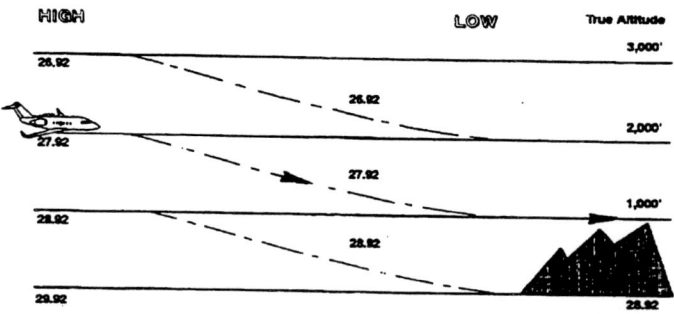

Fig. 6-3: Effect of incorrect altimeter setting. Where altimeter setting goes from 29.92 to 28.92 and pilot fails to reset altimeter, aircraft will be 1,000 feet lower than indicated.

DENSITY ALTITUDE is pressure altitude corrected for nonstandard temperature. To calculate density altitude, use a flight computer or Koch chart. High density altitude increases takeoff distance and reduces the rate of climb. Conditions causing drastic reductions in aircraft performance are high airport elevation, high ambient temperature and high atmospheric humidity.

6.03 Temperature

The Earth's atmosphere is heated from BELOW: the Sun's radiation heats the earth, which in turn heats the lower atmosphere. CONVECTION CURRENTS occur where there is differential heating of the Earth's surface. The normal rate of temperature decrease with altitude is 1.98° C per 1,000 ft (standard lapse rate). Temperature does not always drop with height: an ISOTHERMAL LAYER is when the temperature remains constant with increasing height. An INVERSION is when temperature increases with height. The CELSIUS SCALE is based on water freezing at 0° C and boiling at 100° C. The FAHRENHEIT SCALE is based on water freezing at 32° F and boiling at 212° F. Use a flight computer to convert from °F to °C, or from °F to °C: on CR computers there is a conversion scale. With electronic computers, enter the appropriate °F or °C and press CONV.

EXAMPLE: What is 10° C in °F?

ANSWER: 50°F.

6.04 Moisture and Cooling

CONDENSATION is the change of state from water vapour (gas) to liquid water. RELATIVE HUMIDITY is the percentage of saturation of air for a given temperature. DEWPOINT is the temperature to which air must be cooled to reach 100% saturation. When the dewpoint temperature is equal to the ambient temperature, cloud forms. Most cloud forms when moist air rises, cools and condenses. SUBLIMATION is when moisture is added to the air by the change of state of ice to water vapour. Cooling processes are RADIATION COOLING (Earth radiates heat out to space), NOCTURNAL INVERSIONS (night time inversion caused by night time cooling), CLOUD EFFECT (cooling not as great since cloud will absorb terrestrial radiation), TOPOGRAPHICAL EFFECT (cold air gets trapped in valleys and low areas), MARITIME EFFECT (night-time cooling is less over water than land since water traps heat), ADIABATIC COOLING (rising air expands and cools), EXPANSION COOLING (orographic lift or upslope lift leads to adiabatic cooling), MECHANICAL TURBULENCE (low level air mixed and moved upward leading to expansion cooling), CONVECTION CURRENTS (lift air for adiabatic cooling), EVAPORATION (cooling effect as falling rain partially evaporates), ADVECTIVE COOLING (horizontal movement of air over colder surface) and LARGE SCALE ASCENT (large body of air such as frontal system gradually moves up). One common means of cloud formation is caused by lifting air, yielding convective cloud.

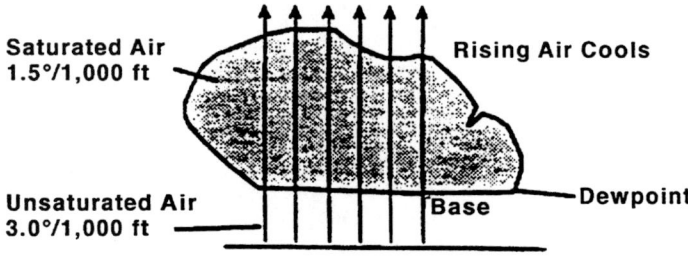

Fig. 6-4: Formation of convective cloud.

The DRY ADIABATIC LAPSE RATE (DALR) is 3.0°C/1000 ft. Use the DALR to find the cloud base. The SATURATED ADIABATIC LAPSE RATE (SALR) is 1.5°C/1000 ft. Use the SALR above the cloud base. The SALR is less than the DALR due to the heat release from condensation.

EXAMPLE: Surface temp. is 15° C, dewpoint is 9° C. Find cloud base and freezing level.

ANSWER: Cloud base = 2,000 ft (15° - 9° = 6°, use DALR). Freezing level = 8,000 ft (use SALR from cloud base for temperature drop of 9° - 0° = 9°).

6.05 Stability and Instability

Stable air resists vertical motion. Stable air has a shallow lapse rate. Unstable air favours vertical motion and has a steep lapse rate. The atmosphere is heated by RADIATION (reflection of Sun's rays on earth's surface), CONDUCTION (warm air contacting cold air), ADVECTION (horizontal movement of air) and CONVECTION (unequal surface heating). Cold air moving over a warm surface will yield unstable conditions, for example, as cold arctic air moves over the Great Lakes in winter, expect vertically developed cloud and snow showers.

The atmosphere is cooled primarily by RADIATION and ADVECTION. Cooling increases stability from below. The lifting processes are CONVECTION, CONVERGENCE (excess air rises when pressure systems meet), MECHANICAL TURBULENCE (friction between air and ground), OROGRAPHIC LIFT (air moves up mountains) or FRONTAL LIFT (advancing cold front). SUBSIDENCE occurs when air sinks, and results in heating and compression, so that the relative humidity decreases and water evaporates.

6.06 Clouds

Clouds are classified as HIGH (above 20,000 feet), MIDDLE (6,500 feet to 20,000 feet), or LOW (having bases below 6,500 feet). The prefix CIRRO means high, and ALTO means middle. There are two basic forms of cloud: LAYER (also called STRATOFORM) or HEAP/CUMULOFORM.

Layer clouds are horizontally developed, and heap clouds are vertically developed. Note the following cloud types:

High
a) Cirrus (Ci).   "Horse tails".
b) Cirrostratus (Cs).   Thin whitish veil; milky sky.
c) Cirrocumulus (Cc).   Small rounded masses.

Middle
a) Altostratus (As). Uniform veil, greyish or bluish.
b) Altocumulus (Ac). Layer or series of flattened rounded masses, long rolls, undulations or waves.

Low
a) Stratus (St).   Low layer of horizontal cloud.
b) Nimbostratus (Ns).   Thick uniform dark gray or black layer from which precipitation falls.
c) Stratocumulus (Sc). Layer or series of patches of cloud with rounded masses of rolling appearance.

Vertically Developed
a) Cumulus (Cu).   "Cotton ball" appearance.
b) Towering Cumulus (TCU).   Cumulus to higher altitudes.
c) Cumulonimbus (Cb).   Thunderstorm cloud.
d) Altocumulus Castellanus (ACC).   Middle level cumulus cloud layer with protruding turrets.

Clouds named with NIMBO produce precipitation. Cumulonimbus cloud produces heavy, intermittent precipitation. Nimbostratus cloud produces heavy, continuous precipitation. Stratoform type cloud produces continuous precipitation, cumuloform cloud produces intermittent precipitation.

VIRGA refers to precipitation that falls below cumuloform type cloud but evaporates before reaching the ground and will appear as visible streamers below the cloud base.

6.07 Turbulence

Turbulence is the result of friction between opposing air currents. Types of turbulence:

a) Convective Turbulence. Unequal surface heating. Also referred to as Thermal Turbulence, referring to updrafts in convective currents.
b) Orographic Turbulence. Friction in air currents in mountainous terrain.
c) Wind Shear Turbulence. Sudden change in wind speed or wind direction. Found near thunderstorms, causing downbursts termed MICROBURSTS. Also found in low level inversions and in frontal zones.
d) Clear Air Turbulence. Found at high altitudes.
e) Mechanical Turbulence. Friction between air and ground as high winds blow over rough, broken terrain.

Fig. 6-5: Mechanical turbulence.

Turbulence is reported as LIGHT (slight changes in attitude or altitude), MODERATE (greater intensity than light turbulence, aircraft remains in positive control), SEVERE (large, abrupt changes in attitude or altitude, aircraft may be momentarily out of control), or EXTREME (aircraft violently tossed about, impossible to control).

The first rule in flying in turbulence is to reduce airspeed. The aircraft flight manual should be referred to determine the appropriate airspeed for penetrating turbulence.

Knowing where to expect turbulence is important to being able to minimize it. Turbulence can be estimated by the cloud type. Cumulus clouds can indicate moderate to severe turbulence.

Cumulonimbus and stacks of lens shaped clouds indicate severe turbulence. Under stratocumulus, light to moderate turbulence may be expected. Stratus clouds mean smooth flying. Cumulus fractus indicates strong, turbulent winds.

You can anticipate turbulence in mountainous areas when the winds blowing across the mountains is greater than 40 knots.

Convective turbulence can be avoided by flying above convective cloud tops.

6.08 Wind

Wind is caused by horizontal pressure differences. Air flows from high to low pressure. PRESSURE GRADIENT is the rate of change of atmospheric pressure in the horizontal. If the pressure gradient is steep, the isobars will be spaced closely together on a weather chart, and the winds will be strong.

Because the Earth rotates, air is deflected to the right in the northern hemisphere. This deflective force is termed CORIOLIS FORCE. It results in clockwise circulation at high pressure systems, and counterclockwise circulation in low pressure systems.

DIURNAL EFFECT is the variation in wind from day to night. The eddying motion caused by friction between the air and the ground often extends 3,000 feet or higher. The result is that the direction and speed of the 3,000 feet winds tends to be transferred to the surface. During the night, the surface wind assumes its normal direction and speed. The 3,000 feet winds can be very strong, although the surface wind is almost calm. To summarized diurnal effects, by day, surface wind veers in increases, by night, surface wind tends to back and decrease. Diurnal variation is least over temperate oceans and greatest over deserts.

Since isobars are normally in a curved pattern, air circulation is generally circular and this involves the CENTRIFUGAL FORCE. The centrifugal force is when winds tend to move to the outside of the circular path described by the isobars. The pressure gradient will act in opposition to the centrifugal force in lows, and with the pressure gradient force in highs. The result is that for the same pressure gradient force, the winds about a low will be less than a high. However, the winds about a low will usually be much stronger than about a high because the pressure gradient force about a low is normally much higher.

At about 3,000 ft above ground level, wind tends to blow parallel to isobars. Below 3,000 feet, friction with the earth results in a change in direction and decrease in wind speed. In a climb, wind direction will VEER (clockwise change in direction) and wind speed will increase. In a descent, wind speed will decrease and wind direction will BACK (counter-clockwise change in direction). Surface friction effects are about 30 degrees over land, with a decrease in velocity, with about 0 - 15 degrees over water, with less of a decrease in velocity.

BUYS-BALLOT'S LAW states that in the northern hemisphere, if you stand with your back to the wind, the area of lower pressure lies to your left.

A GUST is a rapid change of wind speed/direction of brief duration. A SQUALL is a rapid change of wind speed/direction of prolonged duration. A SEA BREEZE occurs at day when wind blows from sea to land, a LAND BREEZE occurs at night when wind blows from land to sea. A KATABATIC WIND blows down valleys at night or down ice-covered slopes by day. An ANABATIC WIND blows up a slope or valley by day.

MOUNTAIN WAVES occur as air crosses mountain ranges in North America, where the prevailing wind is from west to east. In Canada, mountain waves are often over the foothills of the eastern Rockies. If the air is moist, clouds will develop which indicate mountain waves.

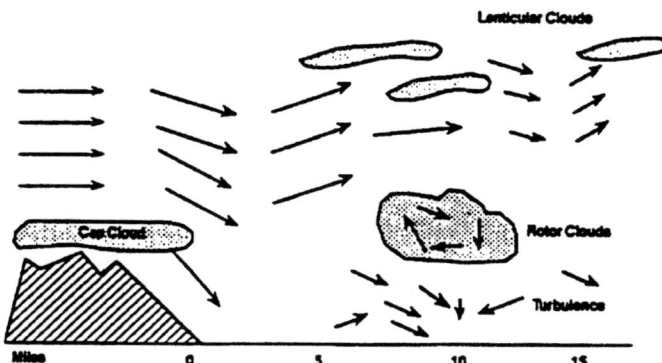

Fig. 6-6: Mountain waves.

A CAP CLOUD lies over the tops of mountain ranges and extends partially down the leeward slopes and indicates an extremely strong downdraft. A ROTOR CLOUD is formed downwind from each wave crest, and is simultaneously being formed and dissipated due to the rotation of the air, and may reach heights of up to 30,000 feet. ROLL CLOUDS resemble a long line of stratocumulus cloud lying parallel to the range. LENTICULAR CLOUD forms in the wave crests aloft, and is hundreds of miles long.

Mountain waves make strong downdrafts of up to 2000 FPM or more. Updraft areas can be used to gain altitude. Turbulence is most significant between the ground and the roll clouds. The altimeter tends to read high at the crests due to the low pressure created by the high winds.

WIND SHEAR occurs when the wind through which an aircraft is flying changes so rapidly that it causes a change in airspeed. Wind shear is either horizontal (low level inversion, changes in headwind or tailwind) or vertical (downbursts from thunderstorms, strong surface winds, passing fronts or mountain waves). Review section 4.16 for flight procedures in wind shear.

6.09 Air Masses

An air mass is a large body of air having uniform properties of moisture and temperature in the horizontal.

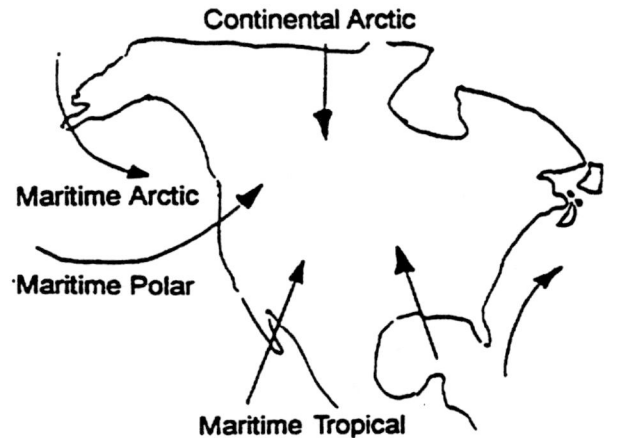

Fig. 6-7: Air masses affecting Canada.

The main air masses in Canada are:

a) Continental Arctic (cA). Dry and cold. Forms over polar cap.
b) Maritime Arctic (mA). Moist and cold. Forms over polar oceans.
c) Maritime Polar (mP). Moist and temperate. Forms over temperate oceans.
d) Maritime Tropical (mT). Moist and warm. Forms over warm, semi-tropical oceans.

Air masses are MODIFIED by taking on the characteristics of their source, such as the cooling of mT as it moves over the north Atlantic in summer, to become mP.

6.10 Fronts

A FRONT is a transition zone between two air masses. A COLD FRONT is an advancing wedge of cold air, that displaces the warm air aloft. A WARM FRONT is an extensive slope of warm air, that overrides the cold air. Frontal weather is determined by the stability and moisture content of the warm air mass.

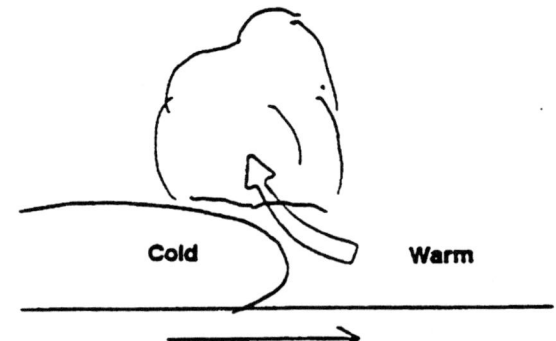

Fig. 6-8: Cross section of Cold Front.

Fig. 6-9: Cross section of Warm Front.

When approaching a warm front from the cold air side, expect high cirrus, cirrostratus, with altostratus or altocumulus at mid level, followed by stratus or nimbostratus at lower levels. If the warm air mass is moist and unstable, Cb cells can be found embedded in the main cloud deck.

FRONTAL SLOPE is the angle a warm front makes with the ground, generally 1:150 or 1:200.

EXAMPLE: If frontal slope is 1:200, what is the cloud height AGL 86 SM forward of the front?

ANSWER: 1/200 X 86 = .43 SM. .43 X 5280 ft/mile = 2,270 feet AGL.

The POLAR FRONT is the theoretical edge between the cold air dome that overlies the northern hemisphere where it meets the tropical air from the south.

In winter, a warm front can have its freezing level at the surface on the cold side, and a higher freezing level on the warm side. The result is shown in Fig. 6-10. Note the areas of freezing rain and ice pellets: if ice pellets are encountered, it is because warm air is above, and freezing rain will be next.

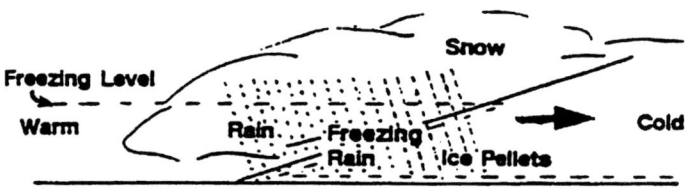

Fig. 6-10: Warm Front in winter. Note freezing level at surface in cold air, much higher in warm air.

A TROWAL is a "Trough of Warm Air Aloft", and is formed when a cold front advances on a warm front, undercutting and forcing the warm air up. A warm or cold OCCLUSION involves similar factors.

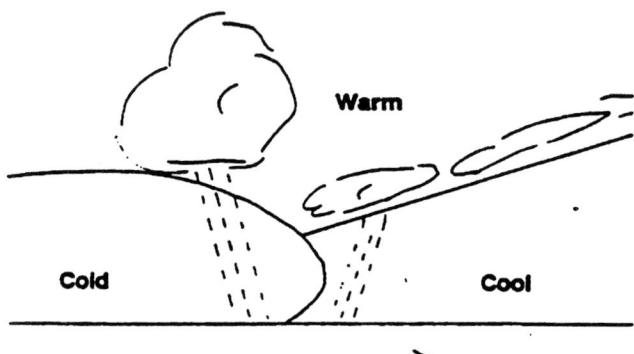

Fig. 6-11: Cold Front Occlusion.

Frontal systems are always associated with a low pressure system, often referred as a FRONTAL WAVE, consisting of a rotating warm front with a cold front in sequence behind and gaining on the warm front. The entire warm front moves parallel to the isobars in the warm air sector.

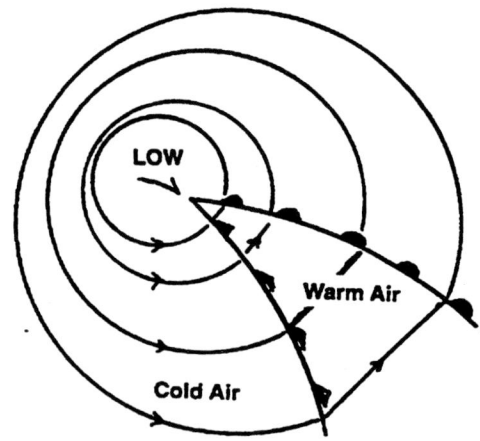

Fig. 6-12: Frontal wave. Movement of this system is northeast.

FRONTOGENESIS is when a front forms, FRONTOLYSIS is when a front dissipates. At a STATIONARY FRONT, the cold air mass does not move, and the winds will blow parallel to the frontal surface.

An UPPER FRONT is produced when very cold air is trapped on the surface. In an UPPER COLD FRONT, advancing cold air advances over a shallow layer of colder, heavier air. The leading edge of the cold front leaves the surface and moves along the top of the colder air as an upper front. In an UPPER WARM FRONT, the heavier cold air is retreating. A station in the colder air does not experience a change of air mass because the front passes overhead. Upper fronts can also be produced where there is substantial lower level heating, resulting in an insufficient temperature contrast for a front to exist at the surface.

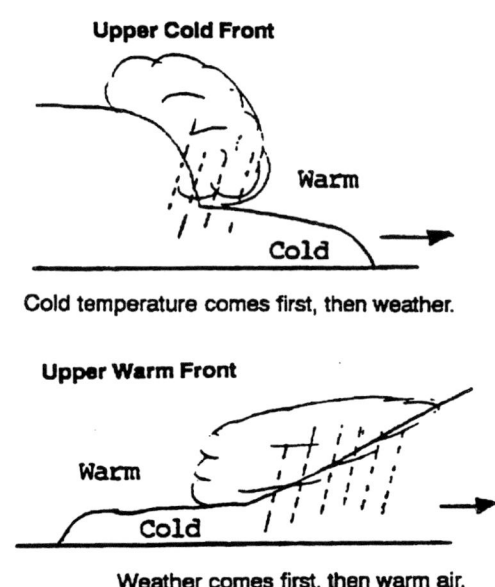

Fig. 6-13: Upper cold front, upper warm front.

DISCONTINUITIES ACROSS A FRONTAL SURFACE refer to differing properties of moisture, temperature, wind and visibility associated with a frontal surface. An example is wind: the alteration of aircraft heading to maintain constant track will always be to the right when flying through a cold or warm front, regardless of direction of flight.

6.11 Complex Low

A COMPLEX LOW is a low frontal depression having a range of overlapping frontal systems and air masses. Understanding the expected weather at a complex low involves a thorough understanding of the principals of air masses, frontal systems, and a close reading of the GFA's and TAFs. The basic structure should be understood as a 3-D image.

Fig. 6-14 is a plan view and cross section of a complex low. Note the air masses involved are Ca, mA, and mP. Cross section lines are drawn from A - B and C - D: note in particular the pocket of warm mP air seen aloft in the A - B cross section.

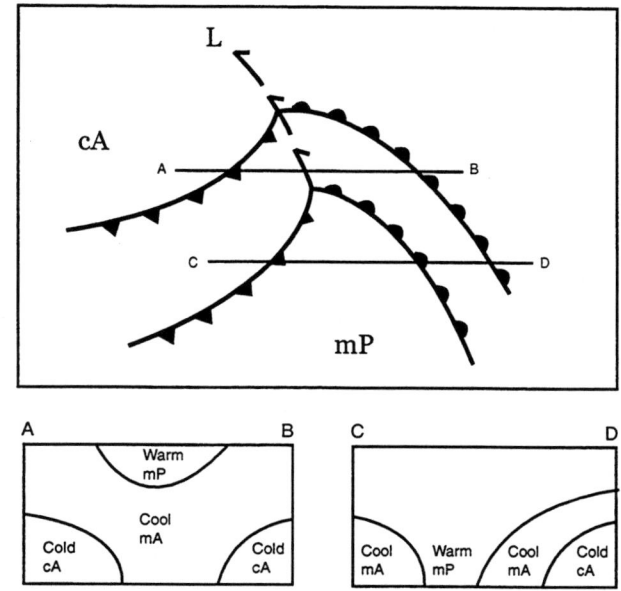

Refer to the complex low surface chart and weather at Fig. 6-15 to complete the exercise. While the weather data provided is based on the old FA (Area Forecast), it still offers a useful exercise in putting together a 3-D image of the overlapping weather systems involved for forecast arriving frontal weather.

Fig. 6-14: Plan and x-section of complex low.

INTENTIONALLY LEFT BLANK

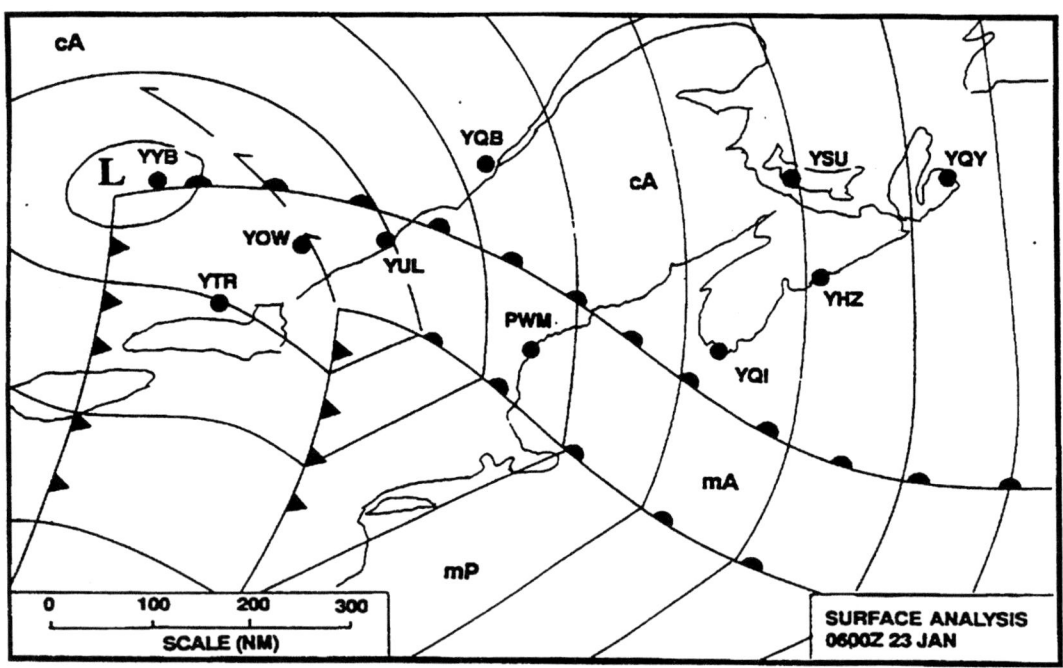

Fig. 6-15: Complex Low (based on old FA).

FACN1 CYHZ 231130Z
12 - 24
HTS ASL UNLESS NOTED

PROG
LO 50 NE NORTH BAY AT 12Z WITH CA WRM FNT SEWD-N QUEBEC
CITY AND 50 NM N YARMOUTH. CREST MA WV NR MONTREAL AT
12Z WITH TROWAL NWWD AND WRM FNT SEWD-N PORTLAND-100
NM S YARMOUTH. AT 24Z CA WRM FNT WILL LIE N SUMMERSIDE-S
SYDNEY WITH MA WRM FNT IN NW-SE LINE 150 SM TO S.

YHZ 1-2-3
W SCOTIA NEW BRUNSWICK STRAITS RGNS
CLDS AND WX. AHD WRM FNTS 20 OVC 40 60 OVC 180 VSBYS 3 -
SN BLSN 150 NM AHD CA FNT BCMG 2 -SN -IP BLSN 100 NM AHD
FNT. CIGS OCNLY 5 HND ABV GND AND VSBYS OCNLY 1/4-1 -SN -
BLSN. CLDS GRDLY LWRG AND MRGG TO 15 OVC 180 VSBYS 2 -
FZDZ FG AT CA FNT BCMG 5 OVC 10 15 OVC 180 VSBYS 2 -RA FG
BHND CA FNT AND 3 OVC 180 VSBYS 1/2 RA FG AT MA FNT. SNW
AND ICE PELLETS CHGG TO FRZG RAIN 50 NM AHD CA FNT.
VSBYS VRBL 1/2-3 IN RAIN AND FOG BHND CA FNT. SFC WIND
11020G30KT INCRG TO 14025G35KT BHND FNT.

ICG. LGT TO MDT RIME ICGIC ABV FRLVL WITH MDT CLR ICG IN
FRZG PCPN AHD CA FNT. FRLVL SFC AHD CA FNT AND 70 BHND
FNT.

TURBC. LGT BCMG MDT LWR LVLS.

BHND MA WRM FNT
CLDS AND WX. 10 OVC 25 80 OVC 130 -RA VSBYS 4-8 SFC WND
20025G40KT.

ICG. LGT RIME ICGIC ABV FRLVL. FRLVL 90.

TURBC. MDT IN LWR LVLS.
CYQI TAF 231045Z 231123 14025G35KT 1SM RA FG OVC005
    FM1700Z 14025G35KT 1/2SM RA FG OVC003 FM1800Z
    20025G40KT 6SM -RA BKN010 OVC080 FROPA
CYHZ TAF 231045Z 231111 11020G35KT 3/4SM -SN -IP -BLSN
    VV008. FM1300Z 11020G35KT 2SM -FZRA OVC020 FM1700Z
    14020G35KT 2SM -RA FG OVC010 FROPA FM2200Z
    14020G35KT 1SM -RA FG OVC005 FM0000Z 14025G35KT 1/
    2SM RA FG OVC003 FM0100Z 21025G40KT 6SM -RA BKN010
    OVC080
CYSU TAF 231045Z 231111 11020KT 3SM -SN OVC020 FM1500Z
    11020G35KT 3/4SM -IP -BLSN VV008 FM1800Z 11020G35 2SM -
    FZRA OVC020 FM2100Z 14025G35KT 2SM -RA FG OVC015
    FROPA. FM0300Z 14024G35KT 1SM -RA FG OVC005 FM0500Z
    14025G35KT 1/2SM RA FG OVC003 FM0600Z 21025G40KT 6SM
    -RA BKN010 OVC080

EXERCISE (refer to Fig. 6-15):
01. When were the surface chart observations made?
02. What kind of air mass lies aloft in the TROWAL over Ottawa (YOW) at 0600Z?
03. What is causing the forecasted precipitation?
04. What is the moisture and stability of the overrunning air mass?
05. What is the amount of the lower cloud layer forecast for Yarmouth (YQI) at 1900Z?
06. What are the overrunning air masses aloft over Summerside (YSU) at 1800Z?
07. What is the maximum height of the cloud top associated with the warm fronts?
08. What are the forecast precipitation as the warm fronts approach and pass a station?
09. What is a positive indication to confirm passage of the cA front at a station?
10. When is the cA front forecast to pass YSU?
11. What is the freezing level over Yarmouth (YQI) at 2400Z?
12. For a flight from YQI at 0600Z on January 24, the appropriate TAF will be issued at what time?
13. Draw a cross section showing the air masses in a straight line from YTR (Trenton) to Quebec (YQB) at 0600 Z.
14. Draw a cross section showing the air masses over YQI at 1200 Z.
15. Where will the mA front lie at 24Z?

ANSWERS:
01. January 23, 0600 UTC. See bottom right edge of chart.
02. mP. See weather chart (0600Z). Note TROWAL symbol over Ottawa, mP symbol is indicated as warm air mass being forced aloft by cold front.
03. Frontal lift. Over-running of warm, moist mA air over cA causes rain, snow and freezing rain.
04. Moist and stable. Moist because of precipitation, stable because there is no indication of TCu, Cb etc. Rain comes from stable air, showers come from unstable air.
05. 5/8 - < 8/8 inclusive. Refer to TAF for YQI. After 18Z the lowest layer is forecast to be C10BKN. This layer is forecasted for the period 18Z to end of FT at 2300Z, 1900Z is within this period.
06. According to chart and FA, there will be three air masses over YSU at 1800Z: cA, mA and mP. The OVERRUNNING air masses are mA and mP.
07. The max height is 18,000. Refer to clds and wx of FA. Note that only the FA provides cloud tops.
08. Snow, ice pellets, freezing rain, rain. See CYSU TAF as all systems will pass this station.
09. Freezing rain changes to rain. The freezing rain requires rain from warm air masses falling into cA. See clds and wx and TAF's.
10. 2100Z. See TAF for YSU: FROPA indicates frontal passage.
11. 9,000 feet. At 2400Z, the mA warm front will lie past CYSU (see PROG). Therefore, refer to ICG, "BHND MA WRM FNT" where FRLVL indicated as 9,000.
12. 0500Z is next issue time. TAFs are issued at 0500, 1100, 1700 and 2300Z.

13.

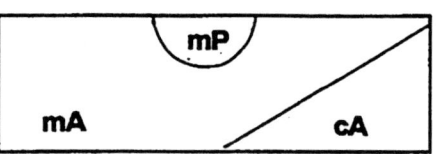

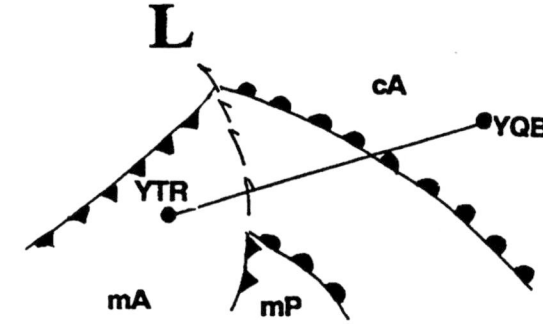

14.

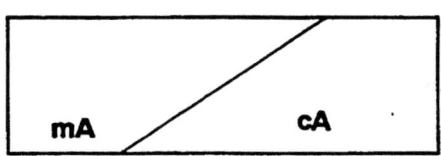

15. 150 NM to south of line from Summerside - south of Sidney. See Prognosis.

6.12 Jet Stream Weather

The key to jet stream weather is the tropopause. The height of the tropopause is higher over the equator than over the poles, and the temperature of the tropopause is colder over the equator. When an air mass moves, it carries the tropopause with it.

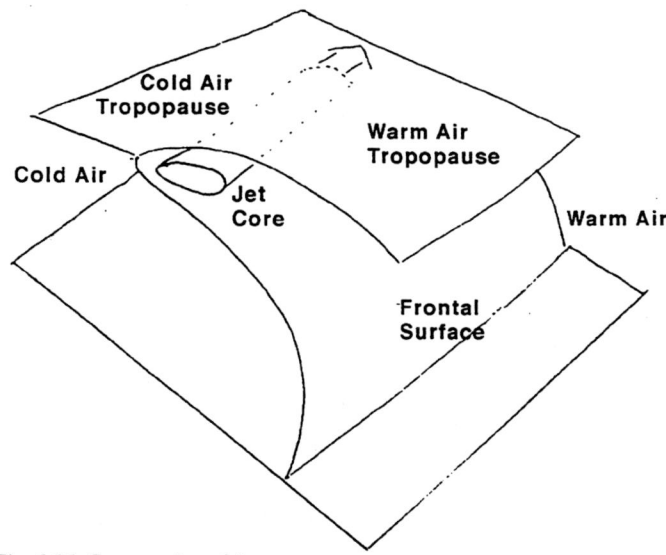

Fig. 6-16: Cross section of Jet stream.

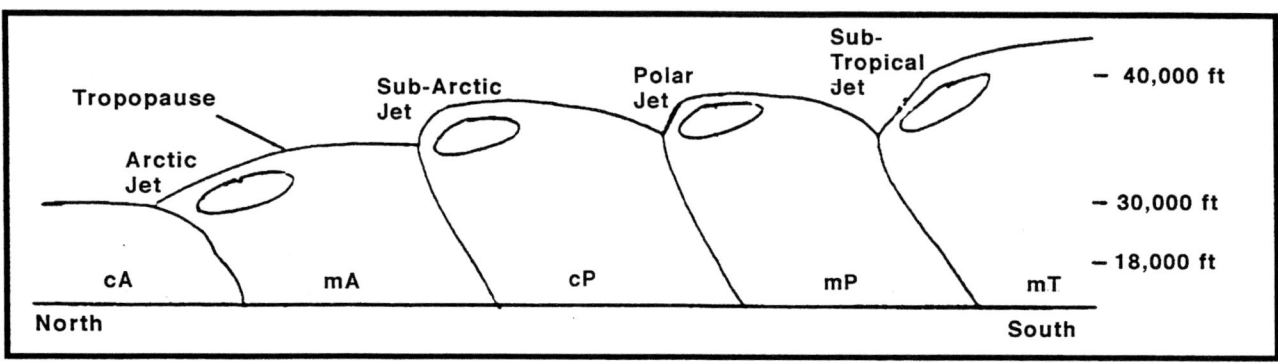

Fig. 6-17: Jet streams in northern hemisphere.

Where there is a strong temperature gradient, there will be a greater pressure gradient and therefore stronger winds. As this continues to high altitudes, the wind gets continually stronger. Upper level winds are caused by the THERMAL WIND COMPONENT, which is dependent on the temperature gradient. JET STREAMS are strong upper winds that develop where there is a strong horizontal temperature gradient. To qualify as a jet stream, the wind speed must be at least 60 kts, and the stream may be thousands of miles long and a few hundred miles wide.

Jet streams lie to the north of frontal systems where the temperature gradient is greatest. The jet stream will take the name of the frontal system causing it. The jet stream is located between the warm air tropopause on one side of the frontal system, and the cold air tropopause on the other and at about the height of the cold air tropopause. For this reason, the height of the jet stream will vary according to the front it is associated with and the season of the year.

Due to the high winds in a jet stream, there is substantial horizontal and vertical wind shear.

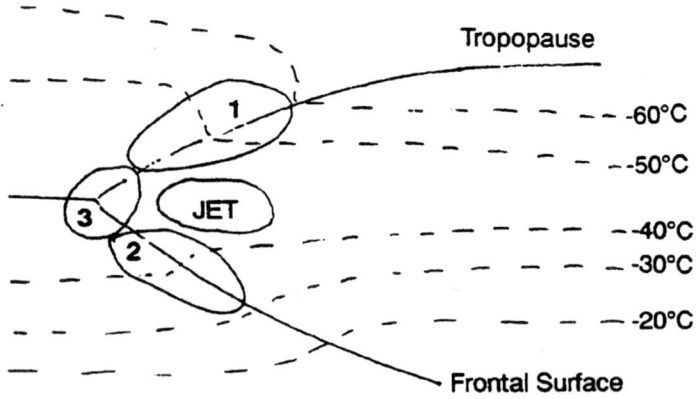

Fig. 6-18: Cross section of jet stream. Turbulence is found in jet streams having more than 110 kts at core in sloping tropopause above core (1), in front below core (2), and on low pressure side of core (3). OAT gauge can be used to minimize CAT when crossing jet stream.

CLEAR AIR TURBULENCE (CAT) is all turbulence not related to convective activity. Strong vertical wind shear is a dominant cause of CAT, and is therefore often found in areas of jet streams. Since jet streams are stronger in winter than in summer, CAT is more common in winter. If CAT is encountered while crossing a jet stream, the note the OAT: if it rises, it is fastest to climb to depart the CAT. If the temperature is constant, it will be just as

fast to climb or descend, and if the temperature drops, descend. If CAT is encountered while flying in a direct tailwind or headwind, generally it is best to turn to the south. Where 30 kt isotachs are spaced closer than 90 NM on the 250 mb upper charts, CAT can be expected from wind shear.

Expect vertical wind shear when 5° C isotherms are spaced closer than 2° latitude (120 NM).

6.13 Airframe Icing

Airframe ice forms when flying in or near cloud with temperatures from +2° C to -40° C. CLEAR ICE is smooth, heavy ice which forms from large, supercooled water droplets having a high rate of catch at temperatures just below 0° C, in cumulus type clouds. Clear ice spreads back as it collects.

RIME ICE forms from small, super-cooled water droplets at temperatures well below 0° C and is opaque and granular. Rime ice forms from small, super-cooled water droplets with a low rate of catch in stable clouds. Rime ice moves forward as it forms and builds up on sharp surfaces such as antennae and masts.

Airframe icing is most severe at temperatures from 0° C to -15° C in cloud thickness from 5,000 feet to 8,000 feet. Icing in towering cumulus clouds can exist at temperatures as low as -25° C. Small, super-cooled water droplets exist at lower temperatures than large, super-cooled water droplets.

The largest droplets of water come from clouds having the most vertical motion. The rate of freezing of super-cooled water droplets on the aircraft depends on the temperature of the aircraft skin and the air temperature. The smaller the size of a super-cooled water droplet, the lower the temperature it can exist at before freezing.

Droplet size affects where ice forms: small droplets form ice at the leading edge, medium droplets extend aft of leading edge, and large droplets extend aft of protected surfaces (de-ice, anti-ice etc.).

Where an aircraft in flight encounters ice crystals, icing conditions decrease because ice crystals build at the expense of water vapour.

The COLLECTION EFFICIENCY of aircraft icing on an aerodynamic surface is inversely proportional to the surface geometry. Thin wings catch more droplets per square inch of leading edge than thick wings, because collection efficiencies vary inversely with the geometry of the collecting surface. For this reason, the collection efficiency is greater for antennae masts, thin leading edges etc. than for canopies, thick wings and other rounded surfaces like fuselages. Rime ice tends to move forward as it accumulates, clear ice tends to move backward as it accumulates. High speed aircraft have higher collection efficiencies than lower speed aircraft.

Icing intensity is reported as TRACE (slight accretions, not hazardous), LIGHT (occasional use of de-icing/anti-icing equipment will remove/prevent accretion), MODERATE (the rate of accretion could be potentially hazardous and use of de-icing/anti-icing equipment necessary) and SEVERE (the rate is so great that de-icing/anti-icing equipment fails to reduce or control the hazard, immediate diversion is necessary).

AERODYNAMIC HEATING refers to the heating effect caused by the compressibility and subsequent heating effect encountered at high speeds. Aerodynamic heating, especially if the surface is dry, can reduce the onset of aircraft icing. As an example, no ice would form on a leading edge at 12,000 feet with a speed of 300 kts at an OAT no warmer than -5° C. To insure no ice at all would collect, speeds of 500 to 600 knots are required.

Cumulus type clouds will yield more severe icing conditions, especially in conditions of high moisture and unstable air. In cumulus cloud, both clear icing and rime icing can be encountered. In layer type cloud, icing will be rime type icing and intensity will be limited to light or moderate.

As shown in Fig. 6-10 above, in areas of freezing rain or freezing drizzle where lower levels and the surface are at temperatures below freezing, very serious areas of clear icing can result.

HOAR FROST is a light, feathery, crystalline deposit that forms on parked aircraft at night due to sublimation. It can also form in flight when an aircraft in very cold air descends into warm, moist air.

Airframe icing impairs aerodynamic efficiency by reducing the airflow, resulting is reduced lift and increased drag. The stall speed will be increased. Icing also seriously impairs aircraft engine performance. Other icing effects include false indications on flight instruments, loss of radio communications, and loss of operation of control surfaces, brakes and landing gear.

Here are some recommended precautions:

a) Before takeoff, check forecasted and actual weather for possible icing areas along the planned flight route. As well, check any PIREPS thate may have been issued.
b) If your aircraft is not equipped with de-icing or anti-icing equipment, avoid areas of icing. Water (clouds or precipitation) must be visible and OAT must be near 0° C for ice to form.
c) Always remove ice or frost from airfoils and control surfaces before attempting takeoff.
d) In cold weather, avoid, when possible, taxiing or taking off through mud, water, or slush. If you have taxied through any of these, make a preflight check to insure freedom of controls.
e) When climbing out through an icing layer, climb at an airspeed a little faster than normal to avoid a stall.
f) Use de-icing or anti-icing equipment when accumulations of ice are not too great. When such equipment becomes less than totally effective, change route or

altitude to get out of the icing as soon as possible.
g) If your aircraft is not equipped with a pitot static system de-icer, be alert for erroneous readings from the ASI, VSI and altimeter.
h) In stratiform clouds, you can likely alleviate icing by changing to a flight level at above freezing temperatures or to one colder than -10° C. An altitude change also may take you out of clouds. Rime icing in stratiform clouds can be very extensive horizontally.
i) In frontal freezing rain, you may be able to climb or descend to a layer warmer than freezing. Temperature is always warmer than freezing at some higher altitude. If you are going to climb, move quickly; procrastination may leave you with too much ice. If you are going to descend, you must know the temperature and terrain below.
j) Avoid cumuliform clouds if at all possible. Clear ice may be encountered anywhere above the freezing level. Most rapid accumulations are usually at temperatures from 0° C to -15° C.
k) Avoid abrupt maneuvers when your aircraft is heavily coated with ice since the aircraft has lost some of its aerodynamic efficiency.
l) When ice has accumulated, fly your landing approach with power.

6.14 Thunderstorms

A thunderstorm will form with an adequate supply of moisture, an unstable lapse rate and lifting agent.

Life Cycle

In the DEVELOPING STAGE, a cumulus type cloud grows and merges with other cumulus clouds. The base is about 4 miles wide and the height is up to about 20,000 feet. In this stage, the cloud consists mainly of updrafts. Early during the cumulus stage, water droplets are quite small but grow to raindrop size as the cloud grows. The upwelling air carries to liquid water above the freezing level creating an icing hazard. As the raindrops grow still heavier, they fall. The cold rain drags air with it creating a cold downdraft coexisting with the updraft at which point the next stage is reached.

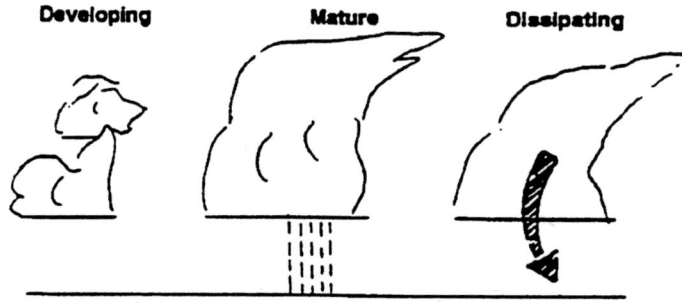

Fig. 6-19: Stages of a thunderstorm.

In the MATURE STAGE, rain or other precipitation such as hail begins to fall. Updrafts have reached their maximum speed. The cloud top reaches the tropopause and an anvil shape forms due to the strong winds at the tropopause. As the rain falls, there are severe downdrafts. When downdrafts hit the ground, air spreads horizontally causing a GUST FRONT, consisting of surface gusts. The gust front produces strong, gusting surface winds, sharp winds, sharp drop in temperature and abrupt rise in pressure. A ROLL CLOUD (also termed SCUD ROLL CLOUD) may form near the main cloud based in the shear area ahead of the cloud between the updraft and downdraft. The speed of the downdraft may be as high as 2,500 FPM.

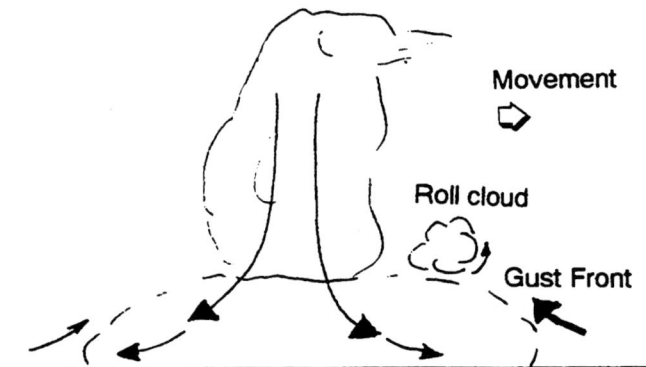

Fig. 6-20: Mature stage detail.

Ahead of the gusts are downdraft areas termed DOWNBURSTS. MICROBURSTS are extreme downdrafts within the downbursts. Microbursts are about 1 NM in diameter at 2,000 feet AGL with a horizontal extent at the surface of about 2 - 2.5 NM. Vertical winds can be as high as 6,000 FPM, and horizontal winds can be as high as 45 kts at the surface (i.e. 90 kts shear). Maximum intensity winds in a microcburst are generally of short duration (from 2 - 4 minutes). Microburst life cycle from initial downburst to dissipation is generally less than 15 minutes.

LIGHTNING occurs in the mature stage due to the movement of water, hail and snow within the cell, creating areas of differential electrical charge. All thunderstorm hazards reach their greatest intensity during the mature stage.

In the DISSIPATING STAGE, the dominant feature is a downdraft and the cloud begins to dissipate. When rain has ended and downdrafts have abated, the dissipating stage is complete.

Classification

The types of thunderstorms are:

a) Orographic. Forms in mountainous regions as air climbs over steep slopes.
b) Convective. Caused by solar heating or radiational cooling. Forms over land or water.
c) Frontal. Caused by frontal lift i.e. fast cold front.
d) Nocturnal. Found in midwest plains areas at night having unusually moist air aloft.

Hazards

Thunderstorms have severe turbulence, heavy showers, hail and icing. As well, some thunderstorms yield tornadoes.

A SQUALL LINE is a non-frontal narrow band of active thunderstorms. Often it develops ahead of a cold front in moist, unstable air, but it may develop in unstable air far removed from any front. The line may be too long to easily detour and too wide and severe to penetrate. It

often contains severe steady state thunderstorms and presents the single most intense weather hazard to aircraft. It usually forms rapidly, generally reaching maximum intensity during the late afternoon and the first few hours of darkness.

Hazardous forms of turbulence are present in all thunderstorms. Strongest turbulence within the cloud occurs with shear between updrafts and downdrafts. Outside the cloud, shear turbulence can be encountered several thousand feet above and 20 miles laterally from a severe storm.

Lightning strikes are more likely in the mature stage where temperatures are from -5° C to +5° C.

Visibility generally is near zero within a thunderstorm cloud. Ceiling and visibility also can become restricted in precipitation, and dust between the cloud base and the ground.

Flight Precautions

It is almost impossible to hold a constant altitude in a thunderstorm, and maneuvering in an attempt to do so greatly increases stresses on the aircraft. Stresses will be least if the aircraft is held in a constant ATTITUDE and allowed to ride the waves. Due to the numerous hazards involved, precision instrument flying is virtually impossible.

Rules for Thunderstorm Avoidance

a) Don't land or take off in the face of an approaching thunderstorm. A sudden wind shift or low level turbulence could cause loss of control.
b) Don't attempt to fly under a thunderstorm even if you can see through to the other side. Turbulence under the storm could be disastrous.
c) Don't try to circumnavigate thunderstorms covering 5/8 of an area or more either visually or with airborne radar.
d) Don't fly without airborne weather radar into a cloud mass with scattered embedded thunderstorms. Scattered thunderstorms not embedded usually can be visually circumnavigated.
e) Avoid by at least 20 NM any thunderstorm identified as severe or giving an intense radar echo. This is especially true under the anvil of a large cumulonimbus.
f) Clear the top of a known or suspected severe thunderstorm by at least 1,000 feet altitude for each 10 knots of wind speed at the cloud top (this would exceed the altitude capability of most aircraft).
g) Remember that vivid and frequent lightning indicates a severe thunderstorm.
h) Regard as severe any thunderstorm with tops 35,000 feet or more whether the top is visually sighted or determined by radar.

Rules for before Penetrating Thunderstorm if Unavoidable

a) Tighten safety belts and shoulder harnesses, secure all loose objects.
b) Plan course to take you through the storm in a minimum time and HOLD it.
c) To avoid the most critical icing, establish a penetration altitude below the freezing level or above the level of -25° C to avoid critical icing areas.

d) Turn on pitot heat and carburetor or jet inlet heat. Icing can be rapid at any altitude and cause almost instantaneous power failure or loss of airspeed indication.

e) Establish power settings for reduced turbulence penetration airspeed as recommended in the aircraft manual - reduced airspeed lessens the structural stresses on the aircraft.

f) Turn up cockpit lights to highest intensity to lessen danger of temporary blindness from lightning.

g) If using automatic pilot, disengage altitude hold mode and speed hold mode. The automatic altitude and speed controls will increase maneuvers of the aircraft thus increasing structural stresses.

h) When using airborne radar, tilt the antenna up and down occasionally. Tilting it up may detect a hail shaft that will reach a point on your course by the time you do. Tilting it down may detect a growing thunderstorm cell that may reach your altitude.

Rules During Thunderstorm Penetration

a) Keep your eyes on your instruments. Looking outside the cockpit can increase danger or temporary blindness from lightning.

b) Don't change power settings; maintain settings for reduced airspeed.

c) Maintain a constant ATTITUDE and let the aircraft ride it out. Maneuvers in trying to maintain constant altitude increase stresses on the aircraft.

d) Do not turn back once you are in the thunderstorm. A straight course through the storm most likely will get you out of the hazards quicker. As well, turning maneuvers increases structural stress on aircraft.

6.15 Hurricanes and Tornadoes

A hurricane is a destructive cyclonic storm that originates in tropical waters. Wind speed is great, and the movement of a hurricane is difficult to predict.

A TORNADO is a rotating tunnel shaped cloud linking the ground with the base of a violent thunderstorm. A WATERSPOUT is like a tornado except it contacts water instead of land.

6.16 Surface Based Layers

A SURFACE BASED LAYER is a layer which contacts the ground, such as fog, haze, smoke, blowing snow, and blowing sand. The requirements for fog are high relative humidity, a cooling process and condensation nuclei. Fog develops when air is cooled below the dewpoint temperature, or moisture is added to air to raise the dewpoint temperature. Fog types are:

a) Radiation Fog. Formed by radiational cooling on clear nights where relative humidity is high. Radiation fog is typically associated with high pressure systems, and can be increased where there is a light wind to enhance the mixing action. Radiation fog is also increased where there is smoke or pollution, since there will be more condensation nuclei for water vapour to collection upon.

b) Advection Fog. Formed by horizontal movement of warm, moist air over a cool surface, such as in coastal areas.

c) Upslope Fog. Moist air moves up rising terrain e.g. as cold air moves west over prairies towards Rocky Mountains.

d) Frontal Fog. Precipitation from warm or cold front falls into colder air below.

e) Arctic Sea Smoke. Forms as arctic air moves over an extensive area of warm water, raising the dewpoint and leading to steam fog. As this air continues to pass over warm water, it leads to the formation of heap cloud.

f) Ice Fog. Forms at very low temperatures when water vapour turns directly to ice crystals (sublimation).

6.17 Meteorological Services for Pilots

Weather information for pilots is available by telephone, computer terminal, weather briefing or radio broadcast. The most common weather service originates from the FSS network. At FSS stations, pilots can obtain all weather information needed for flight. The Aviation Weather Information Service (AWIS) is the basic weather briefing service provided for pilots at FSS stations and is accessible at the station via telephone (collect calls accepted). AWIS is called W2 in the CFS.

The Aviation Weather Briefing Service (AWBS) is a more advanced weather briefing service that presently comes from AES offices and some FSS's. In addition to domestic flights, AWBS is used for long range flights and is called W1 in the CFS.

The Pilots Automatic Telephone Weather Answering Service (PTAWAS) is a recording service that pilots can select various reports and forecasts by pressing different keys on a touchtone telephone. The locations of this service are in the CFS.

Aviation weather is also available from private funded DUATS computer terminals provided by FBO's at some aerodromes. DUATS availability can be found in the CFS, and in addition, DUATS service can be obtained from private companies for home use with a computer and modem.

At several airports, Automatic Terminal Information Service (ATIS) information is available. ATIS is a recorded broadcast providing basic weather information for departure and arrival at an airport. Transcribed Weather Broadcasts (TWB) are recorded weather reports, forecasts and NOTAMS that are broadcast over various NDB stations.

VOLMET service refers to In-Flight Meteorological Information (ICAO). VOLMET provides forecast and actual weather over long distance HF radio for North Atlantic/high arctic weather. For specific HF frequencies and times of operation, refer to the CFS.

6.18 Graphic Area Forecast (GFA)

The GFA replaces the former Area Forecast, and provides a graphical display of forecast weather for a large geographic area. The GFA depicts the most probable weather over a given area below 24,000 feet at a specific time (the valid time).

GFA Domains

The GFA covers one of 7 distinct DOMAINS, that in turn cover all of Canadian airspace. The following domains are provided:

a) PACIFIC - GFACN31;
b) PRAIRIE - GFACN32;
c) ONTARIO-QUEBEC - GFACN33;
d) ATLANTIC - GFACN34;
e) YUKON - GFACN35;
f) NUNAVUT - GFACN36;
g) ARCTIC - GFACN37.

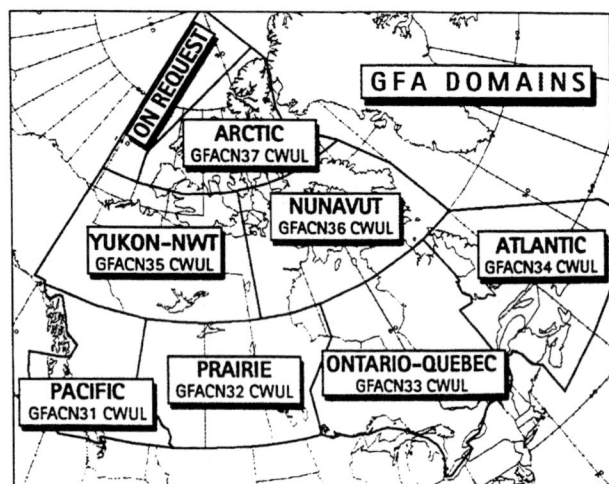

For each domain, 6 charts are produced for every GFA period. Two charts are to be used together for each 6 - hour valid period, one for clouds and weather, and another for icing, turbulence and freezing levels. GFA's are issue 4 times per day, approximately 1/2 hour before the beginning of the forecast period. The GFA issue times are approximately 2330, 0530, 1130 and 1730 UTC and are valid at 0000, 0600, 1200 and 1800 UTC respectively.

In a GFA:

a) Heights are above sea level (ASL), unless noted with AGL abbreviations such as "ABV GND" or "CIGS . . . HND".
b) Heights are given in hundreds of feet.
c) Visibility is given only if it will be 6 SM or less. Visibility over 6 SM is indicated as "P6SM".
d) Wind is given only if expected to be above 20 kts or gusts to 30 kts or more.
e) Cloud heights are given for bases and tops. For example: OVC $\frac{180}{80}$

indicates clouds based at 8,000 feet ASL and topped at 18,000 feet ASL. Scalloped areas indicate organized clouds.
g) All relevant fronts, highs and lows are included. Isobars are depicted at 4 mb intervals. Systems expected to move will be indicated with an arrow for the direction and number designating the speed in knots. QS indicates "quasi-stationary" for a front or pressure system moving at less than 5 kts. For example:

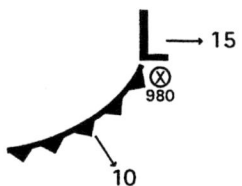

represents a low pressure system moving eastward at 15 kts and cold front moving southeastward at 10 kts.
h) Surface based layers are described using standard abbreviations used in TAFs and METARs including the term OBSCD. Obstructions to vision are only mentioned when the visibility is forecast to be 6 SM or less. For example:

A.

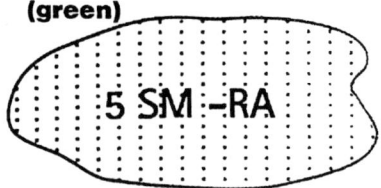

means continuous precipitation, visibility 5 SM in light rain.

B.

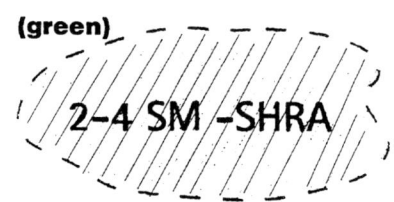

Means showery precipitation, visibility 2 - 4 SM in light rain showers.
i) Icing will be depicted when moderate or severe icing is forecast. Bases, tops and type of icing (RIME, MIXED, CLEAR) are indicated. Icing areas are enclosed with blue dots.
j) Turbulence is depicted when moderate or severe. The type of turbulence if due to mechanical turbulence will be indicated such as MECHANICAL, LEE WAVE, CAT etc.. For example:

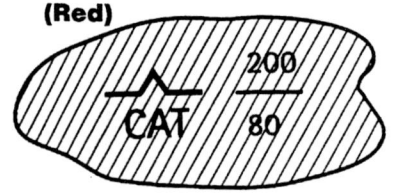

indicates an area of moderate CAT based at 8,000 feet ASL and topped at 20,000 feet ASL.
k) The freezing level is indicated by contours using red dashed lines, starting at the surface and at 2,500 foot levels.

The GFA chart is divided into the following sections:

a) Weather Information Box. This is a graphic representation of the forecast weather, both for Clouds and

Weather, and for Icing, Turbulence, and Freezing Level.

b) Chart name based on GFA domain.
c) Issue time in UTC.
d) Valid time in UTC.
e) Legend. Provides some symbols used on chart and scale in NM.
f) Comments. This is for anything of importance to the forecaster.
g) IFR Outlook. This is a summary of the expected weather for the 12 hr clouds and weather GFA. The three outlook categories used to describe the expected ceiling and visibility for the next 12 hours as follows:

Category	Ceiling	Visibility
IFR	less than 1,000 ft	less than 3 SM
MVFR	1,000 ft - 3,000 ft	3 - 5 SM
VFR	more than 3,000 ft	more than 5 SM

EXERCISE: Refer to the GFA Clouds and Weather chart at Fig. 6-21 (a) to answer the following questions:

1. What GFA domain does this chart cover?
2. What time UTC was this chart issued?
3. For what local time (Eastern Daylight Savings) and date is this forecast based?
4. Why are freezing levels not depicted on this chart?
5. What do the cloud types CB, TCU and ACC imply on this chart?
6. What ceilings can be anticipated in the area marked "1"?
7. What are the lowest visibilities that can be anticipated in the area marked "1"?
8. What is the height of cloud tops associated
9. What kind of frontal system is depicted in the area marked "2"?
10. What visibilities are forecast along the coast of Hudson's Bay?

ANSWERS:
1. Ontario - Quebec region.
2. Issue time is 1736 UTC.
3. Chart is for 1800Z - 4 hrs = 1400 EST for May 6, 2000.
4. Freezing levels are not depicted on this chart because it is the "clouds and weather GFA" - another GFA chart is issued to depict icing/turbulence and this will have freezing levels.
5. The cloud types CB, TCU and ACC all imply significant turbulence and icing, and the cloud type CB implies low level wind shear.
6. Extensive ceilings in stratus could from 200 feet AGL to 800 feet AGL.
7. Visibilities as low as 2 SM in thunderstorms, rain and mist.
8. Thunderstorms to 35,000 feet ASL.
9. Warm front.
10. Visibilities as low as 1/4 SM in fog and mist are forecast along the Hudson Bay Coast (see comments box at A).

EXERCISE: Refer to the GFA Icing, Turbulence and Freezing Level chart at Fig. 6-21 (b) and answer the following questions.

1. What is the valid time for this chart?
2. What degree of icing is forecast for the area marked "1" and for what range of heights?
3. What is the forecast freezing level height at area 2?
4. What degree of turbulence can be expected in area 4 and from what heights?

5. What type of icing is forecast within area 2?
6. What is the approximate speed and direction of movement of the front depicted at area 3?
7. What kind of turbulence can be expected at area 5 from 16,000 to 39,000 feet ASL?
8. What pressure is associated with the centre of the Low?
9. At what time will the next GFA chart be issued for this area?
10. For this chart, what colour would be used to depict the forecast areas of turbulence?

ANSWERS:
1. This chart is valid for 1800Z on May 6, 2000.
2. Severe icing is forecast locally within cloud from the surface to 3000 feet ASL.
3. Freezing level 10,000 feet ASL.
4. Moderate turbulence from 10,000 feet through 24,000 feet ASL.
5. Local moderate rime ice within cloud from freezing level to 14,000 feet ASL.
6. 5 knots in a northeastern direction.
7. Moderate clear air turbulence.
8. 999 mb.
9. At approximately 2330 UTC.
10. Red.

6.19 International Aerodrome Forecasts (TAF)

The TAF is an international weather format for an aerodrome forecast, describing the most probable weather conditions expected for an aerodrome. TAFs give weather within 5 NM of the centre of the runway complex. TAFs are scheduled 4 times daily for 12 or 24 hour periods at 00Z, 06Z, 12Z and 18Z. The information provided is as follows:

a) Message type. TAF or TAF AMD indicates regular or amended forecast. An amended TAF is issued when the current TAF no longer describes the on-going weather.
b) Station identifier. A 4 letter ICAO identifier is used to indicate the airport name.
c) Date and time of issue. This is a 6 digit code: the first two digits are the date, followed by the time of issue. TAFs are normally issued about 30 minutes before their valid period. Time is UTC.
d) Valid period. A 4 digit code represents the valid period. Normally 12 hours, although sometimes 24 hours and for an amended or corrected TAF a different period may be used.
e) Winds. The wind direction and speed is given to the nearest 10 degrees true and knots. Calm is 00000KT. VRB is for variable direction, less than 3 kts. Gusts, if applicable, is given in 2 digits. Where there is significant wind shear, the code WS is used, followed by the height of the shear and forecast wind speed and direction at that height. For example, WS015/27030KT means wind shear at 1500 ft AGL, 270°T at 30 kts.
f) Prevailing visibility in statute miles when 6 SM or less. P6SM means visibility greater than 6 SM.
g) Weather. "-" means light intensity, no sign is moderate, "+" is heavy. There are 7 descriptors which apply to forecast precipitation and/or obstructions to visibility: SH = showers, DR = drifting, FZ = freezing, MI = shallow, BL = blowing, BC = patches. Weather codes:

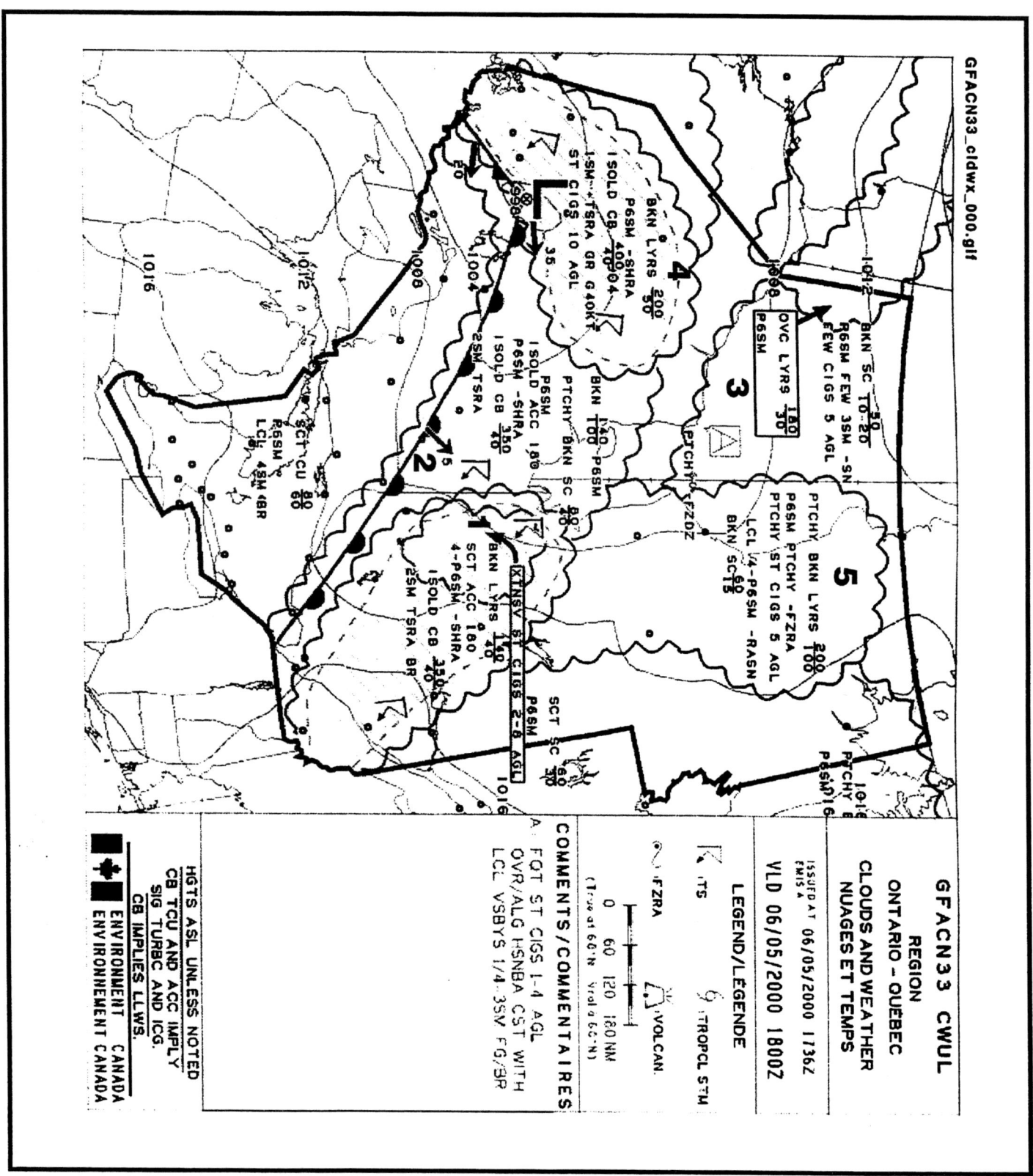

Fig. 6-21 (a): GFA Clouds and Weather Chart.

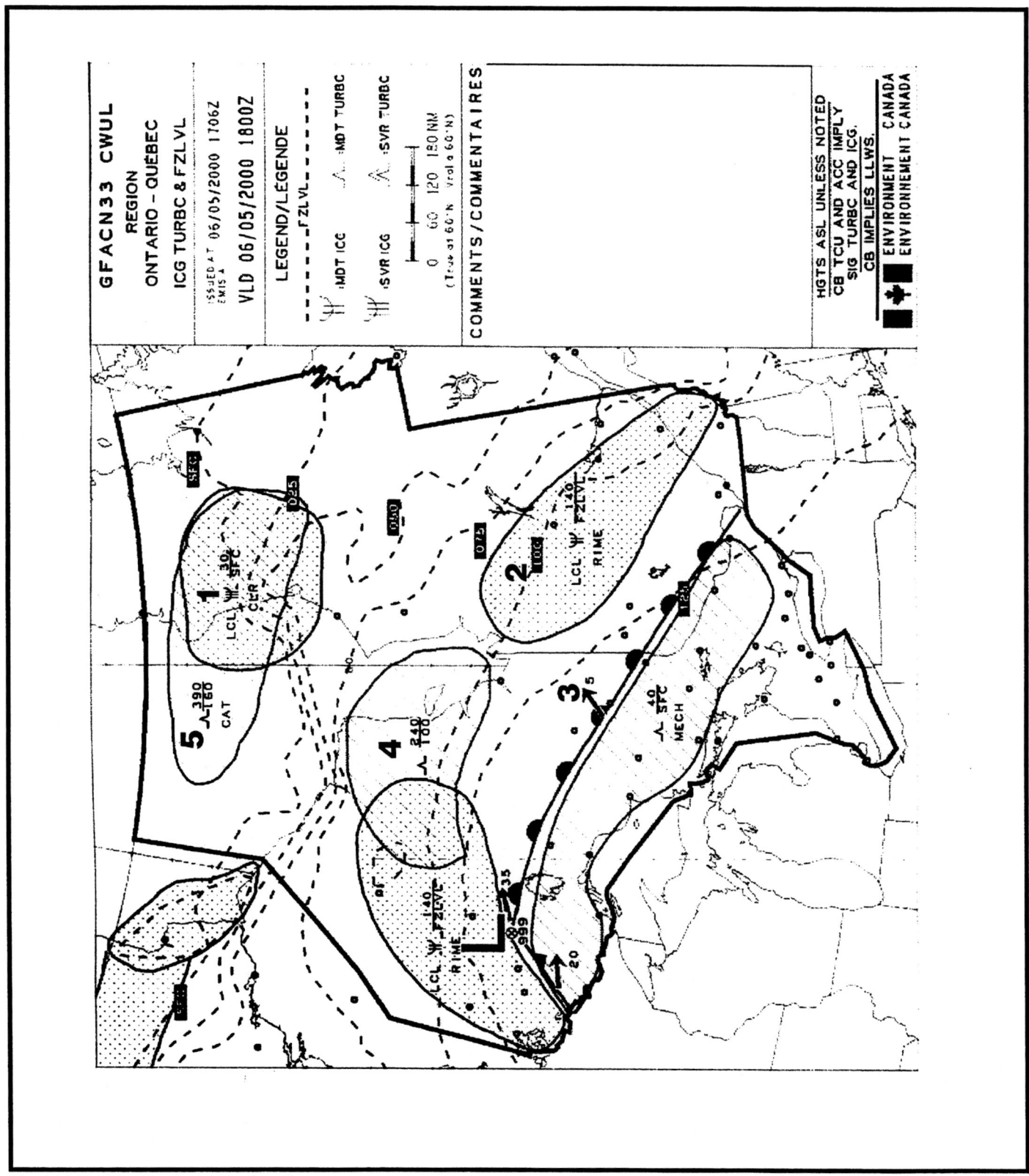

Fig. 6-21 (b): GFA Clouds and Weather Chart.

ABT	About	CVCTV	convective	IMDTLY	Immediately	OCLD	Occlude	ST	Stratus
ABV	Above	CVR	Cover	INCOMP	Incomplete	OCLDG	Occluding	STBL	Stable
AC	Altocumulus	DCRG	Decreasing	INCRG	Increasing	OCLN	Occlusion	STG	Strong
ACC	Altocumulus Castellanus	DEG	Degree	INDEF	Indefinite	OCNL	Occasional	STGTN	Strengthen
		DFUS	Diffuse	INOV	In the Vicinity of	OCNLY	Occasionally	STNRY	Stationary
ACRS	Across	DISTNT	Distant	INSTBY	Instabliity	OFSHR	Offshore	STP	Stop
ACSL	Standing Lenticular Altocumulus	DNS	Dense	INTMT	Intermittent	ONSHR	Onshore	SVR	Severe
		DNSLP	Downslope	INTR	Interior	ORGRHC	Orographic	SVRL	Several
		DP	Deep	INTS	Intense	OT	On Top of Clouds	SW	Southwest
ACTV	Active	DP	Dewpoint	INTSFY	Intensify			SWLY	Southwesterly
ADVY	Advisory	DPNG	Deepening	INVDG	Invading	OTLK	Outlook	SXN	Section
AFT	After	DR	Drifting	IP	Ice Pellets	OTRW	Otherwise	SYS	System
AFTN	Afternoon	DRFTG	Drifting	IS	Island	OVC	Overcast		
AHD	Ahead	DRPG	Dropping	ISLD	Island	OVR	Over	TCU	Towering Cumulus
ALF	Aloft	DSIPTG	Dissipating	ISOL	isolated	OVRNG	Overrunning		
ALG	Along	DTRT	Deteriorate					TEMP	Temperature
ALQDS	All quadrants	DURG	During	JDFSTR	Juan de Fuca Strait	PAC	Pacific	TEMPO	Temporary
ALT	Altitude	DVLPG	Developing			PCPN	Precipitation	THK	Thick
ALTA	Alberta	DZ	Drizzle	JMSBA	James Bay	PD	Period	THKNG	Thickening
AMS	Air Mass					PRECDD	Preceded	THN	Thin
APRCH	Approach	E	East	KT	Knot	PRECDS	Precedes	THNC	Thence
APRCHG	Approaching	ELSW	Elsewhere			PRES	Pressure	THNG	Thinning
ASL	Above Sea Level	ELY	Easterly	LABRDR	Labrador	PROG	Prognostic	THRU	Through
ASSOCD	Associated	EMBD	Embedded	LAT	Latitude	PROG	Prognosis	THRUT	Throughout
ATLC	Atlantic	ENDG	Ending	LCL	Local	PRSTG	Persisting	THSD	Thousand
ATTN	Attention	ENTR	Entire	LFTG	Lifting	PSG	Passage	TIL	Until
AURBO	Aurora Borealis	EVE	Evening	LGT	Light	PSG	Passing	TRML	Terminal
AVG	Average			LIFR	Low IFR	PTCHY	Patchy	TROF	Trough
		FCST	Forecast	LK	Lake	PTLY	Partly	TROWAL	Trough of Warm Air Aloft
BAFDY	Bay of Fundy	FG	Fog	LLWS	Low Level Wind Shear				
BECMG	Becoming	FILG	Filling			QS	Quasi-Stationary	TRRN	Terrain
BFR	Before	FL	Flight level	LMT	Limit	QSTNRY	Quasi-Stationary	TSTM	Thunderstorm
BGN	Begin	FLO	Flow	LN	Line	QUAD	Quadrant	TURBC	Turbulence
BGNG	Beginning	FLRY	Flurry	LND	Land	QUE	Quebec	TWD	Toward
BHND	Behind	FLWD	Followed	LO	Low				
BINOVC	Breaks in Overcast	FLWG	Following	LONG	Longitude	RA	Rain	UNSTBL	Unstable
		FM	From	LTL	Little	RCH	Reach	UNSTDY	Unsteady
BKN	Broken	FNT	Front	LTGIC	Lightning in Cloud	RCKY	Rockies	UPR	Upper
BL	Blowing	FNTGNS	Frontogenesis			RDG	Ridge	UPSLP	Upslope
BLDG	Building	FQT	Frequent	LTNG	Lightning	RFRMG	Reforming	UTC	Co-ordinated Universal Time
BLO	Below	FRLVL	Freezing Level	LTR	Later	RGN	Region		
BLZD	Blizard	FRMG	Forming	LVL	Level	RMK	Remark		
BNDRY	Boundary	FROPA	Frontal Passage	LWR	Lower	RMNG	Remaining	V	Variable
BR	Mist	FROSFC	Frontal Surface	LWRG	Lowering	RPDLY	Rapidly	VC	Vicinity
BRF	Brief	FRQ	Frequent	LYR	Layer	RPT	Repeat	VCNTY	Vicinity
BRFLY	Briefly	FRZG	Freezing			RSG	Rising	VLNT	Violent
BRKS	Breaks	FT	Feet	MAN	Manitoba	RTRN	Return	VLY	Valley
BTN	Between	FTHLS	Foothills	MB	Millibars	RUF	Rough	VR	Veer
		FU	Smoke	MCKNZ	Mackenzie	RVR	River	VRBL	Variable
CASCDS	Cascades	FZ	Freezing	MDFYD	Modified			VRISL	Vancouver Island
CAT	Clear Air Turbulence			MDT	Moderate	S	South	VS	Visual
		GASTR	Strait of Georgia	MI	Mile	SASK	Saskatchewan	VSBY	Visibility
CAVOK	Ceiling and Visibility OK	GGNBA	Georgian Bay	MID	Middle	SC	Stratocumulus		
		GLFALSK	Gulf of Alaska	MOVG	Moving	SCT	Scattered	W	West
CB	Cumulonimbus	GLFST-LAWR	Gulf of St. Lawrence	MPH	Miles per hour	SCTR	Sector	WDLY	Widely
CC	Cirrocumulus			MRNG	Morning	SE	Southeast	WK	weak
CF	Cumlus Fractus			MRTM	Maritime	SELY	Southeasterly	WLY	Westerly
CIG	Ceiling	GND	Ground	MRTMS	Maritimes	SFC	Surface	WND	Wind
CLD	Cloud	GNLY	Generally	MST	Moist	SH	Shower	WNDSHFT	Windshift
CLR	Clear	GRAD	Gradient	MSTR	Moisture	SHFT	Shift	WRM	Warm
CLRG	Clearing	GRDLY	Gradually	MTNS	Mountains	SHFTG	Shifting	WRN	Western
CMA	Continental-Maritime Arctic	GRBNKS	Grand Banks	MVFR	Marginal VFR	SHLW	Shallow	WS	Wind Shear
		GRTLKS	Great Lakes	MXD	Mixed	SHWR	Shower	WV	Wave
CMA	Comma	GVG	Giving	MXG	Mixing	SKC	Sky Clear	WX	Weather
CMPLX	Complex					SLKRS	Selkirks		
CNTR	Centre	HGT	Height	N	North	SLO	Slow	XCP	Except
CNTRD	Centred	HI	High	NE	Northeast	SLOLY	Slowly	XT	Extend
CONDS	Conditions	HIER	Higher	NELY	Northeasterly	SLPG	Sloping	XTGD	Extending
CONTRA	Condensation Trails	HLSTO	Hailstones	NFLD	Newfoundland	SLY	Southerly	XTRM	Extreme
		HLTP	Hilltop	NGT	Night	SMK	Smoke	XTSV	Extensive
CONTG	Continuing	HND	Hundred	NLY	Northerly	SML	Small		
CONTUS	Continuous	HR	Hour	NML	Normal	SN	Snow	YDA	Yesterday
CONTVD	Continental Divide	HNSBA	Hudson Bay	NMRS	Numerous	SNRS	Sunrise	YKN	Yukon
		HVY	Heavy	NR	Near	SNST	Sunset		
CONTG	Continuing			NRLY	Nearly	SNW	Snow	ZULU (Z)	Co-ordinated universal time
COR	Correction	ICG	Icing	NW	Northwest	SPECI	Special		
CST	Coast	ICGIC	Icing in Cloud	NWLY	Northwesterly	SPRDG	Spreading		
CU	Cumulus	ICGIP	Icing in Precip.	OBSC	Obscured	SQL	Squall		

Fig. 6-22: Common GFA abbreviations.

FC = Tornado/Water-
      spout/Funnel cloud
TS = Thunderstorm
DZ = Drizzle
FG = Fog (< 5/8 SM)
BR = Mist (> 5/8 SM)
GS = Small Hail
FU = Smoke
SS = Sandstorm
VA = Volcanic Ash
PO = Dust/Sand whirls

RA = Rain
SG = Snow Grains
PL = Ice Pellets
IC = Ice crystals
SA = Sand
SN = Snow
HZ = Haze
GR = Hail
DU = Dust
SQ = Squall
DS = Duststorm

The code "VC" refers to elements within 5 - 10 NM from the centre of the runway complex.

h) Sky condition. Sky cover amounts are cumulative, so layer amounts include the sum of any layers below. Sky condition is given in eighths or "oktas". SKC = sky clear, no cloud present. FEW = >0 - 2 oktas. SCT = scattered, 3 - 4 octas. BKN = broken, 5 - <8 octas. OVC = sky overcast, 8 octas. A CLOUD CEILING exists at the height of the first layer for which a coverage symbol of BKN or OVC is present. VV = vertical visibility, in hundreds of feet. The presence of a vertical visibility constitutes an OBSCURED CEIL-ING. Cloud height is forecast in three digits in hundreds of feet. CB is added where needed.

i) Change groups. Used when change expected. The following can be used: FM0100Z indicates a change from 0100Z. BECMG 0608 indicates a change between 0600Z and 0800Z. TEMPO 1801 indicates fluctuations between 1800Z and 0100Z. PROB30 2022 indicates a 30% probability of an event between 2000Z and 2200Z.

EXAMPLE:

CYYZ TAF 200438Z 200505 05010KT 5SM BR SCT008 BKN100 TEMPO 0510 2SM BR OVC005 FM1000Z 06010G15KT 3/4SM -RA BR VV003 TEMPO 1420 1/2SM RA FG VV001 FM2200Z 01015G25KT 2SM -RABR OVC008

QUESTIONS BASED ON ABOVE TAF:
1. What date and time was the TAF issued?
2. What is the valid period of the TAF?
3. The lowest visibility is forecast for what time?
4. What is the lowest forecast ceiling for 0700Z?
5. What is the forecast surface wind at 2300Z?
6. What is the forecast visibility at 1000Z?
7. What is the forecast weather at 1000Z?
8. What is the forecast vertical visibility at 1000Z?
9. What time is the forecast lowest ceiling?
10. After what time are the strongest winds forecast?

ANSWERS:
1. 20th day of the month, at 0438Z.
2. From 0500Z on the 20th to 0500Z on the 21st.
3. From 1400Z to 2000Z, the visibility is forecast to fluctuate as low as 1/2 SM in rain and fog.
4. 500 feet AGL overcast.
5. 010° True at 15 kts, gusting to 25 kts.
6. 3/4 SM.
7. Light rain and mist.
8. 300 feet AGL.
9. 100 feet AGL from 1400Z to 2000Z.
10. After 2200Z.

Weather Station Codes

Although the naming of Canadian airports and weather stations can seem confusing, here is a brief explanation. Originally, in the 1930's, Canada used two letters for

identification of a weather reporting station. Additionally, preceding the 2-letter code, was placed a Y (meaning "yes") where the reporting station was co-located with an airport, a W (meaning "without") where the reporting station was not co-located with an airport, and a U where the reporting station was co-located with an NDB. An X was used if the last 2 letters of the code had already been taken by another Canadian ident, and a Z was used if the locator could be confused with a US three letter ident. Vancouver International Airport, for example, was named YVR since there is a weather reporting station at the airport. The ICAO names are in a 4 letter format starting with a C for Canadian airports and a K for US airports. For example: the ICAO identifier for Vancouver International Airport is CYVR, and for Las Vegas, USA the ICAO ident is KLAV. See the CFS for the codes of other airports not having a weather station. Fig. 6-23 provides a summary of Canadian airports having weather reporting stations.

6.20 Upper Level Winds and Temperatures Forecasts (FD)

The FD forecasts wind direction and speed for various altitudes above a selected airport. Altitudes are referenced ASL so at high elevation airports (i.e. Calgary AB) you will not see figures for 3,000 feet. FD's are prepared twice daily. Remember the following:

a) A minus sign ("-") = temperatures below 0°C.
b) Wind direction is in °T, wind speed is in kts.
c) 9900 means "light and variable" winds.
d) "Low" winds are 18,000 and below, "high" winds are above 18,000 feet.
e) Wind speeds from 100 to 199 kts have 50 added to the code. For figures from 51 - 86, subtract 50 to get 010° - 360°. EXAMPLE: 791159 = 290° T at 111 kts, temperature -59° C. Wind speeds over 200 kts have the code 99 and 50 added to the direction. EXAMPLE: 859950 = 350° T at 199 kts or more, temperature -50° C.

EXAMPLE FD:

FDCN1 CWAO 221520

FCST BASED ON 221200 DATA VALID 221800 FOR USE 15-21.

	3000	6000	9000	12000
YVR	2710	2813+03	2921+00	3028-03
YYF	2713	2715+02	2921+00	2927-05
YYE	9900	2708-02	2817-07	2826-12

	18000	39000
YVR	3042-14	851638
YYF	3038-14	851137
YYE	2943-23	820742

QUESTIONS BASED ON ABOVE FD:
1. What is the date and time of issue?
2. What period of time is the FD intended for use?
3. What will the winds be at YVR at 3,000 ft?
4. What will temperature be at YYF at 6,000 ft?
5. What will the wind direction be at 4,500 ft at YVR?
6. What will the wind speed be at 4,500 ft at YYF?
7. What will the temperature be at 4,500 ft at YYF?
8. What are the winds above YVR at 39,000 feet?
9. In a climb from 6,000 feet to 9000 feet above YVR,

Code	Location	Code	Location	Code	Location
XBE	Bearskin Lake, ON	YEK	Arviat, NWT	YKO	Akulivik, QB
XGR	Kangiqsualujjuaq, QB	YEL	Elliot Lake, ON	YKQ	Waskaganish, QB
XKS	Kasabonika, ON	YEM	Manitowaning, ON	YKT	Klemtu, BC
XLB	Lac Brochet, MB	YEN	Estevan, SA	YKU	Chisasibi, QB
XPK	Pukatawagan, MB	YET	Edson, AB	YKX	Kirkland Lake, ON
XQU	Qualicum, BC	YER	Fort Severn, ON	YKY	Kindersley, SA
XSI	South Indian Lake, MB	YEU	Eureka, NWT	YKZ	Toronto - Buttonville, ON
XTL	Tadoule Lake, MB	YEY	Amos, QB	YLA	Aupaluk, QB
YAA	Anahim Lake, BC	YEV	Inuvik, NWT	YLB	Lac La Biche, AB
YAB	Arctic Bay, NWT	YFA	Fort Albany, ON	YLC	Lake Harbour, NWT
YAC	Cat Lake, ON	YFB	Iqaluit, NWT	YLD	Chapleau, ON
YAG	Fort Frances, ON	YFC	Fredericton, NB	YLE	Lac la Martre, NWT
YAH	La Grande-4, QB	YFD	Brantford, ON	YLH	Lansdowne House, ON
YAJ	Lyall Harbour, BC	YFE	Forrestville, QB	YLJ	Meadow Lake, SA
YAM	Sault Ste Marie, ON	YFH	Fort Hope, ON	YLL	Lloydminster, AB
YAQ	Kasabonika, QB	YFJ	La Macaza, QB	YLO	Shilo, MB
YAR	Attawapiskat, ON	YFO	Flin Flon, MB	YLP	Mingan, QB
YAV	Winnipeg, St. Andrews MB	YFR	Ft. Resolution, NWT	YLQ	La Tuque, QB
YAW	Halifax - Shearwater, NS	YFS	Ft. Simpson, NWT	YLR	Leaf Rapids, MB
YAX	Lac du Bonnet, MB	YFT	Makkovik, NF	YLT	Alert, NWT
YAY	St. Anthony, NF	YFX	Fox Harbour, NF	YLU	Kangiqsualujjuaq, QB
YAZ	Tofino, BC	YGB	Gilles Bay, BC	YLW	Kelowna, BC
YBA	Banff, AB	YGD	Goderich, ON	YMA	Mayo, YT
YBB	Pelly Bay, NWT	YGE	Golden, BC	YMD	Mould Bay, NWT
YBC	Baie Comeau, QB	YGG	Ganges Harbor, BC	YME	Metane, QB
YBD	Bella Coola, BC	YGH	Ft. Good Hope, NWT	YMG	Manitouwadge, ON
YBE	Uranium City, SA	YGK	Kingston, ON	YMH	Mary's Harbour, NF
YBG	Bagotville, QB	YGL	La Grande, QB	YMJ	Moose Jaw, SA
YBI	Black Tickle, NF	YGM	Gimli, MB	YML	Charlevoix, QB
YBK	Baker Lake, NWT	YGO	Gods Narrows, MB	YMM	Ft. McMurray, AB
YBL	Campbell River, BC	YGP	Gaspe, QC	YMN	Makkovik, NF
YBN	Borden, ON	YGQ	Geraldton, ON	YMO	Moosonee, ON
YBP	Brooks, AB	YGR	Iles de la Madeleine, QB	YMP	Port McNeil, BC
YBQ	Tadoule Lake, MB	YGS	Germansen Landing, BC	YMQ	Montreal, QB
YBR	Brandon, MB	YGT	Igloolik, NWT	YMT	Chibougamau, QB
YBT	Brochet, MB	YGV	Havre St. Pierre, QB	YMU	Umiujaq, QB
YBU	Nipawin, SA	YGW	Kuujjuarapik, QC	YMW	Maniwaki, QB
YBV	Berens River, MB	YGX	Gillam, MB	YMX	Montreal - Mirabel, QB
YBW	Calgary - Springbank, AB	YGZ	Grise Fiord, NWT	YMY	Ear Falls, ON
YBX	Blanc Sablon, QB	YHA	Quaqtaq, QB	YNA	Natashquan, QB
YCA	Cartwright, NF	YHB	Hudson Bay, SA	YNC	Wemindji, QB
YCB	Cambridge Bay, NWT	YHD	Dryden, ON	YND	Gatineau - Hull, QB
YCC	Cornwal, ON	YHE	Hope, BC	YNE	Norway House, MB
YCD	Nanaimo, BC	YHF	Hearst, ON	YNH	Hudson's Hope, BC
YCE	Centralia-Huron, ON	YHG	Charlottetown, PEI	YNJ	Langley, BC
YCG	Castlegar, BC	YHH	Nemiscau, QB	YNL	Points North Landing, SA
YCH	Chatham, NB	YHI	Holman Island, NWT	YNM	Matagami, QB
YCK	Colville Lake, NWT	YHK	Gioa Haven, NWT	YNO	North Spirit Lake, ON
YCL	Charlo, NB	YHM	Hamilton, ON	YNS	Nemiscau, QB
YCN	Cochrane, ON	YHN	Hornepayne, ON	YOC	Old Crow, YT
YCO	Coppermine, NWT	YHO	Hopedale, NF	YOD	Cold Lake, AB
YCP	Blue River, BC	YHP	Poplar Hill, ON	YOG	Ogoki, ON
YCQ	Chetwind, BC	YHR	Chevery, QB	YOH	Oxford House, MB
YCR	Cross Lake, MB	YHS	Sechelt, BC	YOJ	High Level, AB
YCS	Chesterfield Inlet, NWT	YHU	Montreal - St. Hubert, QB	YOO	Oshawa, ON
YCT	Coronation, AB	YHY	Hay River, NWT	YOP	Rainbow Lake, AB
YCW	Chilliwack, BC	YHZ	Halifax, NS	YOS	Owen Sound, ON
YCX	Gagetown, NB	YIB	Atikokan, ON	YOW	Ottawa, ON
YCY	Clyde River, NWT	YID	Digby, NS	YOY	Valcartier - Lecomte, QB
YCZ	Fairmont Hot Springs, BC	YIF	St-Augustin, QB	YPA	Prince Albert, SA
YDA	Dawson City, YT	YIK	Ivujivik, QC	YPB	Port Alberni, BC
YDB	Burwash, YT	YIO	Pond Inlet, NWT	YPC	Paulatuk, NWT
YDC	Princeton, BC	YIV	Island Lake MB	YPD	Port Hawkesbury, NS
YDE	Paradise River, NF	YJA	Jasper, AB	YPE	Peace River, AB
YDF	Deer Lake, NF	YJF	Fort Liard, NWT	YPG	Portage La Prairie - Southport, MB
YDG	Chute des Passes, QB	YJM	Fort St. James, BC	YPH	Inukjuak, QB
YDI	Davis Inlet, NF	YJP	Fort Providence, NWT	YPI	Port Simpson, BC
YDL	Dease Lake, BC	YJQ	Bella Bella, QB	YPJ	Aupaluk, QB
YDM	Ross River, YT	YJT	Stephenville, NF	YPK	Pitt Meadows, BC
YDN	Dauphin, MB	YKA	Kamloops, BC	YPL	Pickle Lake, ON
YDO	Dolbeau-St-Methode, QB	YKC	Collins Bay, SA	YPM	Pikangikum, ON
YDP	Nain, NF	YKD	Aklavik, NWT	YPN	Port Menier, QB
YDQ	Dawson Creek, BC	YKF	Kitchener - Waterloo-Guelph, ON	YPO	Peawanuck, ON
YDS	Desolation Sound, BC	YKG	Kangiqsujuaq, QC	YPQ	Peterborough, ON
YEA	Empress, AB	YKJ	Key Lake, SA		
YED	Edmonton Nameo, AB	YKK	Kitkatla, BC		
YEE	Midland-Huronia, ON	YKL	Schefferville, QB		
YEG	Edmonton Int'l, AB				

Code	Location	Code	Location	Code	Location
YPR	Prince Rupert, BC	YTL	Big Trout Lake, ON	YZE	Gore Bay, ON
YPS	Pemberton, BC	YTQ	Tasiujuaq, QB	YZF	Yellowknife, NWT
YPT	Pelee Island, ON	YTR	Trenton, ON	YZG	Salluit, QB
YPU	Puntzi Mountain, BC	YTS	Timmins, ON	YZH	Slave Lake, AB
YPW	Powell River, BC	YTZ	Toronto - Toronto Island, ON	YZP	Sandspit, BC
YPX	Povungnituk, QB	YUB	Tuktoyaktuk, NWT	YZR	Sarnia, ON
YPY	Ft. Chipewyan, AB	YUD	Umiujaq, QB	YZS	Coral Harbour, NWT
YPZ	Burns Lake, BC	YUF	Pelly Bay, NWT	YZT	Port Hardy, BC
YQA	Muskoka, ON	YUL	Montreal - Dorval, QC	YZU	Whitecourt, AB
YQB	Quebec, QB	YUS	Shepard Bay, NWT	YZV	Sept-Iles, QB
YQC	Quaqtaq, QB	YUT	Repulse Bay, NWT	YZW	Teslin, YT
YQD	The Pass, MB	YUW	Dewar Lakes, NWT	YZX	Greenwood, NS
YQF	Red Deer, AB	YUX	Hall Beach, NWT	YZY	Mackenzie, BC
YQG	Windsor, ON	YUY	Rouyn - Noranda, QB	ZAC	York Landing, MB
YQH	Watson Lake, YT	YVB	Bonaventure, QB	ZAM	Salmon Arm, BC
YQI	Yarmouth, NS	YVC	La Ronge, SA	ZBA	Burlington Airpark, ON
YQK	Kenora, ON	YVG	Vermilion, AB	ZBB	Boundary Bay, BC
YQL	Lethbridge, AB	YVK	Vernon, BC	ZBD	Ilford, MB
YQM	Moncton, NB	YVM	Broughton Island, NWT	ZBF	Bathurst, NB
YQN	Nakina, ON	YVN	Cape Dyer, NWT	ZEE	Kelsey, MB
YQQ	Comox, BC	YVO	Val D'Or, QB	ZEG	Edmonton Centre, AB
YQR	Regina, SA	YVP	Kuujjuaq, QB	ZEL	Bella Bella, BC
YQS	St. Thomas, ON	YVQ	Norman Wells, NWT	ZEM	Eastmain River, QB
YQT	Thunder Bay, ON	YVR	Vancouver, BC	ZFA	Faro, YT
YQU	Grande Prairie, AB	YVT	Buffalo Narrows, SA	ZFD	Fond du Lac, SA
YQV	Yorkton, SA	YVV	Wiarton, ON	ZFG	Pukatawagan, MB
YQW	North Battleford, SA	YVZ	Deer Lake, ON	ZFM	Fort McPherson, NWT
YQX	Gander, NF	YWA	Petawawa, ON	ZFN	Tulita, NWT
YQY	Sydney, NS	YWB	Kangiqsujuaq, QB	ZGF	Grand Forks, BC
YQZ	Quesnel, BC	YWG	Winnipeg, MB	ZGI	Gods River, MB
YRA	Rae Lakes, NWT	YWH	Victoria Harbour, BC	ZGR	Little Grand Rapids, MB
YRB	Resolute, NWT	YWJ	Deline, NWT	ZGS	Gethsemani, QB
YRC	Chicoutimi, QB	YWK	Wabush, NF	ZHP	High Prairie, AB
YRD	Dean River, BC	YWL	Williams Lake, BC	ZJG	Jenpeg, MB
YRF	Cartwright, NF	YWM	Athabasca, AB	ZJN	Swan River, MB
YRG	Rigolet, NF	YWO	Lupin, NWT	ZKE	Kaschechewan, ON
YRI	Riviere-du-Loup, QB	YWP	Webequie, ON	ZKG	Kegaska, QB
YRJ	Roberval, QC	YWV	Wainwright, AB	ZLQ	Thicket Portage, MB
YRL	Red Lake, ON	YWY	Wrigley, NWT	ZLT	La Tabatiere, QB
YRM	Rocky Mountain House, AB	YXC	Cranbrook, BC	ZMD	Muskrat Dam, ON
YRN	Rivers Inlet, BC	YXD	Edmonton - Municipal, AB	ZML	108 Mile Airport, BC
YRO	Ottawa - Rockcliffe, ON	YXE	Saskatoon, SA	ZMN	Pikwitonei, MB
YRP	Ottawa - Carp, ON	YXH	Medicine Hat, AB	ZMT	Masset, BC
YRQ	Trois-Rivieres, AB	YXJ	Ft. St. John, BC	ZNA	Nanaimo - Harbour, BC
YRR	Stuart Island, BC	YXK	Rimouski, QB	ZNG	Poplar River, MB
YRS	Red Sucker Lake, MB	YXL	Sioux Lookout, ON	ZNL	Nelson, BC
YRT	Rankin Inlet, NWT	YXN	Whale Cove, NWT	ZPB	Sachigo Lake, ON
YRV	Revelstoke, BC	YXP	Pangnirtung, NWT	ZPC	Pincher Creek, AB
YSA	Sable Island, NS	YXQ	Beaver Creek, YT	ZPO	Pinehouse Lake, SA
YSB	Sudbury, ON	YXR	Earlton, ON	ZQM	Moncton Centre, NB
YSC	Sherbrooke, QB	YXS	Prince George, BC	ZQS	Queen Charlotte Is, BC
YSD	Suffield, AB	YXT	Terrace, BC	ZQX	Gander Centre, NF
YSE	Squamish, BC	YXU	London, ON	ZRJ	Round Lake, ON
YSF	Stony Rapids, SA	YXX	Abbotsford, BC	ZSJ	Sandly Lake, ON
YSG	Snowdrift, NWT	YXY	Whitehorse, YT	ZSN	South Indian Lake, MB
YSH	Smiths Falls, ON	YXZ	Wawa, ON	ZST	Stewart, BC
YSJ	Saint John, NB	YYB	North Bay, ON	ZSW	Prince Rupert - Cove, BC
YSK	Sanikilauq, NWT	YYC	Calgary - Int'l, AB	ZTA	Bloodvein River, MB
YSL	St. Leonard, NB	YYD	Smithers, BC	ZTB	Tete a la Baleine, QB
YSM	Ft. Smith, NWT	YYE	Ft. Nelson, BC	ZTM	Shamattawa, MB
YSN	St. Catharines, ON	YYF	Penticton, BC	ZUC	Ignace, ON
YSO	Postville, NF	YYG	Charlottetown, PEI	ZUE	Cape Parry, NWT
YSP	Marathon, ON	YYH	Taloyoak, NWT	ZUL	Montreal Centre, QB
YSQ	Atlin, BC	YYJ	Victoria, BC	ZUM	Churchill Falls, NF
YSR	Nanisivik, NWT	YYL	Lynn Lake, MB	ZVL	Edmonton - Villeneuve, AB
YST	St. Therese Point, MB	YYM	Cowley, AB	ZVR	Vancouver Centre, BC
YSU	Summerside, PEI	YYN	Swift Current, SA	ZWG	Winnipeg Centre, MB
YSW	Sparwood, BC	YYO	Wynyard, SA	ZWH	Lac Brochet, MB
YSY	Sachs Harbour, NWT	YYQ	Churchill, MB	ZWL	Wollaston Lake, SA
YSZ	Ste-Anne-des-Monts, QB	YYR	Goose Bay, NF	ZYZ	Toronto Centre, ON
YTA	Pembroke, ON	YYT	St. Johns, NF		
YTB	Koala, NWT	YYU	Kapuskasing, QB		
YTE	Cape Dorset, NWT	YYW	Armstrong, ON	FVP	St. Pierre, France
YTF	Alma, QB	YYY	Mont Joli, QB	FVM	Miquelon, France
YTG	Sullivan Bay, BC	YYZ	Toronto - Pearson Int'l, ON		
YTH	Thompson, MB	YZD	Toronto - Downsview, ON		
YTJ	Terrace Bay, ON				

Fig. 6-23: Weather Station/Airport Codes.

will the wind direction veer or back?

10. In descent from 12,000 ft to 9,000 ft above YVR, will the wind direction veer or back? Will the wind speed increase or decrease?

ANSWERS:
1. 22nd day of month at 1520Z.
2. 1500Z to 2100Z (6 hours).
3. 270°T at 10 knots.
4. 2° C.
5. 275° T.
6. 14 kts.
7. +4 ° C.
8. 350° T at 116 kts, temperature -38° C.
9. Veer. 280° T at 6,000 feet to 290° T at 9,000 feet = clockwise change.
10. Back (from 300° T to 290° T). Wind speed will decrease, from 28 knots to 21 knots.

6.21 Aviation Routine Weather Reports (METAR)

METARs are an international ICAO format which is now being used by Canada and the USA in the place of the former Hourly Weather Reports (SA).

Each country is permitted to make minor changes to the METAR format, so the coding is not exactly the same worldwide. The order and information presented in the METAR is as follows:

a) Message type. Indicated by METAR (routine observation report) or SPECI (special METAR observation) and official observation time.
b) Station identifier. Indicated by 4 letter ICAO identifier and time of observation in UTC (Z).
c) Wind. The first three digits are wind direction to the nearest 10° true. VRB is for variable direction less than 3 knots. Calm is 00000KT. The next two digits indicate wind speed, then gusts with a two digit maximum speed. Speeds equal to or greater than 100 kts use 3 digits.
d) Prevailing visibility. Reported in SM and fractions. Visibility of 15+ is encoded as 15 SM. Lower visibilities which are half or less of prevailing visibility are reported as supplementary information.
e) Runway Visual Range (RVR). The 10 minute average RVR is reported when prevailing visibility is 1 SM or less, and/or the RVR is less than 6000 ft. D indicates a downward trend, U an upward trend, and N for no change.
f) Present weather. Comprised of weather phenomena (precipitation, obscuration or other) preceded by one or two qualifiers (intensity or proximity to the station and descriptor). "-" refers to light intensity, no sign is moderate, + is heavy. Descriptors which apply to forecast precipitation and/or obstructions to visibility: SH = showers, DR = drifting, FZ = freezing, MI = shallow, BL = blowing, BC = patches.
g) Weather codes:

FC = Tornado/Water-spout/Funnel cloud	RA = Rain
TS = Thunderstorm	SG = Snow Grains
DZ = Drizzle	PL = Ice Pellets
FG = Fog (< 5/8 SM)	IC = Ice crystals
BR = Mist (> 5/8 SM)	SA = Sand
GS = Small Hail	SN = Snow
FU = Smoke	HZ = Haze
SS = Sandstorm	GR = Hail
VA = Volcanic Ash	DU = Dust
	SQ = Squall

PO = Dust/Sand whirls   DS = Duststorm

h) Sky condition. Amounts of sky cover are cumulative. Therefore layer amounts include the sum of any layers below. SKC = sky clear, no cloud present. FEW = >0 - 2/8 summation. SCT = 3/8 - 4/8. BKN = 5/8 - < 8/8. OVC = sky overcast, 8/8. CLR = clear below 10,000 as interpretated by an autostation. W = sky obscured. VV = vertical visibility and is forecast in hundreds of feet. Partially obscured is forecast as SKC or is included in the first layer. Cloud height is forecast in 3 digits in hundreds of feet. CB is added where needed.
i) Temperature and dewpoint. Observed values within 0.5° are rounded up to the next warmer degree. M means a negative temperature.
j) Altimeter setting. A indicates inches of mercury, Q indicates hectopascals.
k) Recent weather. Any recent weather in the period since the last routine report but not at the time of observation.
l) Supplementary information such as windshear (up to 1,600 ft, provided by an aircraft on takeoff or approach path), sector visibilities etc. SLP = Sea Level Pressure in Hpa or mb: add 9 or 10 to first digit and decimal point before last, ie 996 = 999.6 mb, 017 = 1001.7 mb.

EXAMPLE METARS:

CYVR METAR 0500Z 01002KT 30SM CT250 03/01 A3010 RMK SLP200
CYWG METAR 0500Z 36009KT 15SM -SN SCT015 OVC044 M11/M12 A3024 RMK SC2 SC2 SLP240
CYYZ METAR 0500Z 09005KT 5SM BR OVC006 04/ 03 A2966 RMK SC5 SLP150

QUESTIONS BASED ON ABOVE METARS:
1. What time was the METAR at CYVR issued?
2. What are the reported surface winds at CYVR?
3. What is the reported visibility at CYWG at 0500Z?
4. What weather is reported at CYWG?
5. What is the reported ceiling at CYWG?
6. What is the temperature and dewpoint at CYYZ at 0500Z?
7. What is the CYWG temperature and dewpoint?
8. What is the reported weather at CYYZ?
9. What is the reported ceiling at CYYZ?
10. What is the CYVR reported altimeter setting?

ANSWERS:
1. 0500Z.
2. 010° T at 2 kts.
3. 15 SM or more.
4. Light snow.
5. Ceiling overcast, 4,400 feet AGL.
6. Temperature +4° C, dewpoint +3° C.
7. Temperature -11° C, dewpoint -12° C.
8. Mist.
9. Ceiling overcast 600 feet AGL
10. 30.10 inches hg

6.22 SIGMET

A SIGMET is a serious weather condition reported as a warning to aircraft in flight. Examples are active thunderstorms, squall lines, heavy hail, serious turbulence or icing, mountain waves, hurricanes, volcanic ash, wind shear etc.

EXAMPLE SIGMET:

WSCN CYYZ 171800
171800 - 172200
SIGMET A5. LN OF TSTMS OBSERVED ON WA
RADAR NORTH BAY - MUSKOKA - LONDON TOPS TO
30 THSD MOVG E 20.

6.23 PIREP and AIREP

PIREP
PIREP means "pilot report". Pilots are encouraged to contact FSS and report significant observations made in flight. UACN10 = PIREP. UACN01 = Urgent PIREP. OV = over, referenced to 3 letter navaid or fix. TM = time. FL = altitude or flight level in hundreds of feet. TP = aircraft type. SK = cloud layers. WX = weather and visibility. TA = temp °C. TB = turbulence. IC = icing. RM = remarks.

EXAMPLE PIREP:

UACN10 OV VQC 045043 1342 FL090/TP BE80/SK
OVC 075/TA - 18/IC LGT RIME BLO - 075 RM WIND
COMP HEAD 010 MH 065 TAS 210

MEANING OF EXAMPLE:
Location - Stirling VORTAC 045° Radial, 43 NM at 1342Z at 9,000 ft, type of aircraft - Beech Queen Air; Sky Condition - overcast, tops 7,500 ft; air temperature -18° C; light rime icing below 7,500 ft; Remarks - headwind component 10 kts, magnetic heading 065, true airspeed 210 kts.

AIREP
An AIREP is a meteorological report made on a routine basis by international air carrier aircraft transitioning Canadian domestic airspace north of 60° N and east of 80° W. As well, aircraft operating in the Gander Oceanic area should use the AIREP format.

6.24 AIRMET

An AIRMET is a "Pilot's Meteorological Advisory" and is a short term weather advisory intended for aircraft in flight, to notify pilots of potentially hazardous weather conditions not described in the current GFA and not requiring a SIGMET. Examples are freezing precipitation, moderate icing, moderate turbulence, surface winds over 20 knots etc.

EXAMPLE AIRMET:

WACN1 CWHX 200700

AIRMET A1 ISSUED AT 0700Z AT CWHZ-AMEND
GFACN34 CWHX 200600 ISSUE

DC9 RPRTD MDT RIME ICG IN ZL ALG THE S CST
OF NB AT 07Z. ZL EXPCD TO CONT UNCHGED TO
14Z.

MEANING OF EXAMPLE:
AIRMET from Maritime Weather Centre, time 0700 on 20th day of month. AIRMET A1 issued at 0700 Z at Maritime Weather Centre which amends the Graphic Area Forecast GFACN34 issued at 0600Z. A DC 9 aircraft reported moderate rime icing in freezing drizzle along the south coast of New Brunswick at 0700Z. The freezing drizzle is expected to continue unchanged until 1400Z.

6.25 Not Allocated

This subject reference is currently not allocated.

6.26 CAVOK

CAVOK means "ceiling and visibility OK" and refers to the following simultaneous conditions:

a) No cloud below 5,000 feet.
b) No cumulonimbus clouds.
c) Visibility at least 6 SM or more.
d) No precipitation, thunderstorms, shallow fog or low drifting snow.

CAVOK is often used in ATIS broadcasts or when indicating generally good VFR weather prevails. CAVOK provides no indication of prevailing winds.
CAVOK is not used in TAFs and METARS.

6.27 Surface Weather Charts

Surface weather charts (also called surface weather maps) are made from reports of surface weather from various reporting stations and are written by means of numbers and symbols. With these symbols, lower level weather conditions such as pressure distribution, air masses, fronts, precipitation and obstructions to vision can be seen. Some upper features such as TROWALS and upper fronts are shown, thereby giving an excellent representation for flight planning purposes. When the surface charts are based on actual data from existing reports, they are referred to as SURFACE ANALYSIS CHARTS. A SURFACE PROGNOSTIC CHART shows future expected pressure patterns and frontal positions for a specific future time.

Prognostic Surface Charts
Prognostic surface charts are issued 48 and 36 hours before valid times 00Z and 12Z. Remember that the weather may be considerably different as flight time will usually not coincide with chart valid time. Review the symbols below used to represent weather on surface charts at Fig. 6-24.

EXERCISE: Using Fig. 6-24, convert the following surface weather chart symbols to plain language.

1. _____

2. _____

3. _____

4. _____

5. _____

SYMBOL IN COLOUR	SYMBOL IN BLACK AND WHITE	FRONTAL PHENOMENON	DESCRIPTION
continuous blue line		cold front	leading edge of advancing portion of cold air mass
continuous red line		warm front	trailing edge of retreating portion of cold air mass
alternately red and blue line		stationary front	edge of cold air mass is neither advancing nor retreating
continuous purple line		occluded front	formed by cold front overtaking warm front (frontal depressions only)
blue ... red		trowal (trough of warm air aloft)	line on ground above which warm air is closest to the ground
broken blue line / broken red line		upper cold front / upper warm front	line along which a frontal surface abruptly steepens, or a front riding over colder air
blue or red		upper cold or upper warm front becoming surface	
blue or red		cold / warm frontogenesis	front forming
/////// blue or red or purple		cold / warm / occluded frontolysis	front dissipating
purple		line squall	

REPRESENTATION	PHENOMENON
continuous green shading	continuous rain
green hatching	intermittent rain
continuous green shading or green hatching with following symbols superimposed	continuous or intermittent
in green	drizzle
in green	snow
in green	ice pellets
in green with dot or comma	freezing rain, drizzle
inverted green triangle dot above in green	rain shower
inverted green triangle star above in green	snow shower
inverted green triangle symbol above in green	hail shower
in green	thunderstorm
continuous yellow shading	fog
continuous brown shading	smoke, dust

Fig. 6-24: Surface chart symbols.

ANSWERS:
1. Freezing drizzle.
2. Mixed intermittent rain and snow.
3. Intermittent rain.
4. Heavy drifting snow.
5. Continuous rain.

Refer to the surface prognostic chart at Fig. 6-26 and complete the questions that follow.

Surface Analysis Charts

Surface analysis charts are considered to represent weather from the surface up to 3,000 feet, and weather visible from the surface at any level.

Since surface analysis charts are made from data at a certain time in the past, the information will already be a few hours old when it first comes out. They are based on observation times of 00Z, 06Z, 12Z and 18Z, and each chart is issued 2 - 3 hours after observations, therefore, the weather may be very different at the time of departure.

The STATION MODEL is used to enter data from the hourly reports onto the surface weather chart. The station model can be used to obtain information such as winds, temperatures, cloud cover, humidity etc.

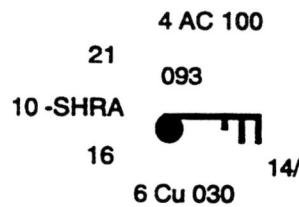

Fig. 6-25: Station model used in surface analysis charts.

The above station model shows the following information:

1. Total sky cover: overcast (by darkened circle).
2. Temperature: 21° C
3. Dewpoint: 16° C.
4. Wind: from 090° T at 25 kts. Shaft of pennant is direction toward station, each barb is 10 kts, each half barb is 5 knots. A second circle around the station means calm.
5. Present visibility and weather is 10 SM in light rain showers.
6. Middle cloud is 4/8 Altocumulus at 10,000 ft.
7. Sea level pressure 1009.3 mb. Use same rules to convert as in METAR.
8. Pressure change in past 3 hours: increased steadily 1.4 mb.
9. Low cloud: 6/8 Cumulus at 3,000 ft.

Refer to the surface analysis chart at Fig. 6-27 and answer the questions that follow.

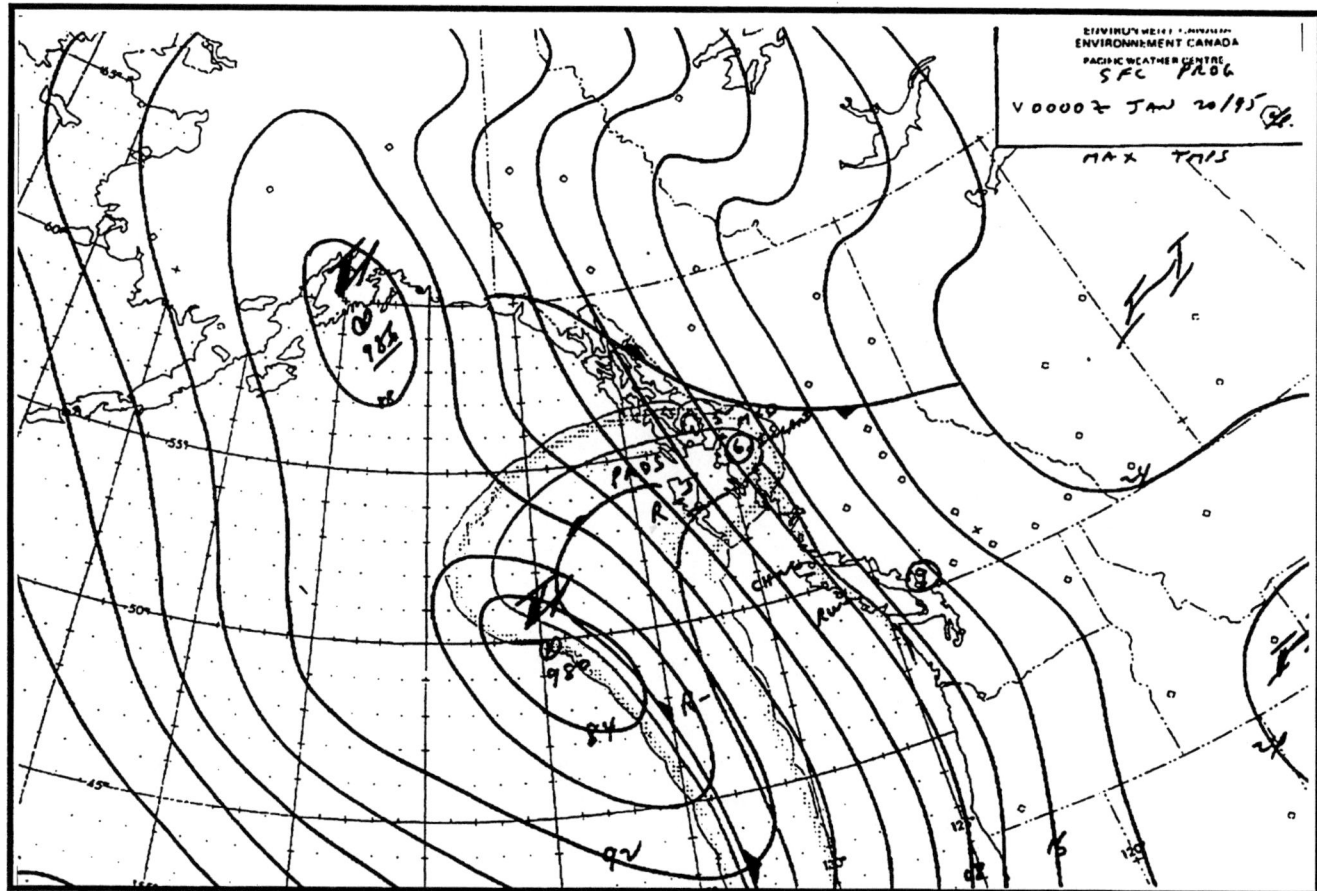

Fig. 6-26: Surface Prognostic Chart.

EXERCISE: Answer the following questions based on the surface prognostic chart above.

1. Is the chart above giving actual weather or predicted weather?
2. What time is the chart effective?
3. Assume the scalloped area off the coast is shaded green. What type of weather is anticipated in the shaded area?
4. What is the barometric pressure at the centre of the low?
5. What kind of frontal system is forecast to lie over mid to northern BC at 00Z?
6. Would you expect greater winds over the BC coast or over Saskatchewan, and why?
7. What is the general 3,000 ft wind direction forecast over southern Alberta at the US border and why?
8. What kind of frontal system lies over the Pacific ocean?

ANSWERS:

1. Predicted weather, because it is named "prognostic".
2. Chart is an estimate of conditions anticipated at 00Z, January 20, 1995.
3. Periods of continuous light rain and light mixed snow inland. Green shading areas indicate continuous precipitation.
4. 998.0 mb. Add 9 or 10, whichever is closer to 1000 mb.
5. Stationary front.
6. Faster winds over BC coast due to closer spacing of isobars.
7. Easterly. Winds circulate parallel to isobars, winds circulate clockwise about High located in middle of Sask.
8. Stationary.

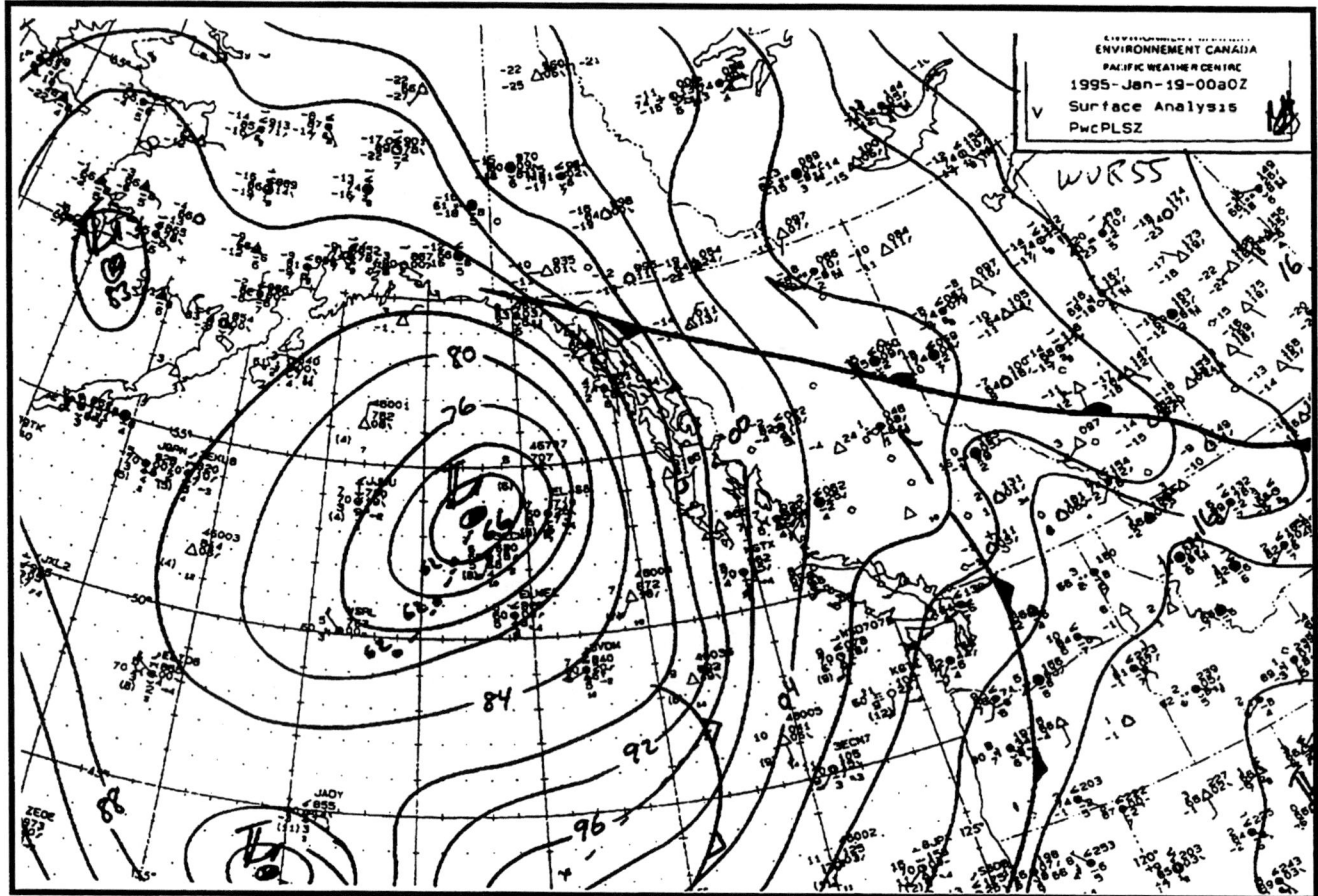

Fig. 6-27: Surface Analysis Chart.

EXERCISE: Based on the above surface analysis chart, answer the following questions.

1. Is the above chart based on forecast weather or actual weather?
2. What is the pressure at the centre of Low in the Pacific at about latitude 54°?
3. With reference to the low in question 2, in the last 24 hours, has it been increasing or decreasing in pressure, and what has its movement been?
4. Wind will be more or less parallel to the isobars shown on this chart at what altitude?
5. What time were the observations made upon which this chart is based?
6. What kind of a frontal systems lie over BC at 00Z?
7. Where will winds be faster, over west coast of BC or northern Alberta and why?
8. What are the approximate winds at 3,000 ft over the north coast of BC?

ANSWERS:

1. Actual weather - because it is called an ANALYSIS chart.
2. 966 mb (add 9 or 10 to 66, whichever is closer to 1,000 mb).
3. Increasing. Note the history of movement of the system is indicated by the 62 - 62 - 62 dots and trail to the lower left of the low, showing the pressure and movement. This cell is increasing in pressure and moving northeast. Each position represents the last 6 hours.
4. 3,000 ft.
5. 00Z on Jan 19, 1995.
6. Stationary front in northern BC, TROWAL in north south line and cold front at southern part at US border.
7. Winds will be faster over north coast of BC because isobars are spaced closer together.
8. Southerly - see isobars, lows circulate counterclockwise.

6.28 Upper Air Charts

Upper weather charts are either PROGNOSTIC (describing future, forecast weather) or ANALYSIS (describing existing conditions at a given time).

Upper charts are often named in reference to the particular pressure level which they are based on. Remember that standard MSL pressure is 1013.2 mb. Therefore, upper charts will be indicated by a pressure lower than this. The altitude of a particular pressure level will vary widely. Some representative heights for pressure levels are:

Pressure	Approximate height
850 mb	5,000 ft ASL
700 mb	10,000 ft ASL
500 mb	18,000 ft ASL
400 mb	24,000 ft ASL
300 mb	30,000 ft ASL
250 mb	34,000 ft ASL
200 mb	40,000 ft ASL
150 mb	45,000 ft ASL
100 mb	53,000 ft ASL

The benefit of an upper air chart is that it provides a 3-D picture of weather above the surface.

The new term for pressure is "hectopascals" or hPa, which are the same thing as Millibars (mb). The millibar terminology is still widely used by pilots and forecasters even though technically speaking is not the current terminology.

CONTOUR LINES on upper air charts represent the pressure level on which the chart is based (not points of equal MSL pressure as on surface charts). Pressure levels will be indicated in decametres, and a decametre is 10 metres (multiply metres by 3.28 to convert to feet). The contour lines can be considered the same basic thing as isobars: wind will blow parallel to the contour lines, and closely spaced contour lines indicate high wind speed. As well, centres of high and low pressure are indicated the same way as with surface charts. Since temperatures are colder as you get closer to the north pole, the pressure levels are lower as well toward the north.

Since an upper air chart is above the effects of surface friction, winds can be considered for practical purposes to blow parallel to the contours.

Contour patterns on an upper chart that are formed by pressure contours can also be described as MACROSCALE WAVES. In other words, the wavy, curved pattern of the pressure contours shows pressure troughs and crests. Poor weather and turbulence will be found at Macroscale troughs, while good weather will generally be found at Macroscale ridges (although CAT may still exist). Macroscale waves having shorter lengths will move faster than longer ones and will lead to surface low frontal weather systems.

Broken or dashed lines on upper analyses charts indicate ISOTHERMS, which show points of equal temperature. By inspecting isotherms carefully, you can determine if your flight will be toward colder or warmer air. Upper Level Analysis Charts

Upper air analysis charts are produced from information coming from weather stations scattered around the continent. This data is used to produce the upper air analysis charts at 00Z and 12Z each day. Be familiar with the following charts:

a) 850 MB ANALYSIS CHART (see Fig. 6-28). At about 5,000 feet. Includes contours at 60 meter intervals (solid lines) starting from the 1500 metre level, with isotherms at 5° C spacing (dashed lines) and low and high heights. Usually the fronts are shown in surface position.

b) 700 MB ANALYSIS CHART (see Fig. 6-29). At about 10,000 feet. Same information as 850 mb chart, but fronts not indicated. Contours spaced at 60 meter intervals.

c) 500 MB ANALYSIS CHART (see Fig. 6-30). Used for flight planning near 18,000 feet. Contours are space at 60 metre intervals.

d) 250 MB ANALYSIS CHART (see Fig. 6-31). Used for flight planning approximately 34,000 feet. On these charts you will see ISOTACHS, which are drawn at 30 kt intervals. For winds from 60 kts to 90 kts the area is shaded, heavier shading is used for 90 - 120 kts, and a clear area is used to indicate 120 kts to 150 kts, and the area is shaded again for 150 kts plus.

Wind speed is also indicated on upper charts by a system of pennants and feathers. The pennant is aligned with the feathers in the direction from which the wind is blowing. Each feather represents 10 kts, a half feather represents 5 kts, and a small darkened triangle represent 50 kts.

EXAMPLE: What winds are indicated by the following symbol on an upper level analysis chart?

ANSWER: Winds are from 270° True at 115 kts.

Complete the exercises following each of the 850 mb, 700 mb, 500 mb and 250 mb charts at Figs 6-28 to 6-31.

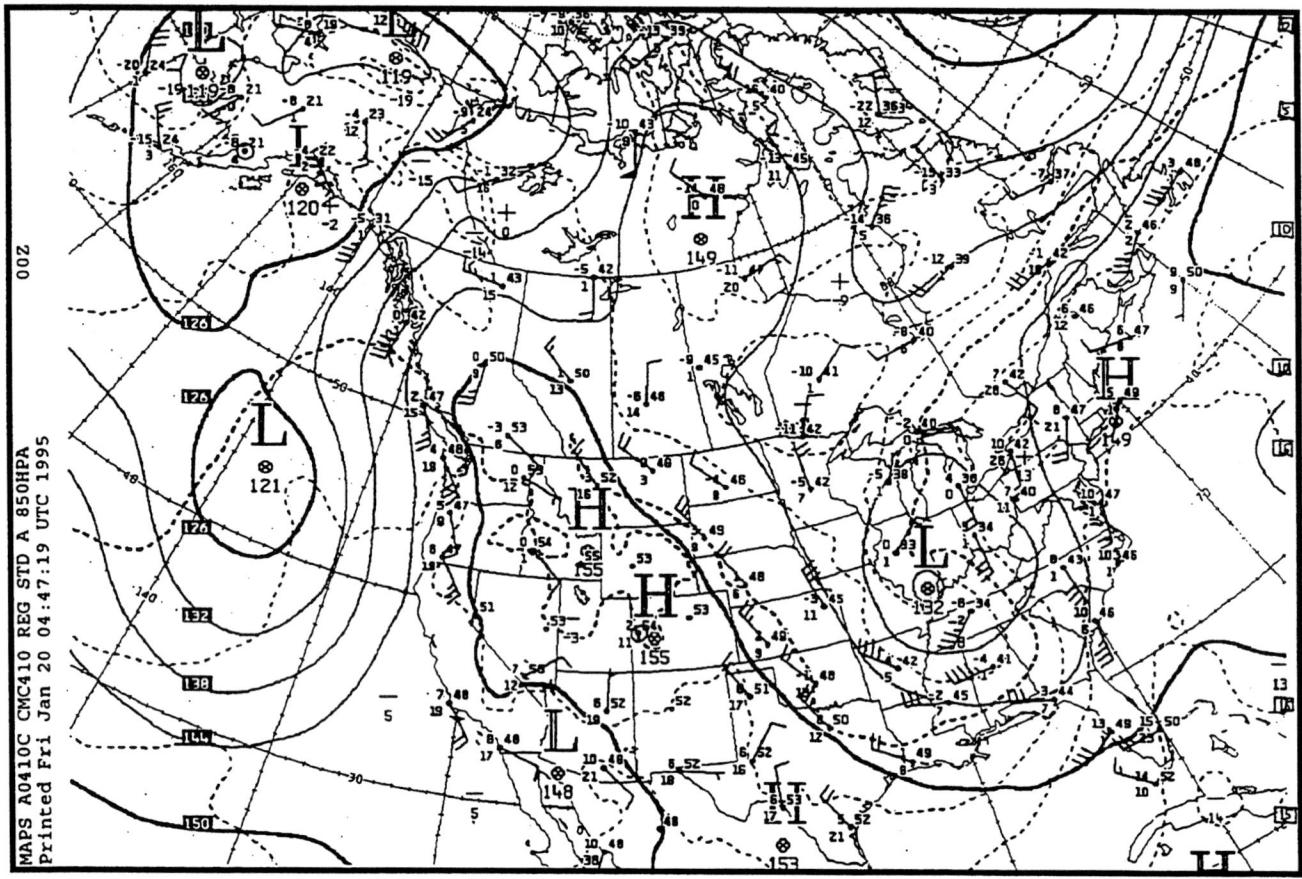

Fig. 6-28: 850 mb Analysis chart.

EXERCISE BASED ON 850 MB ANALYSIS ABOVE:

1. What kind of pressure systems is centered to the south of the great lakes?
2. What does the number 121 indicate beneath the low pressure cell in the Pacific Ocean?
3. Why is the contour line from Florida through Alberta and back to Southern California darker than the other contour lines?
4. What is the height of the 850 mb contour level over Lake Winnipeg?
5. What is the temperature of the 850 mb pressure level over Lake Winnipeg?
6. What is the wind direction and wind speed at the 850 mb pressure level in central Saskatchewan?
7. What is the temperature and dewpoint of the 850 mb pressure level over central Saskatchewan?
8. What is the height of the 850 mb pressure level over central Saskatchewan?

ANSWERS:

1. Low pressure.
2. Height of 850 mb pressure level at centre of low = 1210 metres or 121 decametres.
3. The darkened contour line represents the 1500 metre level of the 850 mb contour and is where the contour lines start on this chart. See the "150" in the box on the western edge of this contour.
4. 1440 metres.
5. -10° C. See isotherms - dashed lines. 0° C is darker than the rest, then other isotherms are marked at 5 degree intervals.
6. Winds are from the north at 10 knots - see vertical pennant with one barb.
7. Temperature -6° C, dewpoint -20° C. See the figures - 6 and 14: -6 is temperature in °C, 14 is the dewpoint depression, or how many degrees colder the dewpoint is than the temperature.
8. 1460 metres. See the 46 figure.

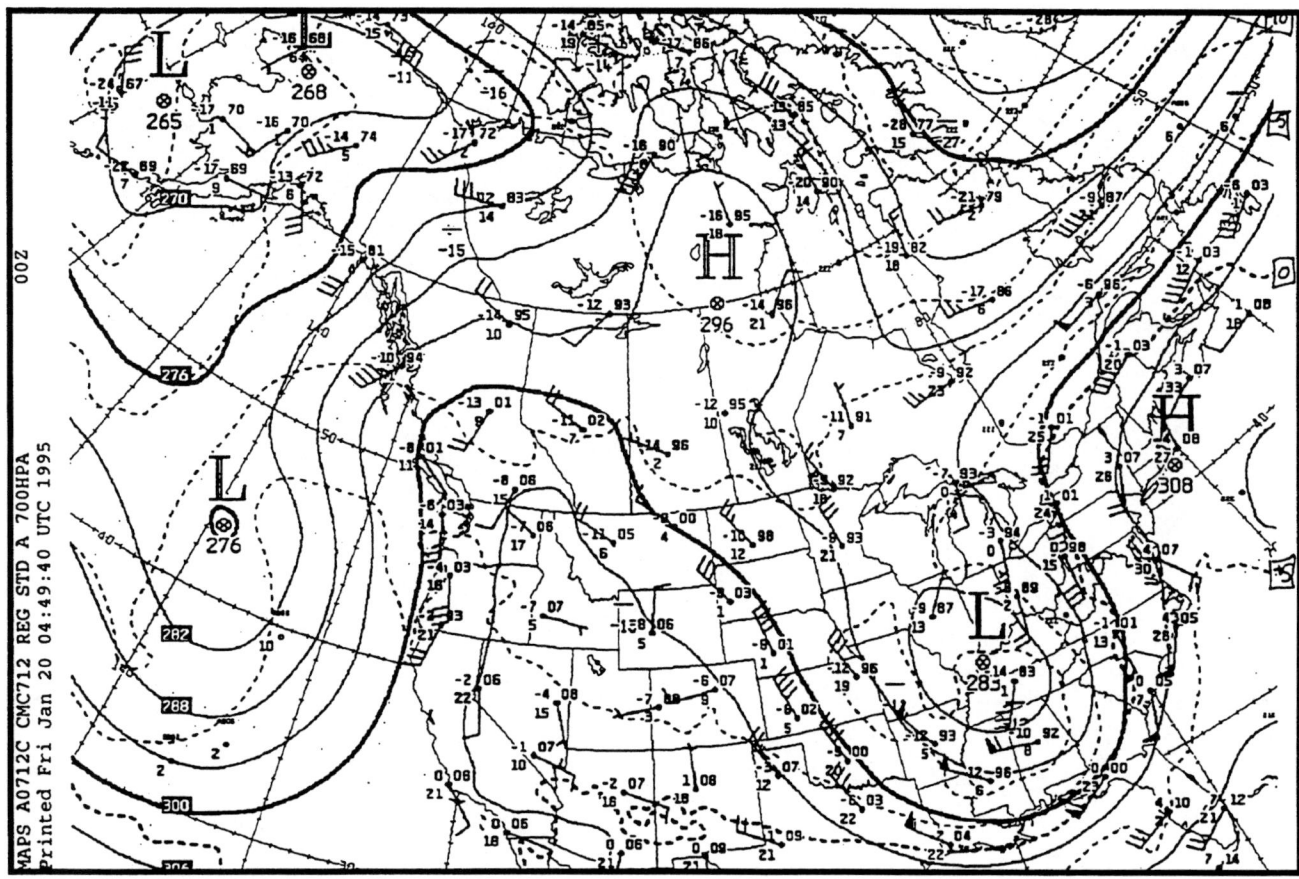

Fig. 6-29: 700 mb Analysis Chart.

EXERCISE BASED ON 700 MB ANALYSIS ABOVE:

1. When is the information presented on this chart based on, and is it based on actual or forecast weather?
2. What is the height of the 700 mb pressure level in feet for central Newfoundland?
3. What is the wind direction and speed at the 700 mb level in eastern Newfoundland?
4. What is the temperature and dewpoint at the 700 mb level for eastern Newfoundland?
5. What is the height of the 700 mb pressure level at eastern Newfoundland?
6. The 700 MB analysis chart shows the winds for what level in the atmosphere in terms of feet ASL?
7. What are the approximate 700 mb pressure level winds over the Gaspe?

ANSWERS:

1. 00Z (see left edge of chart). It is based on actual reported weather.
2. 9,840 ft. 3,000 metres is indicated by 300 symbol, multiply by 3.28 ft per metre.
3. Westerly winds of 40 knots. See pennant aligned from west in degrees true and 4 feathers for speed.
4. Temperature -6° C, dewpoint -7° C. See number -6 (temperature) and 1 (dewpoint depression).
5. 3030 Metres. See 03 just off 300 contour.
6. Approximately 10,000 feet ASL.
7. About 270°T at 45 kts, based on 50 kts from station to north and 40 knots from station to south and interpolating between the two.

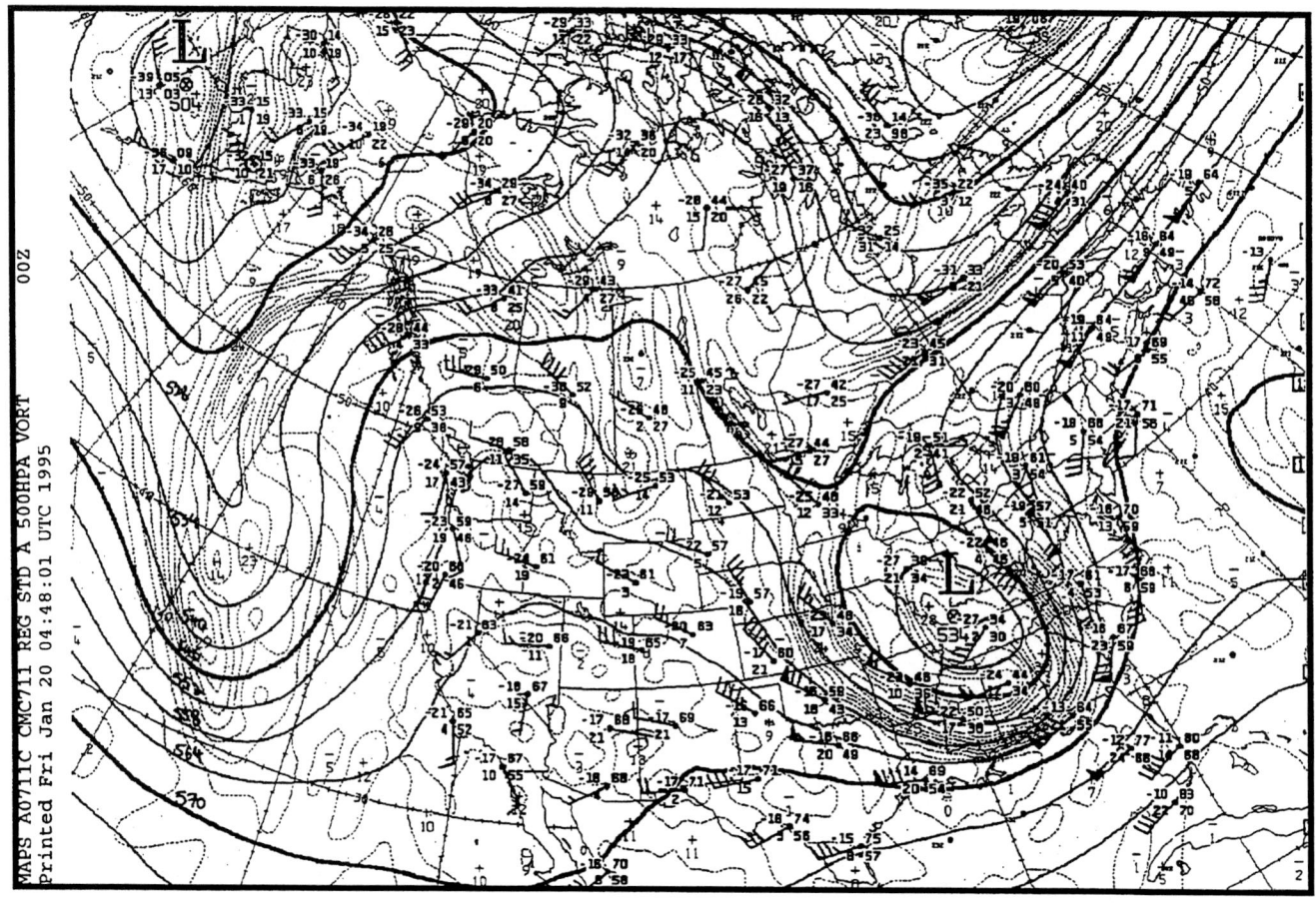

Fig. 6-30: 500 mb Analysis Chart.

EXERCISE BASED ON 500 MB ANALYSIS ABOVE:
1. What approximate ASL height is this weather intended for?
2. What does the abbreviation "HPA" on the edge of the chart refer to?
3. What is the height of the 500 MB pressure level over Winnipeg?
4. Where are the 500 MB level winds faster, over Vancouver Island, or Newfoundland, and by how much faster are they?
5. What is the 500 MB upper temperature over northern Vancouver Island?

ANSWERS:
1. 18,000 ft ASL.
2. Hectopascals. Millibars and hectopascals are equivalent units, therefore this chart is equally the 500 MB or 500 HPA chart, depending on which unit you prefer.
3. 5445 Metres. See 45 on station model.
4. Over Newfoundland, by 25 knots. By referencing wind symbol, winds are about 50 knots over Newfoundland, and 25 kts over Vancouver Island.
5. -26° C.

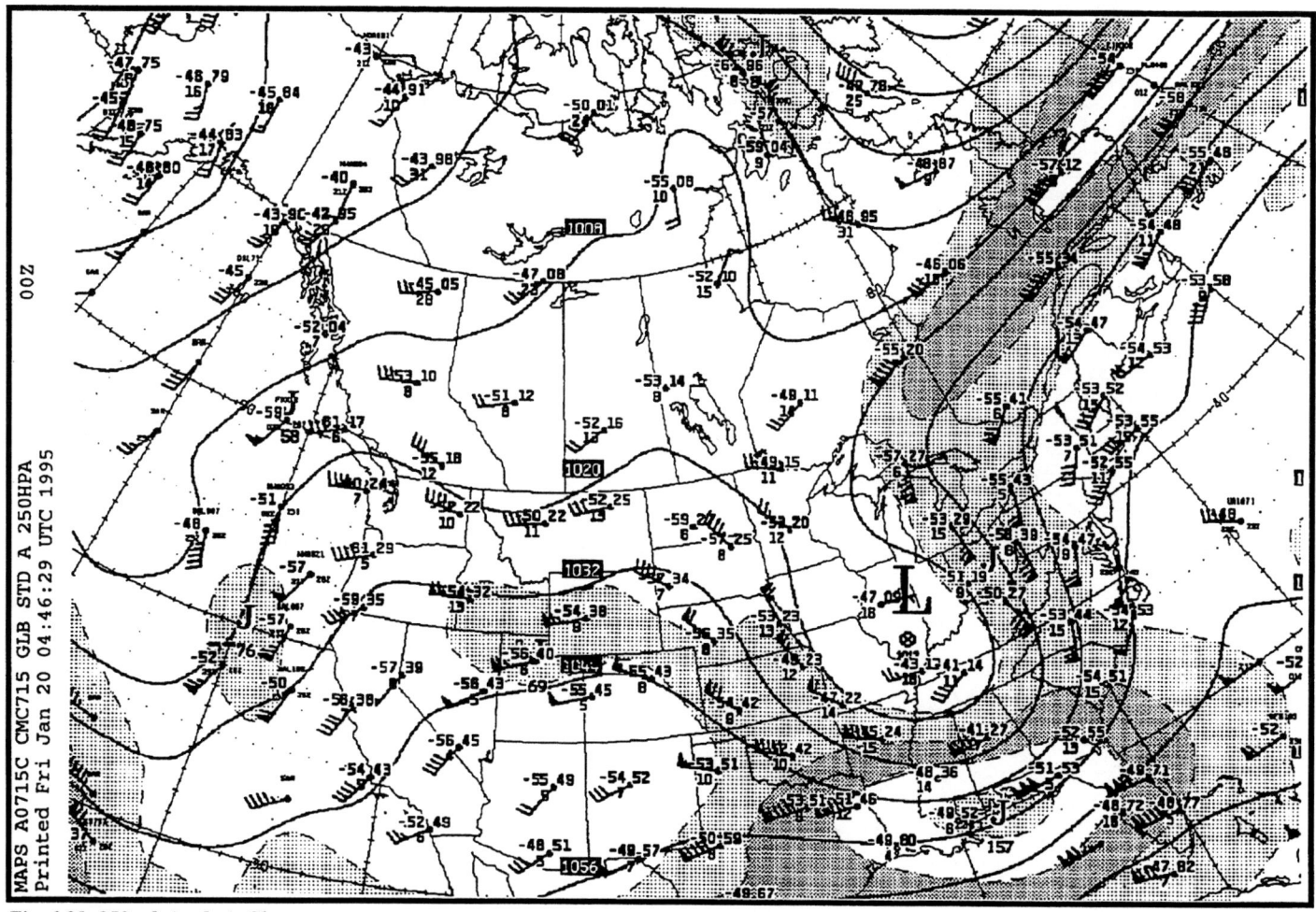

Fig. 6-31: 250 mb Analysis Chart.

EXERCISE BASED ON 250 MB ANALYSIS ABOVE:
1. What approximate ASL height is this chart intended for?
2. What is the height of the 250 hPa contour over Saskatoon, SASK?
3. What is the temperature over Saskatoon, SASK at the 250 mb pressure level?
4. What are the winds over south eastern Ontario at the 250 mb pressure level?
5. What are the maximum winds depicted on this chart?

ANSWERS:
1. 34,000 feet ASL.
2. 10,160 metres or 33,325 feet.
3. -52° C.
4. 60 - 90 kts (see shading).
5. 150 kts +. See jet core (J) over southern Mississippi/Alabama USA - note shaded area within a clear area.

Upper Level Prognostic Charts: Winds and Temperatures Aloft

Upper level forecast winds and temperatures charts are available for FL 240, FL 340 and FL 450. These charts provide coverage for extensive areas and are also known as WORLD AREA FORECAST SYSTEM (WAFS) charts.

The valid period of each chart will be shown on the data on the side panel. In these charts the forecast temperature for the flight level of the chart will be indicated in a circle, and the temperature is assumed to be negative so no "-" sign will appear.

Wind speeds are indicated with the pennant and shaft system, where the shaft shows where the wind is from. The small number at the end of the barb is the nearest ten degrees where the wind is from. First determine the wind to the nearest 10 degrees by looking at the shaft, then use the small number to determine the nearest 10 degrees.

EXAMPLE: What wind speed and direction would be represented by the following symbol on an upper level winds and temperatures prog chart?

ANSWER: Winds from 270° T at 115 kts.

Review the sample charts that are provided and complete the exercises that follow.

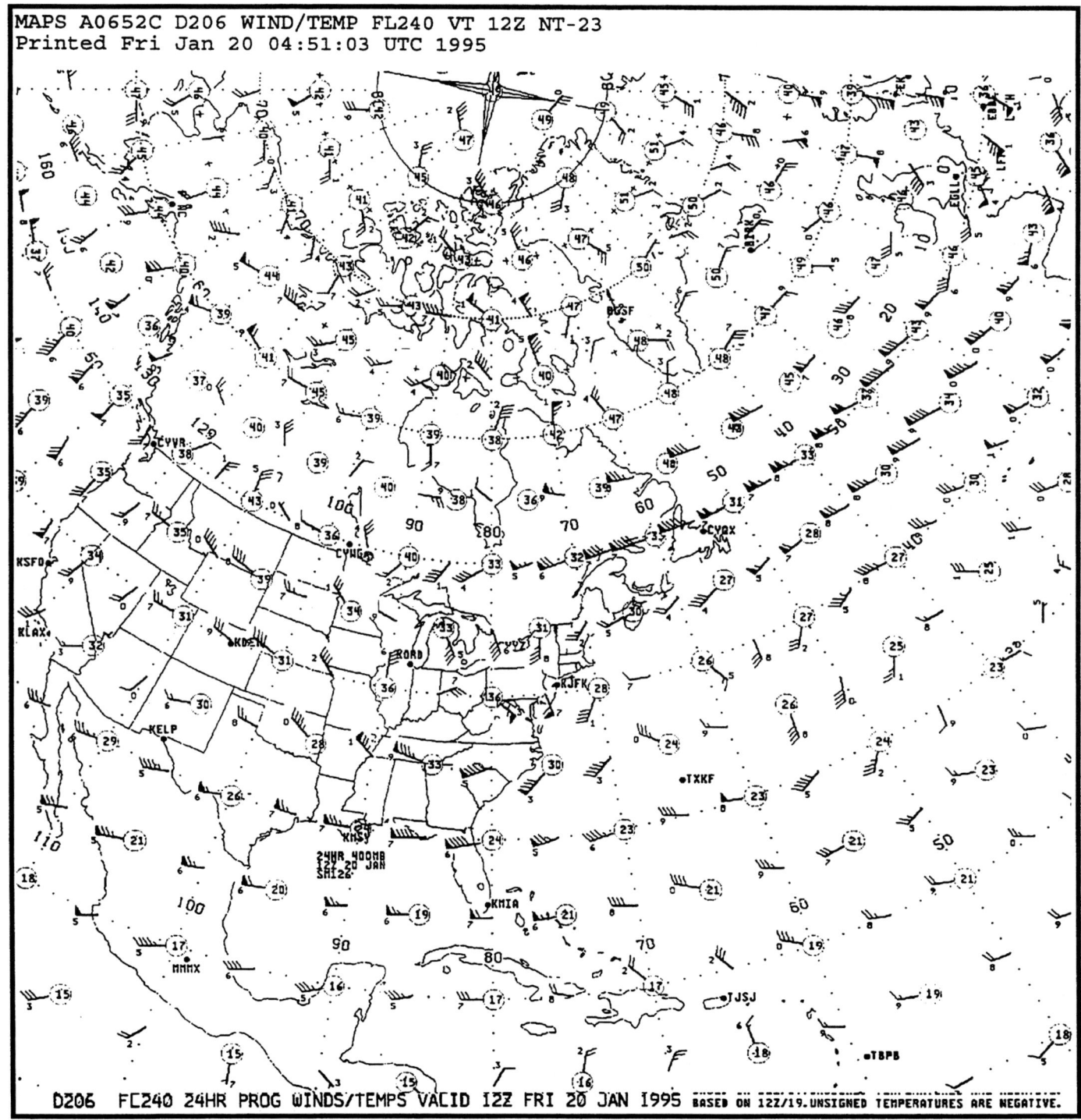

Fig. 6-32: FL 240 Upper level winds and temperatures Prog chart.

EXERCISE BASED ON FL 240 WAFS CHART:
1. What is the chart valid period?
2. What kind of chart projection is the above chart based on?
3. What is the wind speed, wind direction and temperature above Winnipeg at FL 240?
4. What is the average flight planned temperature for a flight from Vancouver to Winnipeg?

ANSWERS:
1. Valid 24 hrs commencing 12Z Jan 20.
2. Polar Stereographic.
3. 340° T, 30 kts, -36° C.
4. -39° C.

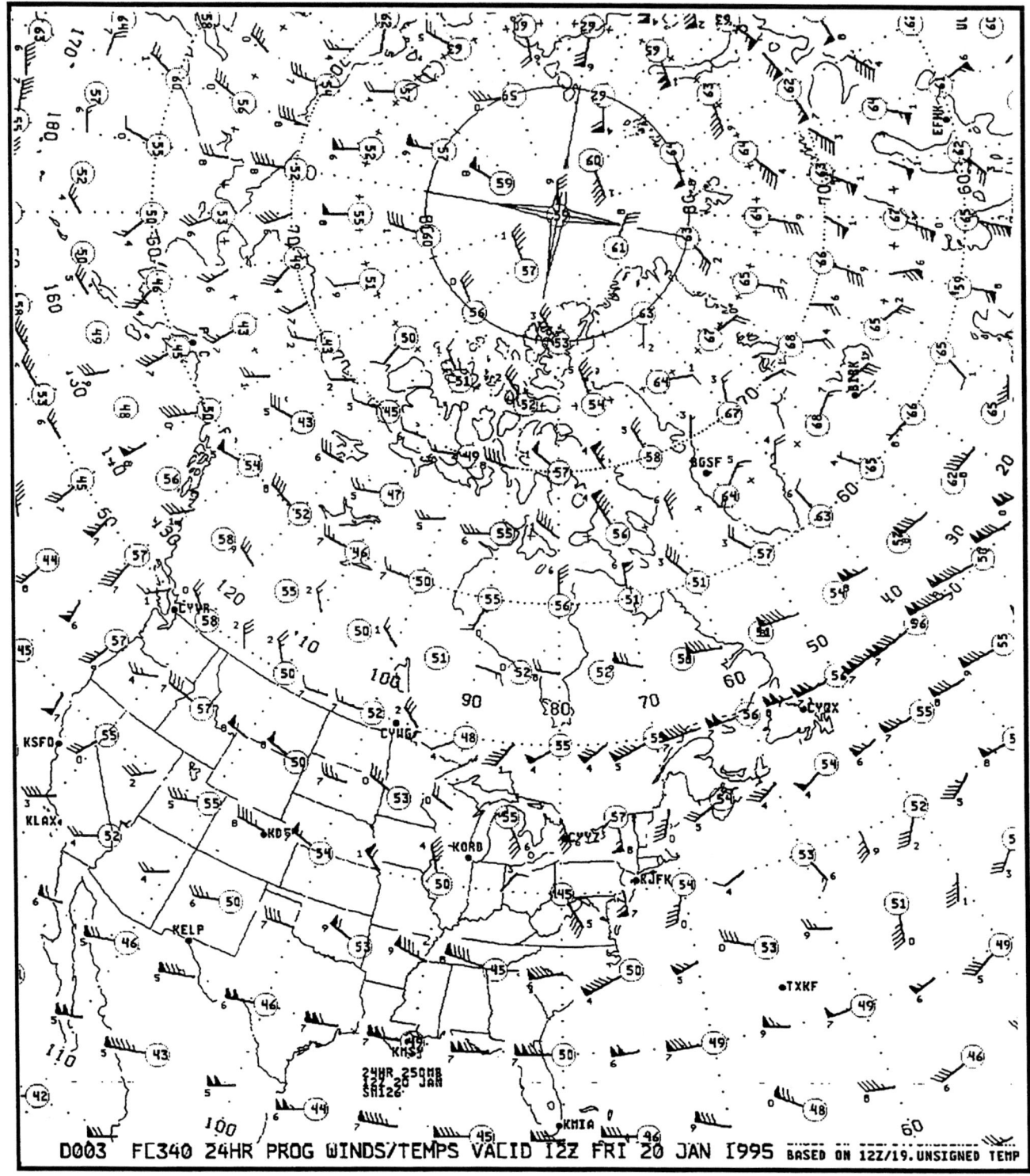

Fig. 6-33: FL 340 Upper level winds and temperatures Prog chart.

EXERCISE BASED ON FL340 WAFS CHART:
1. What is the valid period of this chart?
2. What type of projection is the chart based on?
3. What is the forecast Toronto wind speed, wind direction and temperature for FL 340?

ANSWERS:
1. 24 hrs commencing 12Z Jan 20.
2. Polar stereographic.
3. Winds from 160° T at 45 kts, temperature -57.

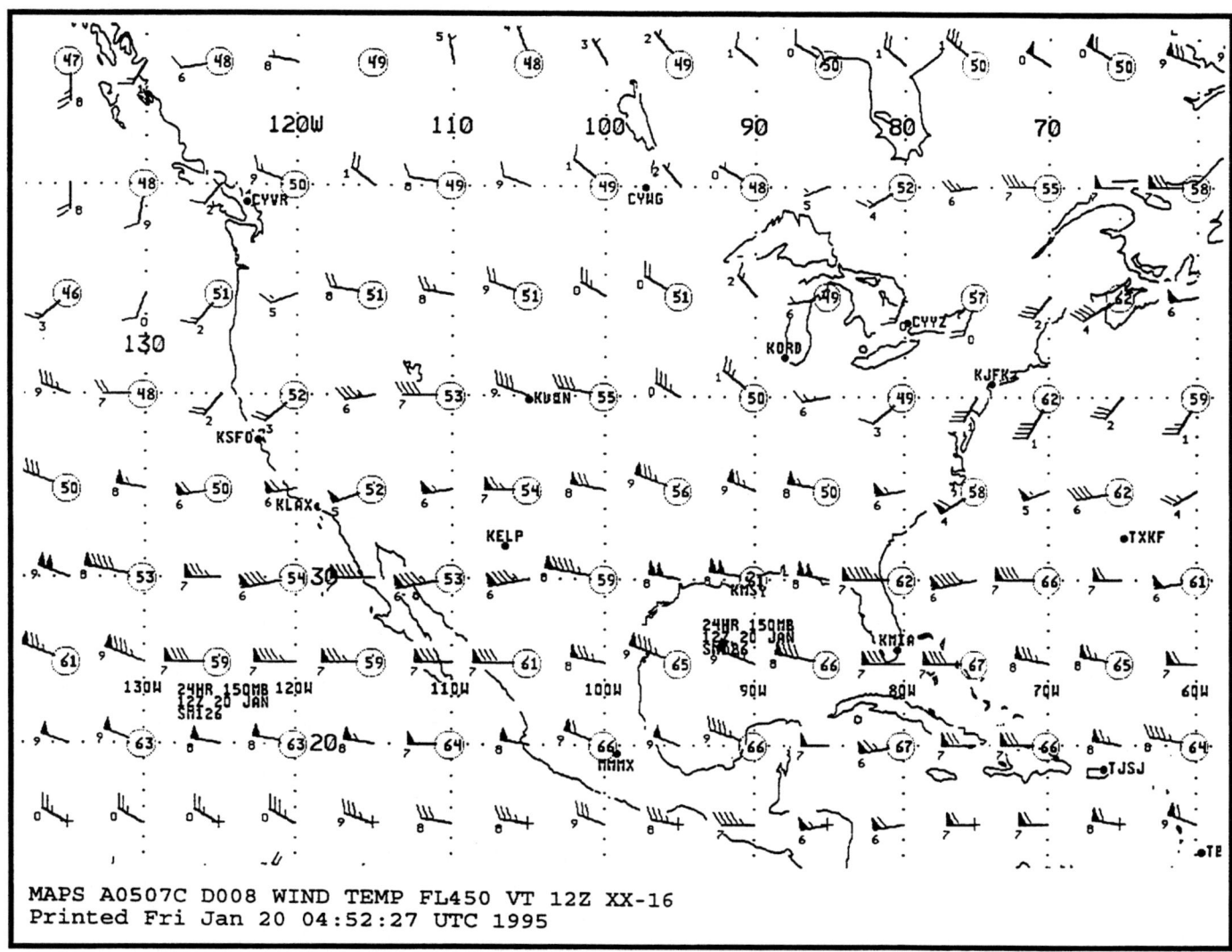

Fig. 6-34: FL 450 upper level winds and temperatures prog chart.

EXERCISE BASED ON FL 450 WAFS CHART:
1. What is the valid period of the chart?
2. What type of chart projection is used?
3. At the level of the chart, what is the temperature deviation from standard over Vancouver?
4. What are the forecast winds and temperatures over 50° N 90° W?

ANSWERS:
1. 12 hours from 12Z on Jan 20.
2. Mercator projection.
3. Temperature is 6.5° C warmer than ISA.
4. 300° T at 5 kts, -48° C.

Upper Level Significant Weather Prognostic Charts, CMC and RAFC

CMC Charts

The Canadian Meteorological Centre (CMC) provides significant weather prognostic charts for FL 100 to FL 250 and are based on the pressure levels 700 mb - 400 mb (see Fig. 6-36).

Upper level prog charts are useful to determine areas of forecast moderate to severe icing, significant cloud layers, mountain waves, freezing level at 10,000 feet, and surface position of highs and lows.

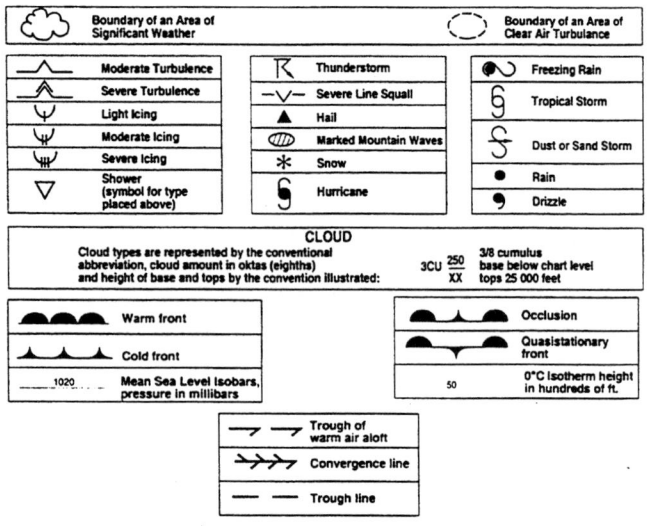

Fig. 6-35: Significant Weather Symbols.

Complete the exercise questions that follow the sample CMC chart at Fig. 6-36.

RAFC

Significant Weather Prognostic Charts - (RAFC) charts are issued to show weather conditions considered to be of concern to aircraft operations from FL 250 to FL 600. The charts are produced for 00Z, 06Z, 12Z and 18Z every day. Remember the following points:

a) Active thunderstorms are indicated as ISOL (for individual CB's, less than 1/8 cover), OCNL (for well-separated CB's, 1/8 to 4/8 cover), and FRQ (for CB's with little or no separation, 5/8 to 8/8 cover).
b) Cloud other than CB's is indicated in oktas (eighths). SCT = 1 - 4 oktas, BKN = 5 - 7 oktas, OVC = 8 oktas.
c) XXX is used to indicate upper or lower limits are above or below the limits of the chart (FL 250 or FL 600).
d) Tropopause heights are indicated as flight levels and are enclosed in a rectangular box in hundreds of feet.
e) Jet streams are indicated if they have a core speed of 80 kts or more. When indicated on the chart, the jet streams are orientated to true north, with arrows and pennants, and sufficiently spaced to show speed or height changes: Where a double hatched line crosses a jet stream, this indicates a change in speed of 20 kts at 100 kts, 120 kts, 140 kts or higher.

Note how jet stream depicted above is at 80 kts at FL 370, then, it increases to 100 kts (see double hatched line).

f) Standard SIGMET symbols are used for turbulence (see Fig. 6-35).
g) Squall lines are indicated as follows:

h) Duststorms are indicated as follows:

i) Tropical cyclones are named and indicated as follows:

j) Frontal systems are indicated using standard symbols, showing SURFACE POSITION.
k) Well defined convergence zones are indicated as follows:

Refer to the RAFC chart at Fig. 6-37 and complete the exercise that follows.

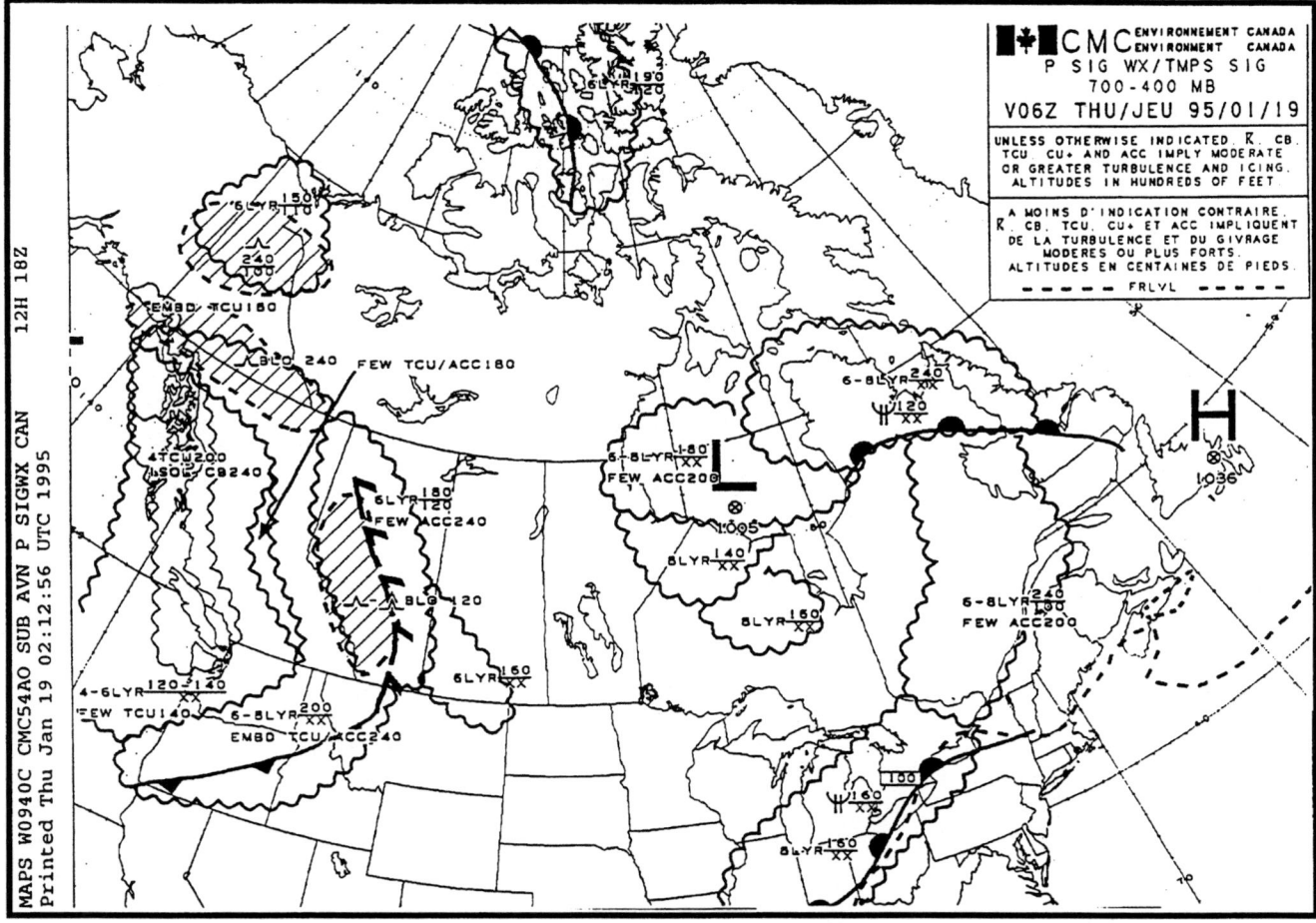

Fig. 6-36: CMC Significant Weather Prognostic Chart.

EXERCISE BASED ON CMC 700 - 400 SIG WX:

1. Does this chart represent forecast or actual weather?
2. What approximate flight levels is this chart intended for?
3. What kind of turbulence can be anticipated over western Alberta?
4. What does the hatching over areas of western Alberta warn of?
5. What kind of icing is forecast for the eastern great lakes area?
6. What is the amount and type of cloud cover forecast for mid Quebec?
7. What is the forecast height of the freezing level over Nova Scotia?

ANSWERS:

1. Forecast weather from 06Z - 18Z.
2. From FL 100 - FL 240.
3. Moderate to severe below 12,000 ft.
4. Areas of mountain wave activity.
5. Moderate icing from below the level of the chart (FL 100) to 16,000 ft.
6. 6/8 to 8/8 layers from 10,000 to 24,000 feet with a few Alto Cumulus Castellanus to 20,000 feet.
7. 10,000 feet. See dashed lines.

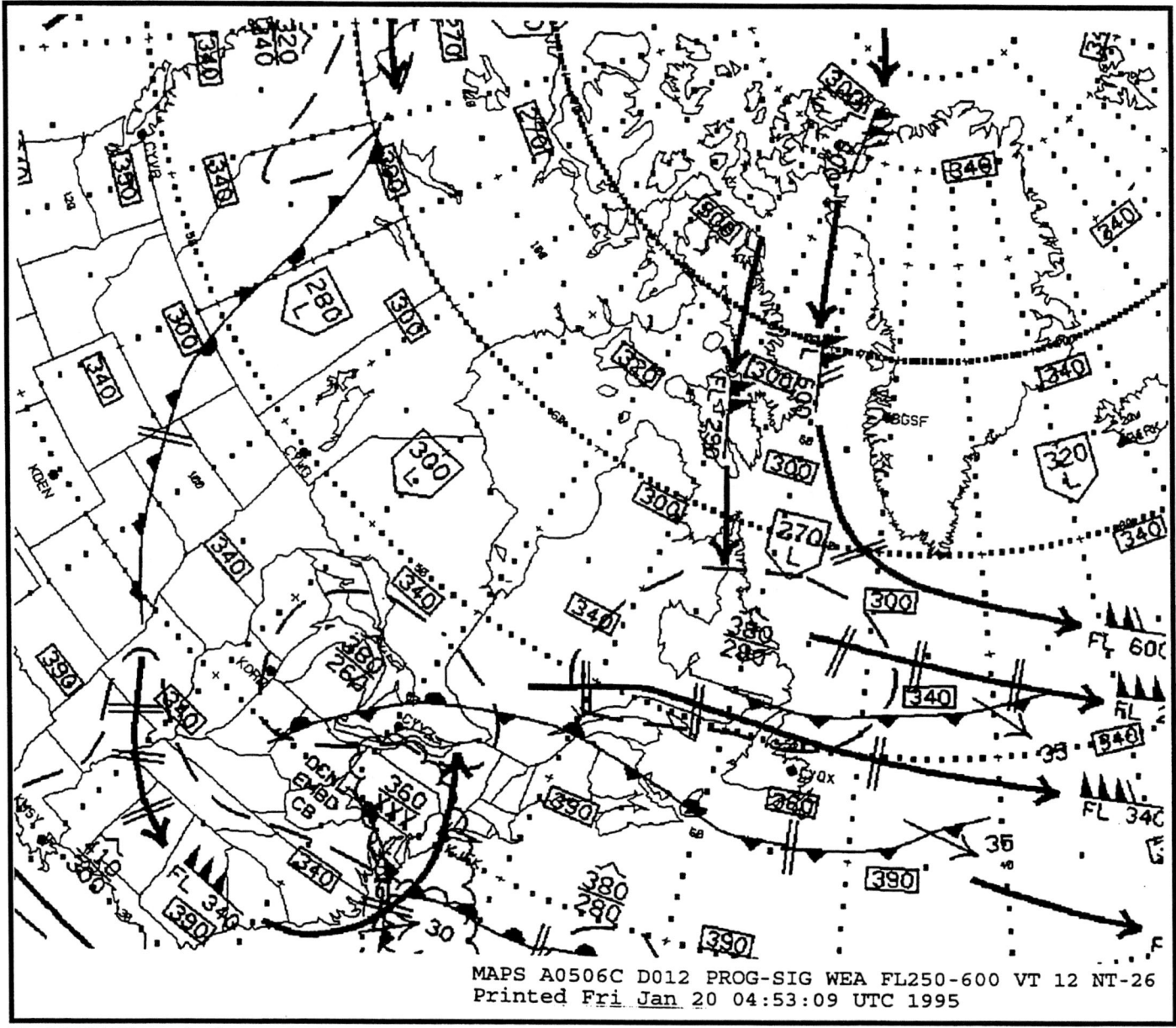

MAPS A0506C D012 PROG-SIG WEA FL250-600 VT 12 NT-26
Printed Fri Jan 20 04:53:09 UTC 1995

Fig. 6-37: RAFC upper level prognostic chart.

EXERCISE BASED ON RAFC CHART ABOVE:

1. What are the vertical limits for which the chart is provided?
2. What is the height of the tropopause over New Brunswick?
3. What turbulence is forecast over Labrador?
4. What is the jet stream direction, height and speed over northern Newfoundland, and is it strengthening, weakening or remaining constant?
5. What significant weather lies just to the south of Toronto and what are the vertical limits of the significant weather?
6. What is the direction of movement of the cold front at the northern tip of Newfoundland?

ANSWERS:

1. From FL 250 - FL 600.
2. 39,000 feet.
3. Moderate from FL 280 to FL 380.
4. From 270° T at 160 kts, FL 340, weakening (by pair of vertical bars that cross jet stream symbol).
5. Occasional CB embedded in cloud from below FL 250 to FL 360.
6. Southeast (see arrow).

Notes:

Notes:

Radio Communications, Aids to Navigation

7.01 Radio Wave Theory

ALTERNATING CURRENT (AC) means the current flow is constantly reversing in a wave pattern. The peak value of the current is the AMPLITUDE. The number of cycles per second is the FREQUENCY, expressed in HERTZ (Hz).

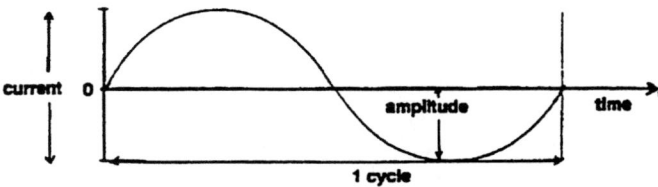

A radio wave that can be transmitted through space is an ELECTROMAGNETIC FIELD. Aviation transmissions are expressed in KHz (1,000 Hz) or MHz (1,000,000 Hz).

Freq & Range	Characteristics
VLF 3 - 30 KHz	Strong surface waves Long range.
LF 30 - 300 KHz	Strong surface waves Long range.
MF 300 - 3,000 KHz	Long range. Static problems.
HF 3,000 - 30,000 KHz	Extreme long range. Static problems.
VHF 30 - 300 MHz	Short range. Static free. Line of site.
UHF 300 - 3,000 MHz	Pulse signals. Nav equip, Radar. DME.

PHASE is the stage of a cycle at an instant, from 0° to 360°.

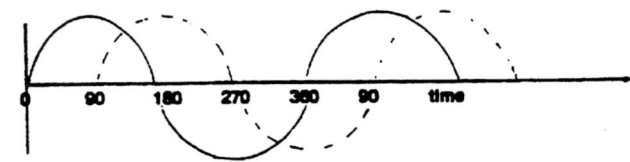

Fig 7-2: Radio wave 90° out of phase. The VOR utilizes this principle to establish radials.

GROUND WAVES are sent from the transmitter as DIRECT WAVES, GROUND REFLECTED WAVES or SURFACE WAVES (remain close to surface and transmit over hundreds of miles). Ground waves are VLF and higher.

SKY WAVES are reflected from the IONOSPHERE in the upper atmosphere. Radio waves are absorbed by the ionosphere at day and are reflected at night, so reception is better at night. Sky waves permit radio reception up to 8,000 miles. Sky waves are HF. SKIP ZONE is the distance between the last ground wave received and the first sky wave returned. SKIP DISTANCE is the distance between the transmitter and the point where the first sky wave returns to Earth.

ATTENUATION is weakening of radio waves. Ground wave attenuation is directly proportional to the frequency: lower frequencies have lower attenuation. REFRACTION is when radio waves are re-directed after reflecting off other objects.

MODULATION is when two frequencies are combined to enable hearing an audible tone. The radio signal carrying the audio frequency is the CARRIER WAVE. AMPLITUDE MODULATION (AM) is where the frequency of the carrier wave remains constant and its amplitude varies. FREQUENCY MODULATION (FM) is where the amplitude of the carrier wave remains constant and the frequency varies. Most civilian aircraft radios are AM.

VHF transmissions are relatively clear of static. HF signals have high radio noise.

BANDWIDTH is the spread between the carrier wave and audio wave frequency. AM signal quality on HF radio can be improved by suppressing one side of the bandwidth at the transmitter. These transmissions are called SINGLE SIDEBAND (SSB). One side of the signal is UPPER SIDEBAND (USB) and the other is LOWER SIDEBAND (LSB). SSB transmissions have greater strength, less interference and greater range.

DATALINK is an advanced transponder system based on MODE S that automatically transmits aircraft speed, rate of climb or clearance acknowledgements direct to the controller.

7.02 HF Communications

Beyond about 200 NM of an ATC unit or station, HF communications are required unless within range of a relay station. Reports may be relayed by other aircraft.

Generally, the higher HF frequencies (10 - 30 MHz) are used by day, or at greater range. The lower HF frequencies (3 - 10 MHz) are used at night, or at closer range.

Often, even with secondary stations, radio contact cannot be established over any frequency. Static noise may be so strong that pilots must shout their radio calls. The same HF frequency may be in use for different stations and several can be heard simultaneously.

As a result of these difficulties, the SELECTIVE CALLING system (SELCAL) has been established. SELCAL is a voice calling feature where a ground station calls an aircraft in flight with a series of audio tones or chimes. The tones are generated in the ground station coder and are received by a decoder in the airborne receiver. Receiving a SELCAL code results in a light or chime signal being activated in the aircraft.

7.03 VOR

VOR means "Very High Frequency Omni Directional Radio Range" and is a navigation system based on VHF transmissions. VHF signals are "line of sight", which means they will be blocked by mountains and high terrain. The VOR frequency range is from 112.1 MHz to 117.9 MHz. The aircraft equipment consists of a radio tuner and navigation indicator.

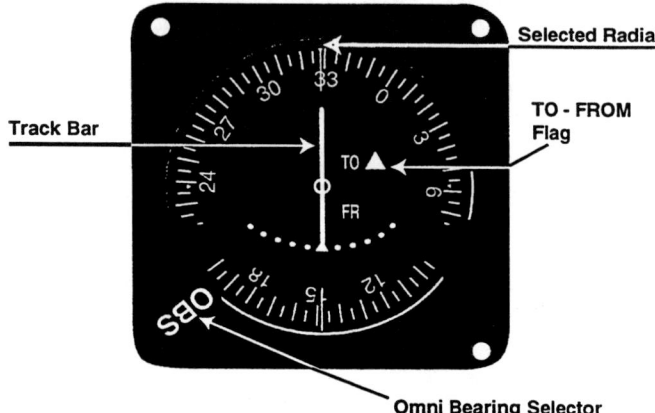

Fig. 7-3: VOR equipment: navigation indicator and radio tuner.

Aircraft VOR equipment electronically measures and displays the phase difference between two signals that are transmitted by a ground station. The VARIABLE PHASE signal is radiated at 1,800 RPM at a constant rate, and the REFERENCE SIGNAL is sent out at 360 degrees at a constant phase. Both signals are in phase with magnetic north, and the phase difference between the signals provides radial information e.g. 45 degrees out of phase would equate to the 045° radial.

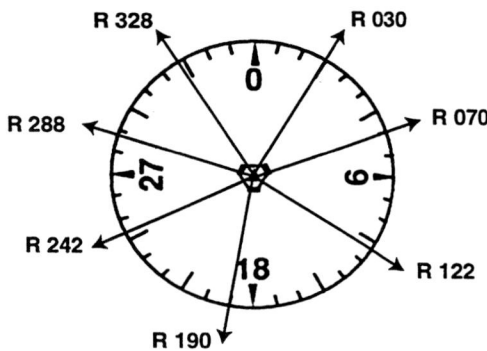

Fig. 7-4: VOR Radials.

There are 360 radials sent out from a VOR station, each representing a given track. The operational advantage to the VOR is that the pilot can verify position and track to or from a VOR station on any selected radial/track.

Before navigating with a given VOR station, the proper frequency should always be verified by listening to the morse code identifier. Some VOR stations have a voice feature which means ATC can send voice messages over the VOR frequency.

VOR reception range, assuming a flat surface, can be estimated using this formula:

VOR Recep. dist. (NM) = 1.23 X $\sqrt{\text{Height AGL}}$

EXAMPLE: What is the VOR reception range for a station when flying at 7,000 ft AGL?

ANSWER: 102.9 NM.

On aviation charts, VOR stations have a COMPASS ROSE, to identify radials. The 360 radial aligns with magnetic north. Radials designed as airways are given a "Victor" designation, such as "V-308". Accuracy of course alignment for published VOR radials is +/- 3°.

Remember that, for a selected radial, there are various quadrants or envelopes. The envelopes are the "TO-FROM" envelope, and the "TRACK BAR" envelope (needle left, needle right). The envelopes are always oriented on the basis of the selected radial. When over the selected radial, the track bar needle is centered, with a FROM flag. When over the reciprocal radial to the selected radial, the track bar needle is centered with a TO flag. 90° either side of the selected radial is the CONE OF AMBIGUITY, where the TO-FROM flag indicates OFF. The OFF indication means the VOR indication should not be relied upon (the TO-FROM sensor cannot properly sense whether the aircraft is in the TO or FROM area).

The TRACK BAR needle senses whether the aircraft is within 10° of the selected radial. Each dot on the face of the instrument represents 2°. Full scale deflection indicates that the aircraft is more than 10° away from the selected radial.

VOR indications depend upon the location of the aircraft in relation to the radial selected (not in relation to the aircraft's heading). It is exemely important to rememeber that VOR indications are based upon orienting the various envelopes to the selected radial. If you can't figure out a VOR indication, visualize the selected radial, then reconstruct the TO-FROM and track bar envelopes.

VOR Receiver Checks - VOT: OMNITEST
Two omnitest transmitters send out low power signals to test the accuracy of VOR equipment at selected airports. To test the VOR, first tune in the VOT frequency (usually 114.8). You should hear the ATIS broadcast over this frequency (now is a good time to get the ATIS copied so you don't have to re-select later!). Next, move the OBS to confirm the following indications (if you forget these, think of a "C-182 = centered, 180°, TO"):

	OBS	CDI Needle	TO-FROM	Max Error
1.	000	Centered	FROM	+/- 4°
2.	090	Full Right	OFF	N/A
3.	180	Centered	TO	+/- 4°
4.	270	Full Left	OFF	N/A

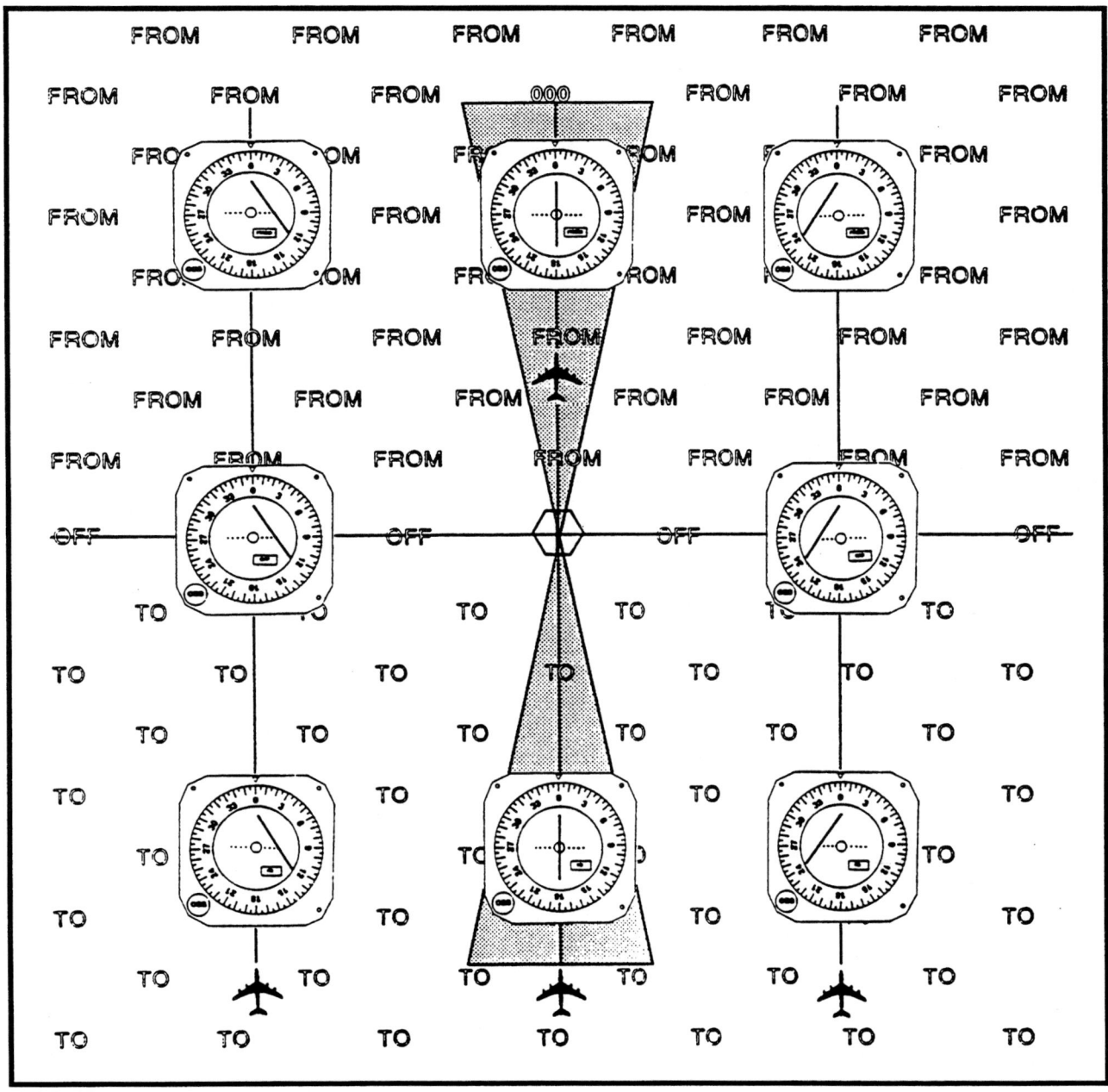

Fig. 7-5: VOR indications for 000 radial. Note TO-FROM areas: FROM is indicated when on the same side of the station as the selected radial (000) and TO is indicated when on opposite side of station. Cone of Ambiguity is 90° either side of selected radial, TO-FROM will show OFF. Track Bar or Course Deviation Indicator needle will show left when on right side of radial and right when on left side of radial WHEN ORIENTED TO THE SELECTED RADIAL. If other radials are selected on OBS, simply re-orient the above diagram to new radial and all VOR indications will follow the same pattern. If you remember these fundamentals, you will never be confused by the VOR.

Airborne Check

Aircraft VOR equipment may be checked while airborne by flying over a landmark located on a published radial and noting the indicated radial. Equipment which varies by more than +/- 6° from the published radial should not be used for IFR navigation.

VOR Check Point

VOR check point signs indicate a location on the aerodrome surface where there is a sufficiently strong VOR signal to check VOR equipment against a designated radial. The indicated radial should be within +/- 4° of the posted radial and the DME should be within 0.5 NM of the posted distance.

Aircraft Dual VOR Check

If neither a VOT test signal nor a designated check point on the surface is available and an aircraft is equipped with dual VORs (units independent of each other except for the antenna) the equipment may be checked against each other by tuning both sets to the same VOR facility and noting the indicated bearings to that station. A difference greater than 4° between the two VOR receivers indicate that one of the aircraft's receivers may be beyond acceptable tolerances.

EXERCISE: Based on aircraft A - G, determine VOR indications for selected radials.

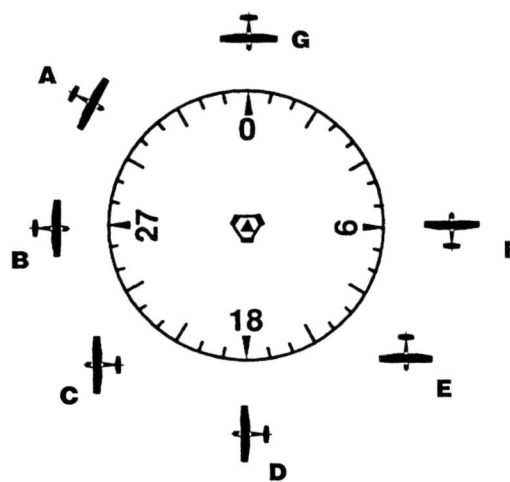

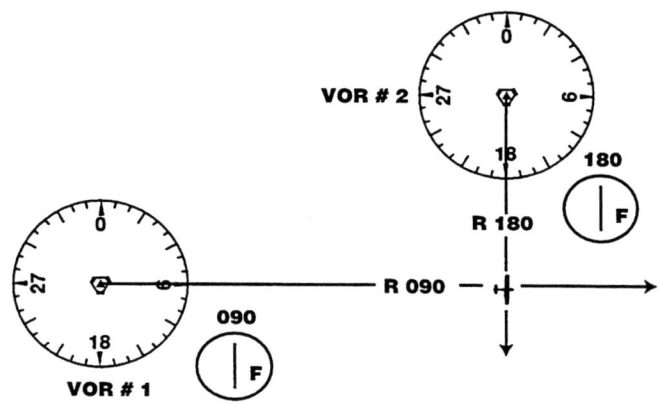

Fig. 7-6: Position Fix using two VOR stations. Centered needle with FROM flag indicates radial from each station.

	Position	Selected Radial	Track Bar	TO-FROM
1.	A	360	_____	_____
2.	B	270	_____	_____
3.	C	180	_____	_____
4.	D	360	_____	_____
5.	E	270	_____	_____
6.	F	_____	Right	OFF
7.	B	_____	Centered	FROM
8.	G	_____	Centered	TO
9.	F	_____	Centered	FROM
10.	D	_____	Left	OFF

ANSWERS:
1. Right, FROM
2. Centered, FROM
3. Left, FROM
4. Centered, TO
5. Right, TO
6. 180
7. 270
8. 180
9. 090
10. 090

One useful application of VOR principles is taking a POSITION FIX. A position can be verified by finding the intersection point of two LINES OF POSITION (LOP). An LOP is also known as a BEARING. A VOR radial represents a magnetic track (or magnetic bearing) away from a station. When the track bar is centered, with a FROM flag, the selected radial will be a line of position from the VOR station. To take a fix, take two VOR indications with a centered Track Bar and FROM indication, and draw in the identified radials. Use the VOR Compass Rose as a reference. Where the two radials intersect will be your position.

Resolving Ambiguity: Malfunctioning TO-FROM Flag

Should an aircraft TO-FROM flag malfunction, a method to resolve TO-FROM side ambiguity is as follows. Centre the track bar and note the selected radial, then select another radial that is offset by 10 degrees. If on the FROM side, the track bar will deflect toward the radial, while if on the TO side, the track bar will deflect opposite the radial.

EXAMPLE: The TO-FROM flag of a VOR is malfunctioning. You have selected 050 on the OBS and have a centered Track Bar, but you are not sure if you are on the TO or FROM side of the station. How can this TO-FROM ambiguity be resolved?

ANSWER: Set 060 on the OBS. If on the FROM side, the Track Bar will deflect RIGHT, while if on the TO side, the Track Bar will deflect LEFT.

VOR: Airway Tracking Procedures, Position Fix, Orientation

Refer to the LO chart segment and legend at Fig. A-2 in the Chart Appendix.

VOR Airway Tracking
When en route NW from WYLDE on V-301, the following procedure would be used:

1. Tune, identify YZU, 112.5.
2. Confirm centered CDI with a TO flag, OBS 293 (recip. of 113).
3. After station passage at YZU, select 278 on OBS, maintain tracking with centered T/B and FROM flag.

VOR Position Fix on LO Chart
To check position on an LO Chart:

1. Center the CDI with a FROM flag for station #1, note OBS.
2. Center the CDI with a FROM flag for station #2, note OBS.
3. Each OBS indication will be the magnetic track for each station. Draw these tracks as lines coming away from each stations; use a nearby compass rose as a reference.
4. Intersection point is position.

EXAMPLE: You wish to take a position fix using the VOR. You select YEG, and the CDI needle centers on 322, FROM. Next you select YZU and the CDI needle centers on 057, FROM. What is your position?

ANSWER: Track 322 from YEG is already drawn in as a BFS (so compass rose from YEG is not required), and this meets track 061 from YZU at AROUK intersection (in black).

Interpretation of CDI Needle While Tracking

To identify reaching a position while tracking on an airway, and to ensure you have not yet passed it:

1. Select the approximate radial to determine fix from VOR station.
2. When over fix, CDI needle will be centered.
3. BEFORE fix, needle will be left of center, where VOR station is to left of airway and right of center, where VOR station is to right.
4. AFTER fix, needle will be right of center, where VOR station is to left of airway, and left of center, where VOR station is to right.

EXERCISE: You are tracking North on V-112 from YEG:

1. Assuming your aircraft #1 VOR is for tracking V-112, and #2 VOR is for cross fix from YZU, what VOR indications would positively identify reaching the Northern boundary of the Edmonton TCA on V-112?
2. What VOR indication would you expect from your # 2 VOR prior to reaching the Edmonton TCA boundary?
3. What VOR indication would you expect from your #2 VOR to confirm passing the Edmonton TCA boundary?

ANSWERS:
1. #1 VOR: select YEG, freq. 117.6, OBS 002, FROM, centered T/B. #2 VOR: select YZU, freq. 112.5, OBS 065, FROM, centered T/B.
2. #2 VOR: CDI needle left.
3. #2 VOR: CDI needle right.

VOR Inbound Intercept

Follow these steps for an inbound intercept:

1. Tune, identify VOR station by frequency and morse code.
2. Rotate OBS to reciprocal of desired radial.
3. Note TO-FROM flag. [If FROM or OFF, you are on the wrong side of the station; proceed with off-side procedure].
4. Note T/B. Needle right, add 90° to OBS. Needle left, subtract 90° from OBS.
5. Turn shortest direction to intercept heading.
6. When T/B needle starts to move (when within 10°), reduce intercept by 45°.
7. As needle approaches center, turn to OBS indication, keeping needle centered and making corrections for wind drift [use bracketing technique].

VOR Outbound Intercept

Follow these steps for an outbound intercept:
1. Tune, identify station by frequency and morse code.
2. Rotate OBS to desired radial.
3. Note TO-FROM flag. [If TO or OFF, aircraft is on the wrong side of the station; proceed with off-side procedure].

4. Note T/B. Needle right, add 45° OBS. Needle left, subtract 45° from OBS.
5. Turn shortest direction to intercept heading.
6. As T/B needle approaches center, turn to OBS indication, keeping needle centered and making corrections for drift [use bracketing technique].

Offside Procedure

If an aircraft is on the opposite side of the VOR station to the desired radial, utilize this procedure:

1. Turn to a heading identical to DESIRED RADIAL.
2. Upon TO-FROM flag giving proper indication, wait a further 2 minutes (to get well into proper side).
3. Proceed with normal intercept procedure.

Bracketing Technique: Wind drift Correction

Bracketing is used to find a heading to compensate for wind and maintain track. Procedure:

1. Turn aircraft to zero wind heading.
2. Note drift direction.
3. Return aircraft over track, apply drift correction larger than necessary, ie 30°.
4. Return to track, then cut drift correction in half, ie 15°.
5. Continue halfing drift correction until precise wind drift correction heading is determined.

Time and Distance Remaining to Station

A VOR can be used to obtain an estimate of the time and distance remaining while travelling inbound to a VOR (assuming you don't have functioning DME equipment on board!). To determine the time remaining, follow this procedure:

a) Note present radial.
b) Turn 90° from present radial.
c) Note the time taken to cross a given number of radials. Time in seconds.
d) Time to station (minutes) =

$$\frac{\textbf{Time in seconds}}{\textbf{\# of radials crossed}}$$

To determine the distance to the station, once you have the time, all that is needed is a basic time-speed-distance calculation. Although the groundspeed is best, the TAS can also be used for an approximation. Use this formula to determine the distance remaining:

Distance to station =

$$\textbf{G/S (or TAS)} \ \textbf{X} \ \frac{\textbf{Time to Station}}{\textbf{60}}$$

EXAMPLE: Refer to the LO chart segment. You are enroute from Slave Lake to the Whitecourt VOR (YZU) along V-371. TAS is 120 kts. You wish to determine the time and distance to YZU. You turn the aircraft 90° from R005, and note that it takes 4 minutes and 30 seconds to cross 10 radials (to the 015R). What is your time and distance to the YZU VOR?

ANSWER: 27 minutes to station, 54 NM remaining.

7.04 ADF and RMI Systems

The ADF frequency range is 190 to 1750 KHz. The signal is not line of sight, so is useful where terrain interferes with VOR signals. An "L" NDB has a power of under 50 Watts, an "M" NDB has a power of 50 - 2,000 Watts, and an "H" NDB has a power of 2,000 Watts or more. NDB's are flight checked to +/- 5° for approach, +/- 10° enroute. The ADF is subject to these reception errors:

a) STATION OVERLAP. Where more than one station has been assigned the same frequency.
b) ELECTRICAL STORMS. ADF points to center of storm.
c) NIGHT EFFECT. ADF swings erratically. Strongest just after sunset and just before sunrise.
d) QUADRANTAL ERROR. Radio waves are deflected when they strike portions of the aircraft, such as engines, before striking loop antenna. Also caused by radio energy within aircraft.
e) MOUNTAIN EFFECT. Radio reflections off mountains may cause erroneous indications.
f) COASTAL REFRACTION. Radio waves can be distorted when passing from land to sea or when moving parallel to the coast. Most pronounced at angles less than 30° to coastline.

The most common ADF navigating error is failure to recognize station passage.

ADF/RMI Navigation

On a fixed card ADF the zero index on the ADF is at the top of the instrument (aligned to nose of aircraft). On an RMI (radio magnetic indicator) the top of the instrument is the aircraft heading, and the RMI needle points to the bearing to or from the station.

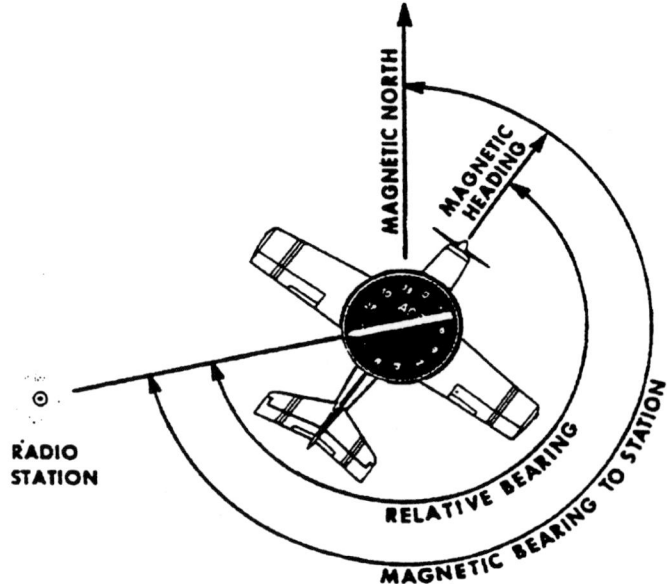

Fig. 7-7: Basic ADF concepts.

Know these definitions:

1. MAGNETIC HEADING. Angular displacement from aircraft longitudinal axis to magnetic north.
2. RELATIVE BEARING. The angular displacement from the longitudinal axis of the aircraft to the NDB sta-

tion. The same as the indication read off the face of a fixed card ADF.
3. MAGNETIC TRACK/BEARING TO STATION. Angular difference between position and NDB.
4. MAGNETIC TRACK/BEARING FROM STATION. The reciprocal of the magnetic track to the station.

From Fig. 7-7, for the fixed card ADF, note that MH + RB = BTS (090 + 050 = 140). Remember this formula. If you know the magnetic heading and the relative bearing, you can determine the bearing to the station. Similarly, if you know the bearing to the station and the magnetic heading, you can determine the relative bearing using this formula: BTS - MH = RB. The reciprocal of the bearing to the station is the bearing from the station.

ADF Position Fix

To determine position using a fixed card ADF, note the RB to each station, then add to heading to determine TTS. Find reciprocals and plot tracks from station using VOR as a reference.

EXAMPLE: Refer to the LO chart segment at Fig. A-2 in the Chart Appendix. You wish to take a position fix using the Slave Lake NDB (YZH) and McLeod NDB (FH). The magnetic heading is 300. When you select YZH, the fixed card ADF shows 340. When you select FH, the fixed card ADF shows 286. What is the approximate position?

ANSWER: Athabasca. BTS YZH = 300 + 340 = 280. Draw BFS 100. BTS FH = 300 + 286 = 226. Draw BFS 046.

ADF Tracking

To intercept a desired track to or from an NDB:

1. Determine the present track in relation to NDB.
2. Note where desired track is in relation to present track.
3. From present track, intercept heading is determined by adding or subtracting intercept angle to desired track.

EXAMPLE: Aircraft magnetic heading is 090, relative bearing on ADF is 250. What heading is required to intercept an inbound track of 270 to the NDB with a 90° intercept?

ANSWER: Present track is 340 TO NDB, therefore aircraft is located to SE of station. To intercept inbound track of 270, using a 90° intercept, aircraft should be flown 270 + 090 = 360.

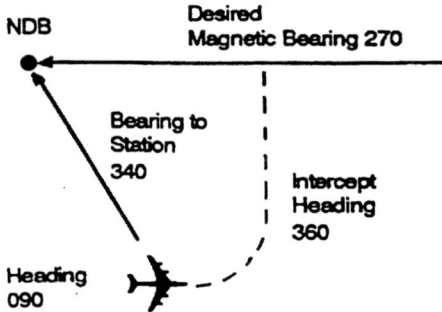

To determine when the aircraft has reached the desired track, the relative bearing can be determined using the formula RB = BTS - MH.

EXAMPLE: For the aircraft above, heading 360, when will the pilot know the aircraft has reached the desired track of 270?

ANSWER: RB = BTS - MH, therefore, aircraft will reach desired track when ADF indicates a relative bearing of 270 - 360 = 270. NOTE: in practice, the aircraft is turned to intercept the desired track by a 5 - 10° lead.

ADF Tracking in Wind

To maintain a constant track in winds, when tracking inbound, the ADF needle will point to the winds and corrections should be applied on the side of needle displacement, resulting in the ADF needle pointing to the OPPOSITE of the wind side once a wind corrected heading has been established.

When tracking outbound, in winds, the ADF needle initially points to the side of the winds, and the drift corrected heading will result in the ADF pointer indicating on the same side as the winds.

EXAMPLE: An aircraft is heading 090 with winds from the north. What should the pilot do to maintain a track to an NDB of 090?

ANSWER: Maintain a drift corrected heading to the left of 090, depending on wind strength.

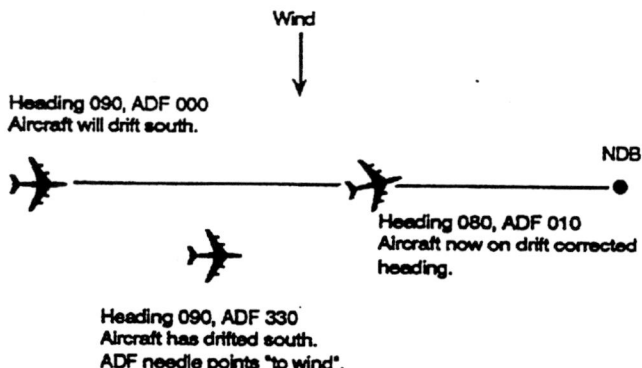

EXAMPLE: Refer to the LO chart segment in the appendix. You are tracking on LF airway A2 from Whitecourt (ZU). Your track is 089 magnetic. If the magnetic heading to maintain track is 095, what will the ADF pointer indicate? Is the wind from the left, or right? Is the drift left, or right?

ANSWER: BTS - MH = RB. 269 - 095 = 174. Needle tail points to 356. Wind is from the right. Drift is to the left, 6 degrees.

Time To Station

To calculate the time remaining to an NDB station, turn the aircraft 90° from the inbound track and take the time (in seconds) for a small change in relative bearing. Then use this formula:

Time to Station in minutes =

$$\frac{\text{Time for bearing change (seconds)}}{\text{Change in relative bearing}}$$

EXAMPLE: A pilot wishes to determine the time remaining to an NDB Station. The pilot turns the aircraft 90° from the inbound track, and notes that it takes 2 minutes and 20 seconds for a change in relative bearing of 10° on the ADF. What is the time to the station?

ANSWER: Convert time to seconds, = 140. Divide by 10° = 14 minutes.

Determination of Expected ADF Indication Over Intersection

When tracking on an LF airway, the ADF can be used to confirm when an aircraft has reached an intersection. Use the formula RB (desired) = TTS - MH.

EXAMPLE: Refer to the LO chart segment in the chart appendix. You are heading 290° M to track 296 on LF airway A-7. You have not yet reached the ACKIN compulsory reporting point. What will the fixed card ADF pointer indicate when you have reached ACKIN?

ANSWER: 095. Study the ACKIN intersection. It is made of two lines of position. The first line is the airway track of 296 for A7, and the second is a 025 bearing to the Slave Lake NDB (YZH). The 025 is the BTS. Applying the formula RB = BTS - MH gives a RB of 095 i.e. RB = (360 + 025) - 290 = 095.

Radio Magnetic Indicator (RMI)

The RMI displays the present bearing to or from an NDB or VOR station and the present heading. When an ADF is selected, the pointer head displays the track to the station and the tail displays the track from the station. If a VOR is selected, the tail pointer shows the radial which the aircraft is over.

Fig. 7-08: Face of RMI.

The RMI indication above shows VOR #2, the wide needle, with a bearing to the station of 315 and a bearing from the station of 135. The aircraft is therefore on the 135 radial on a heading of 300.

EXAMPLE: With the above RMI indication, a left turn of 90° change in direction would cause the interception of which radial - 180, 270, 360 or 090?

ANSWER: 180 radial.

ADF/RMI Indication on Interception of Airway

To intercept a desired track or airway, the pilot must determine the required intercept heading and find the expected ADF indication to confirm the interception. To find the required intercept heading, either add or subtract the intercept angle to the desired track. While any intercept angle can be used, use 90° on inbound intercepts and 45° on outbound intercepts to keep it simple. To confirm the airway or track has been reached, note the airway track as indicated on the chart, then add or subtract the aircraft heading to determine the expected relative bearing. When the fixed card ADF points to that relative bearing, the aircraft is over track. With an RMI, simply wait for the pointer to indicate the desired track.

EXAMPLE: An aircraft is to the south of an NDB with a heading of 070 and a relative bearing on a fixed card ADF of 330. The pilot wishes to intercept an inbound track of 360 to the NDB. What heading should be flown for a 45° intercept?

ANSWER: The present position will be on an outbound track of 220° from the NDB. To track inbound on a track of 360, the track is to the right, and the aircraft should be flown 360 + 045 = 045. The aircraft is over track when a relative bearing of 315° appears on the card of the ADF, or 360 on the RMI pointer.

7.05 TACAN Systems

TACAN means "Tactical Air Navigation" and is for military operations. It is often integrated into the civilian VOR system ("VORTAC"). TACAN operates in the UHF frequency range. Conventional VOR equipment is unable to utilize TACAN signals. Components of a VORTAC operate simultaneously on "paired" frequencies so that aircraft DME receivers, when selected to the VOR frequency, will obtain distance information from the DME component of the TACAN.

7.06 DME

Distance Measuring Equipment operates in the UHF frequency range from 962 MHz to 1213 MHz and provides a distance readout and in some cases groundspeed to the station. DME is normally co-located with a VOR, and has a maximum range of about 300 NM. DME is secondary RADAR, and functions by measuring the time difference between the transmitted and received pulse signal (the aircraft receives the signal from the station at a frequency slightly different than transmitted signal). Because the system is based upon SLANT RANGE, and not true horizontal distance, there will be inaccuracies. For example, an aircraft directly above the station at 35,000 feet indicates 6.6 NM. At low altitudes and at far distances from the station, slant range error is insignificant. There are 126 DME channels, each channel consisting of two frequencies spaced 63 MHz apart, one

for air to ground interrogation and the other for ground to air response. Channel 1 air-to-ground frequency is 1025 MHz, ground-to-air is 962 MHz. DME equipment can be utilized at TACAN, VORTAC or DME facilities. The DME frequency is automatically tuned when the frequency of the collocated VOR is selected.

Actual ground distance can be computed with the following formula:

$$D = \sqrt{(S^2 - A^2)}$$

Where   D = Actual Ground Distance
S = Slant Range DME Readout
A = Altitude in NM (divide feet by 6075)

EXAMPLE: What is the actual ground distance for DME indication of 36.9 NM at 36,000 feet?

ANSWER: 36.4 NM.

7.07 RNAV

RNAV (Area Navigation) permits direct navigation on any desired flight path based on availability of navaids and self contained navigation systems. Current ground based systems with RNAV capability include VOR/DME, DME, LORAN-C, OMEGA and OMEGA/VLF. Self contained systems include INS (Inertial Navigation System) which utilizes accelerometers, a computer and gyro-stabilized platform to gives instantaneous position information. An INS determines the position of the aircraft by measuring the acceleration forces on the aircraft and integrating them. An INS system must be first aligned with the grid coordinates from the departure airport, referred as GYROCOMPASSING. Another self contained system is based on Doppler radar.

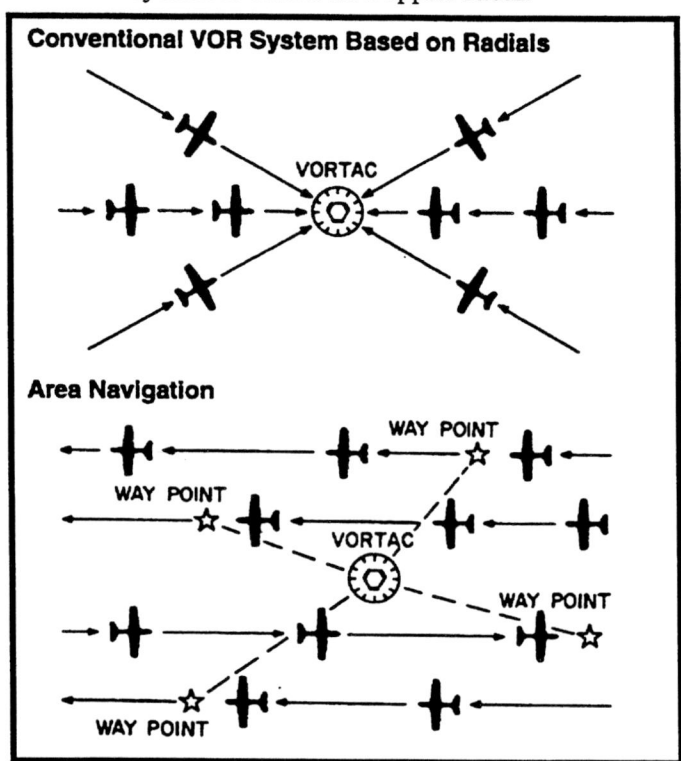

Fig. 7-9: Area Navigation is based on waypoints.

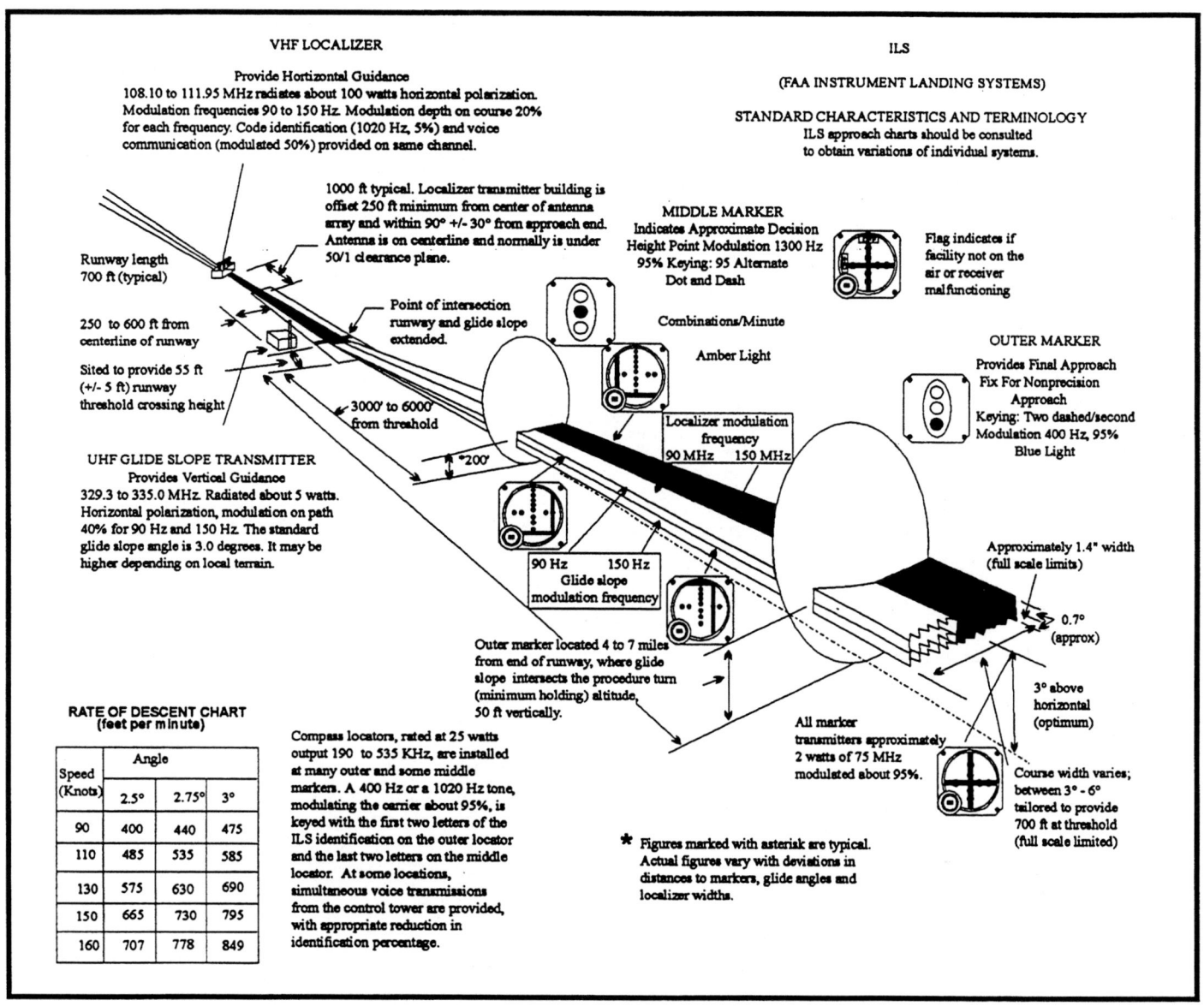

Fig. 7-10: Conventional ILS installation.

In the HYPERBOLIC mode of RNAV operation, an LOP is based on the same relative signal phase or time difference from two ground stations, such as LORAN and OMEGA. Using three stations will provide two LOP's for position identification. The advantages of this mode are no requirement for expensive high precision time reference in the receiver, improved dynamic performance, long term accuracy, and freedom from phase related errors.

In DIRECT RANGING mode RNAV systems, position is defined by measuring the signal phase from two or more stations, such as two DME stations (RHO-RHO). A high precision oscillator reference is required if only two stations are used.

All RNAV systems involve inputting WAYPOINTS. RNAV provides CONSTANT COURSE WIDTH, so a 1° deflection represents 1/2 NM (enroute mode).

7.08 ILS

The Instrument Landing System is a precision approach system allowing landings in conditions of low visibility and ceiling. The ILS consists of a VHF localizer transmitter for horizontal guidance, and a 3° UHF glideslope for vertical guidance.

Category I ILS System

An Operational Category I ILS is for operation down to a minima of 200 feet decision height and RVR 2,600 feet, with a high probability of success. When RVR is not available, 1/2 SM visibility is intended. The ground equipment consists of a localizer and glide path transmitter, marker beacon (outer and middle) and/or a non-directional beacon (NDB) along the approach path. A DME fix may replace the marker beacons and/or NDB (an NDB is not mandatory for an ILS approach).

Category II ILS System

An Operational Category II ILS is for operation down to a minima below 200 feet decision height and RVR 2600, to as low as 100 feet decision height and RVR 1200, with a high probability of success. The following systems must be fully serviceable to meet Category II standards:

a) Airport Lighting: approach and threshold lights, touchdown zone lights, center-line lighting, runway edge lights and runway end lights.
b) ILS components: localizer, glidepath, middle marker.
c) RVR Equipment: two transmissometers, one located adjacent to the runway threshold and one located adjacent to the runway midpoint.
d) Power Source: airport emergency power as primary electrical source for all essential system elements, with commercial power available within one second as backup.

NOTE 1: Category II operations are restricted to aircraft and pilots specifically approved for such operations by Transport Canada. NOTE 2: Middle Markers have been removed from all ILS installations in Canada.

FAN MARKER beacons are placed about 7 miles from the runway (outer marker) and about 1/2 mile from the runway (middle marker). Marker beacons have been removed from Canadian ILS's.

The localizer is a VHF transmitter which provides azimuth information and is 4 times as sensitive as the VOR (full scale deflection = 2.5° either side of CDI). The ILS beam is 3° above the horizontal, and the beam width is about 1.4°.

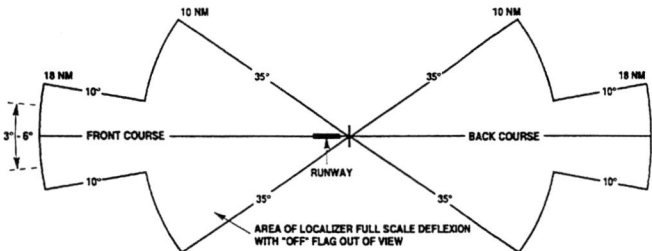

Fig. 7-11: Coverage zones of ILS localizer.

The localizer signal is considered reliable when within 35° of course up to 10 NM, and when within 10° up to 18 NM from station. Outside these areas, there will be false localizer course indications.

A TRACK GUIDANCE LOCALIZER is used in mountainous terrain to provide track guidance information to arriving aircraft (LDA), and in STARs (Standard Terminal Arrival Routings). An "X" localizer identification means that the glidepath is inoperative, and localizer alignment exceeds 3° of the runway heading. Localizers with an alignment of less than 3° of the runway heading will have an "I" as the first letter of the identifier.

The effect of a sudden change in wind velocity on an aircraft's rate of descent while maintaining a steady 3° glide path at constant speed and power will be as follows: STRONG HEADWIND - decrease in groundspeed, decrease in rate of descent required to maintain glideslope. STRONG TAILWIND - increase in groundspeed, increase in rate of descent required to maintain glideslope.

7.09 Microwave Landing System (MLS)

The time referenced scanning beam Microwave Landing System (MLS) was hoped to be adopted by ICAO as the standard precision approach system to replace ILS at international airports. However, with the advent of GPS a widespread transition to MLS has not occurred.

The MLS has several advantages over ILS limitations such as terrain, possibility of FM radio station interference and limited channels available. Aircraft can approach at multiple approach paths and can utilize curved, offset and segmented approach paths at varying glide paths.

With MLS, ground equipment normally consists of azimuth and elevation transmitters to provide horizontal and vertical guidance, and a precision DME transponder to provide range information. Approach signal accuracy is +/- 40° either side of centre line to at least 20 NM and 20, 000 feet.

The back azimuth at the opposite end of the runway is +/- 20°, 5 NM to 5,000 feet. The maximum signal angle is to 15° above the horizontal (both front and back azimuth).

7.10 LORAN

LORAN is a pulsed hyperbolic navigation system operating in the 90 - 100 KHz frequency band which is used primarily for marine navigation. The system is only approved as a secondary navigational system. LORAN is based upon the measurement of the time difference in the arrival of signal pulses from a group or chain of stations. The chain of stations consists of a Master station linked with 4 secondary stations. To provide a position fix, the LORAN measures the time difference between the Master and at least two of the secondary stations.

LORAN stations send ground waves and sky waves. The ground waves are used for navigation. Signal coverage ranges from 120 NM to 900 NM, depending upon ground signal strength. Accuracy is generally +/- 1/2 NM. In the United States, some non-precision approaches have been approved for LORAN.

Caution must be exercised when using LORAN, because errors can occur if the receiver locks on to sky waves or other erroneous signals. There are "dead zones" where signal reception is impossible due to lack of stations. Signals are best over salt water at day.

7.11 Omega and VLF Navigation

OMEGA is a Very Low Frequency (VLF) long range navigation system based upon eight transmitting stations strategically located around the world. The stations transmit on 4 OMEGA frequencies - 10.2 kHz, 11.05 kHz, 11.3 kHz, and 13.6 kHz. Airborne receivers process signals in either the hyperbolic or direct ranging mode. Accuracy of the system depends upon the geometry of the signal pattern and signal propagation, but generally is accurate to +/- 2 NM worldwide. Caution should be exercised when utilizing OMEGA because the stations can be shut down or the frequencies changed without advance warning.

The VLF navigation system uses 9 VLF ground stations located throughout the world, operating from 15 - 30 kHz. These stations are not dedicated to navigation purposes and can be shut down without advance warning. The VLF systems processes signals in the direct ranging mode. The VLF system has not been approved by Transport Canada as a primary navigation aid.

The OMEGA-VLF system is an OMEGA receiver which has the capability of switching over to VLF is signal propagation is poor.

OMEGA or OMEGA-VLF has been approved as a prime means of navigation for IFR flight in the NAT MNPS airspace and CMNPS airspace.

7.12 RADAR Systems

RADAR means "Radio Direction and Ranging" and is based on the time difference between transmission of a radio signal and its reflection by an object. The reflection is illuminated on a cathode ray tube. The reflectivity of a radar echo decreases with the distance of the aircraft. The lag between the radar signal and its reflection provides distance information.

Doppler RADAR Navigation System

Doppler radar is an aircraft based radar system. The Doppler radar sends out four symmetrical beams, two forward (left and right) and two rear (left and right). The DOPPLER EFFECT states that when a sound source moves, the frequency of the sound heard differs from the frequency of the vibrating source from which it originates. The frequency shift in the microwave echo is a measure of the relative motion of the aircraft and reflecting object (ground). The returning frequency is at a higher frequency than the transmitted beam.

A Doppler radar system has three components: Doppler radar, compass and computer. The Doppler radar provides data for groundspeed and drift information, and the compass system provides a heading reference. The computer processes this information into position and guidance data.

The limitations of the system are that it is entirely an aircraft based system, so error is cumulative with time and must be corrected by correlating position fixed with other systems. Another primary source of error is the magnetic compass system, since heading data is correlated into the computer's navigational output.

Airport Surveillance Radar (ASR)

ASR refers to primary radar systems in use by ATC to control traffic. The following types of ASR are used:

a) AASR-1. A medium to long range primary radar for distance of 180 to 200 NM.
b) ASR-5. A short range primary radar for distance of up to 60 NM.
c) ASR-803. A short to medium range radar for up to 80 NM.

Precision Approach RADAR

PAR is a short range, high definition primary radar for conducting a ground instructed approach. There are two radar scopes: search/surveillance and final approach. The first scope provides lateral information, and the second provides vertical information. The controller instructs the pilot orally on which directions to fly in order to maintain a precision approach path to minimums.

Secondary Surveillance RADAR (SSR)

SSR requires a transponder. A transponder receives an INTERROGATION pulse from a ground based radar unit and sends a coded RESPONSE pulse. The result is that the controller sees a more reliable, solid return. By squawking "IDENT", the signal becomes momentarily brilliant, allowing the controller to positively identify an aircraft. With SSR, aircraft can be identified at greater distances than with convectional radar, and targets can be identified quickly and positively. A conventional transponder is MODE A. MODE C transponders provide the controller with an altitude indication. A MODE C transponder requires an ENCODING ALTIMETER, which sends altitude information to the aircraft transponder. There are two main types of encoding altimeter: the PNEUMATIC type is the simplest, where a conventional pressure altimeter is connected to an internal glass slide which sends an electrical impulse to the transponder for the indicated altitude. The SERVO type encoding altimeter requires an external power source but is highly accurate and sends altitude information not only to the transponder but also to other instruments and systems such as altitude alert, Ground Proximity Warning System (GPWS) and Airborne Collision Avoidance System (ACAS). With ACAS, SSR signals from other aircraft provide the pilot of potential conflicting traffic - the system can only "see" SSR equipped aircraft and operates independently of ground based equipment.

The encoding altimeter sends pressure altitude to ATC, regardless of the aircraft altimeter setting. An ATC computer corrects the received pressure altitude for local pressure, so that all targets have a consistent readout.

Enroute and Terminal Control Radar

SSR is the main source of enroute information. SSR is a long range radar in the +200 NM range. In general, SSR is complemented by short range PSR for terminal operations. Terminal Surveillance Radar (TSR) consists of PSR (short range) to 80 NM and SSR to 250 NM. Independent SSR (ISSR) is another long range radar system (250 NM).

Weather RADAR

Airborne weather radar is a primary radar system that consists of an antenna enclosed within a fibreglass RADOME (normally located in the nose of the aircraft), a receiver-transmitter, and a radar scope. The radar scope shows water droplets within clouds. Water droplets indicate potential turbulence, heavy rain and hail. The larger the raindrops, the stronger the return. The closer the aircraft is to a storm cell, the brighter the return. Hooks, fingers, scalloped edges and nodules are usually associated with heavy turbulence and hail.

Because weather radar only indicates water droplets, using weather radar while in moderate rain makes it difficult to locate thunderstorms. Attenuation occurs when there is a rainfall area between the aircraft and a target cell. The brightest echoes are caused by large water droplets (rather than by many small droplets). The reflective activity is greatest in wet snow, and least in dry snow.

The GAIN control on a weather radar unit is used to adjust the sensitivity of the receiver. The TILT control adjusts the pitch of the antenna beam relative to the horizon. Weather radar ATTENUATION or weakening of returns is caused by precipitation i.e. in heavy rain, a radar signal may be partially or totally absorbed by the target (rain) in the foreground and targets beyond it cannot be seen. Mountainous terrain can also block weather radar returns.

Ice or heavy rain on or near the radome results in radar attenuation. RAIN GRADIENT is the change in rainfall density versus distance and is a reliable indication of storm cell severity. The CONTOUR MODE is selected to identify areas of heavy precipitation and steep rain gradients. Returning signals which exceed a pre-determined receiver strength are blanked for the display and appear as a black hole which indicates an area of heavy precipitation. A steep rain gradient is displayed as a narrow band between the black hole and the edge of the return when the contour mode is selected. The steeper the gradient, the greater the associated turbulence and possibility of hail. Use the radar in normal mode while searching for weather returns, then switch to contour mode momentarily to examine a return, and go back to normal to search for additional returns. If the radar is left in contour mode, areas of heavy precipitation may contour completely off the scope.

Mapping Radar

Although civilian aircraft generally use primary radar only for weather radar purposes, in the military, various uses are made of primary aircraft radar for mapping purposes. In such cases, a normal aircraft radar return is used by a military navigator to determine various ranges and positions during enroute navigation. Mapping radar techniques at best are only a secondary means of navigation.

7.13 Global Positioning System (GPS)

GPS is a satellite based navigation system designed by the US military. It has many advantages, such as avoiding line of sight difficulties and high accuracy. GPS operation is based on ranging and triangulation from a group of satellites in space which act as precise reference points. There are about 24 satellites dedicated for GPS.

Precise timing is the key to GPS. Each satellite has four atomic clocks on board, guaranteeing an acuracy of one billionth of a second. Each atomic clock broadcasts a digital code that appears to be selected randomly but that is actually repeated every milisecond; this is called the PSEUDORANDOM NOISE (PRN) code. Receivers start generating the same code at the same time. Code matching establishes how long ago the signal left the satellite. By knowing the speed of light and the average effects of the ionosphere and atmosphere, the receiver calculates how far it is from the satellite; this is called a

PSEUDORANGE.

In theory, the receiver would need three pseudoranges to calculate a 3-D position, but, in practice, four are required because the receiver clock can never be as accurate as the satellite clocks: the fourth satellite allows calculation of the time difference. GPS also gives a direct reading of velocity, both speed and direction of motion, by measuring the rate of change of the pseudoranges.

GPS accuracy depends on the accuracy of satellite clocks and broadcast orbits, but, more importantly, it depends on the delay as the signal passes through the ionosphere (a zone of charged particles several hundred kilometres above the Earth). This delay varies from day to night and is affected by solar activity. Current receivers contain a model of average day/night delay, but this model does not completely remove ionospheric error.

Another key to GPS accuracy is the relative position of satellites in the sky, or SATELLITE GEOMETRY. When satellites are widely spread, geometry and accuracy are best. If satellites are clustered in a small area of the sky, geometry and accuracy are worst. On average, GPS horizontal and vertical positions are accurate to within about 10 metres and 15 metres, respectively.

An important advantage of GPS is that it permits direct route navigation based on WAYPOINTS which can be pre-programmed into the units.

Subject to Transport Canada installation requirements and certifications, GPS can be used for enroute, terminal and non-precision approach phases of flight. GPS OVERLAY APPROACHES are based on using GPS to supplement a standard non-precision approach and all non-GPS components must be fully operational. Once approved, the approach may be certified as a GPS STAND ALONE APPROACH. The advantage to a GPS stand alone approach is that the procedure turn can usually be eliminated (by a series of waypoints in a "T" pattern) and the approach will have greater accuracy meaning reduced minima.

GPS units for use in IFR must meet certain international standards for ACCURACY, INTEGRITY, CONTINUITY, and AVAILABILITY.

ACCURACY is the measure of position error. GNSS position error is the difference between the estimated and the actual position.

INTEGRITY is the measure of the trust that can be placed in the correctness of the information supplied by the total system. Integrity is by far the most critical factor in air navigation. Integrity includes the system's ability to tell the user in a timely fashion when the system must not be used for the intended operation. GPS integrity resides in the global ground-monitoring and control system that constantly evaluates satellite performance and relays corrections when necessary. This system can also shut down a faulty satellite within minutes, if necessary. The main integrity component in SatNav is the RAIM (Receiver Autonomous Integrity Monitoring) function of IFR approved receivers. RAIM works on the premise that four satellites are required for a navigation solution. When the receiver sees more than four satellites, the avionics can calculate and compare navigation solution results from each group of four and estimate its confidence in the accuracy of the aircraft position. It will warn the pilot when alert limits for the phase of flight (2 NM en route, 1 NM terminal and 0.3 NM on non-precision approach) are exceeded. The pilot must then revert to an alternate navigation source, or overshoot if this happens during a GPS approach.

CONTINUITY is the capability of the system to perform its function without non-scheduled interruptions during the intended operation. This is expressed as a probability. For example, there must be a high probability that the service remains available throughout a full instrument approach procedure.

AVAILABILITY is the portion of time during which the system is delivering the required accuracy, integrity and continuity.

7.14 Lightning Detection Equipment

A STORMSCOPE is lightning detection equipment that is based on an antenna and CRT that displays electrical energy associated with static electrical discharges. It is NOT a radar system. The stormscope display shows ranges of 25, 50, 100 and 200 NM in a typical display.

A stormscope offers the advantage that it will show areas of thunderstorm activity regardless of whether rain is present as with weather radar. As well, stormscopes are not subject to line of sight limitations.

7.15 Emergency Locator Transmitter

ELT's operate on 121.5 MHz, and send a distinctive siren-like sound and signal to locate downed aircraft. The ELT battery should be checked periodically. There are 5 kinds of ELT: automatic ejectable, fixed, automatic portable, personnel and water activated. ELT's can be tested only during the first 5 minutes of any hour and for not more than 5 seconds duration. To test an ELT,

monitor 121.5 MHz and activate the ELT.
Refer to s. 1.43 for details on regulations concerning when an ELT is required to be carried on board.
All accidental ELT activations should be reported to the nearest ATS unit. If the activation was the result of a malfunction of the ELT, it should be reported to the Regional Manager, Aviation Licensing with the model of ELT, position of function switch and description of occurrence. If a pilot hears an ELT signal in flight, they should notify the nearest ATS unit of position, altitude, time first signal heard, and ELT signal strength. The position, altitude and time when contact is lost should also be given. Pilots should not attempt a SAR operation. If the ELT signal remains constant, it may be the pilot's own aircraft.

7.16 DF Direction Finder

Many control towers have a VHF direction finder, by which controllers can give pilots a bearing to fly direct to the tower based on VHF transmissions from the VHF transceiver on the aircraft. DF steering service can also be provided to military aircraft utilizing UHF radios. Consult the CFS for information on whether a particular airport has DF service available.

7.17 TCAS Systems

TCAS refers to Traffic Alert and Collision Avoidance System and is based on aircraft transponder signals. A TCAS computer on the aircraft monitors bearing, range and azimuth information from other aircraft based on transponder interrogation and reply signals, and uses this information to determine whether a potential for collision exists and display appropriate visual/auditory warnings. Refer to s. 1.44 of this text for further information on TCAS requirements, use of TCAS, and pilot-controller actions and interchanges. It is important to remember that TCAS will only function where other aircraft have TCAS equipment installed.

Notes:

Notes:

Notes:

Notes:

Notes:

Transport Canada Subject Codes

INSTRUCTIONS: Transport Canada indicates the subject material for study and review in preparation for the ATPL in the publication *Study and Reference Guide Airline Transport Pilot Licence - Aeroplane, Nineteenth Edition October 2002 - TP 690E*. Following your completion of each written (SARON and SAMRA), you will receive on your scoring results sheet a listing of any subjects scored incorrectly based on the following Transport Canada subjects. The number in brackets following each Transport Canada subject refers to the corresponding subject reference number used in this course. Applicants for the ATPL are expected to have already mastered all materials at the Commerial Pilot Licence level. Where we have used the term "CPL" below, this refers to a subject that we consider to be pre-existing commercial pilot knowledge.

1. Air Law and Procedures

CARs PART I - GENERAL PROVISIONS

101 - Interpretation
101.01 - Interpretation [1.07]

103 - Administration and Compliance
103.02 - Inspection of Aircraft, Requests for Production of Documents and Prohibitions [1.07]
103.03 - Return of Canadian Aviation Documents [1.07]
103.04 - Record Keeping [1.07]

CARs PART III - AERODROMES AND AIRPORTS

300 - Interpretation
300.01 - Interpretation [CPL]

301 - Aerodromes
301.01 - Application [CPL]
301.04 - Markers and Markings [CPL]
301.06 - Wind Direction Indicator [CPL]
301.07 - Lighting [1.14]
301.08 - Prohibitions [CPL]
301.09 - Fire Prevention [CPL]

302 - Airports
302.10 - Prohibitions [CPL]
302.11 - Fire Prevention [CPL]

CARs PART IV - PERSONNEL LICENSING AND TRAINING

400 - General
400.01 - Interpretation [1.03]

401 - Flight Crew Permits, Licences and Ratings

401.03 - Requirements to Hold a Flight Crew Permit, Licence or Rating [1.03]
401.04 - Flight Crew Members of Aircraft Registered in Contracting States other than Canada [1.03]
401.05 - Recency Requirements [1.03]
401.08 - Personal Logs [1.03]
401.10 - Crediting of Flight Time Acquired by a Co-pilot [1.03]
401.34 - ATPL Privileges - Aeroplane [1.03]
401.47 - Instrument Rating Privileges [1.06]
401.48 - Instrument Rating Period of Validity [1.06]
401.52 - Second Officer Rating [1.05]
401.53 - Second Officer Privileges [1.05]
401.61 - Flight Instructor Ratings [1.03]

404 - Medical Requirements
404.03 - Requirement to Hold a Medical Certificate (MC) [1.02]
404.04 - Issuance, Renewal and Validity Period of MC [1.02]
404.06 - Prohibitions Regarding Exercise of Privileges [1.02]
404.10 - MC Requirements for Personnel Licences [1.02]
404.18 - Permission to Continue to Exercise the Privileges of a Licence or Rating [1.02]

CARs PART VI - GENERAL OPERATING AND FLIGHT RULES

600 - Interpretation
600.01 - Interpretation [1.20]

601 - Airspace Structure, Classification and Use
601.01 - Airspace Structure [1.20]
601.02 - Airspace Classification [1.20]
601.03 - Transponder Airspace [1.17]
601.04 - IFR or VFR Flight in Class F Special Use Restricted Airspace or Class F Special Use Advisory Airspace [1.20]
601.05 - IFR Flight in Class A, B, C, D or E Airspace or Class F Special Use Restricted or Advisory Controlled Airspace [1.20]
601.06 - VFR Flight in Class A Airspace [1.20]
601.07 - VFR Flight in Class B Airspace [1.20]
601.08 - VFR Flight in Class C Airspace [1.20]
601.09 - VFR Flight in Class D Airspace [1.20]
601.14 - Interpretation [1.20]

601.15 - Forest Fire Aircraft Operating Restrictions [1.20]
601.16 - Issuance of NOTAM for Forest Fire Aircraft Operating Restrictions [1.20]
601.17 - Exceptions [1.20]

602 - Operating and Flight Rules

GENERAL
602.01 - Reckless or Negligent Operation of Aircraft [1.27]
602.02 - Fitness of Flight Crew Members [1.27]
602.03 - Alcohol or Drugs - Crew Members [1.27]
602.04 - Alcohol or Drugs - Passengers [1.27]
602.06 - Smoking [1.27]
602.07 - Aircraft Operating Limitations [1.27]
602.08 - Portable Electronic Devices [1.27]
602.09 - Fuelling with Engines Running [1.27]
602.10 - Starting and Ground Running of Aircraft Engines [1.27]
602.11 - Aircraft Icing [1.27]
602.12 - Overflight of Built-up Areas or Open-Air Assemblies of Persons during Take-offs, Approaches and Landings [1.27]
602.13 - Take-offs, Approaches and Landings within Built-up Areas of Cities and Towns [1.27]
602.14 - Minimum Altitudes and Distances [1.27]
602.15 - Permissible Low Altitude Flight [1.27]
602.17 - Carriage of Persons during low Altitude Flight [1.27]
602.19 - Right of Way - General [CPL]
602.20 - Right of Way - Aircraft Maneuvering on Water [CPL]
602.21 - Avoidance of Collision [CPL]
602.22 - Towing [CPL]
602.23 - Dropping of Objects [CPL]
602.24 - Formation Flight [CPL]
602.25 - Entering or Leaving and Aircraft in Flight [CPL]
602.26 - Parachute Descents [CPL]
602.27 - Aerobatic Manoeuvres - Prohibited Areas and Flight Conditions [CPL]
602.28 - Aerobatic Manoeuvres with Passengers [CPL]
602.30 - Fuel Dumping [1.27]
602.31 - Compliance with Air Traffic Control

Instructions and Clearances [1.22]
602.32 - Airspeed Limitations [1.29]
602.33 - Supersonic Flight [1.27]
602.34 - Cruising Altitudes and Cruising Flight levels [1.23]
602.35 - Altimeter Setting and Operating Procedures in the Altimeter-Setting Region [1.15]
602.36 - Altimeter Setting and Operating Procedures in the Standard Pressure Region [1.15]
602.37 - Altimeter Setting and Operating Procedures in Transition Between Regions [1.15]
602.38 - Flight over the High Seas [1.27]
602.39 - Transoceanic Flight [1.27]
602.40 - Landing at or Take-off from an Aerodrome at Night [CPL]

OPERATIONAL AND EMERGENCY EQUIPMENT REQUIREMENTS
602.58 - Prohibition [1.33]
602.59 - Equipment Standards [1.33]
602.60 - Requirements for Power-Driven Aircraft [1.33]
602.61 - Survival Equipment - Flights over Land [1.33]
602.62 - Life Preservers and Personal Flotation Devices [1.33]
602.63 - Life Rafts and Survival Equipment - Flights over Water [1.33]

FLIGHT PREPARATION, FLIGHT PLANS & FLIGHT ITINERARIES
602.70 - Interpretation [1.18]
602.71 - Pre-Flight Information [1.18]
602.72 - Weather Information [1.18]
602.73 - Requirements to file a Flight Plan or Flight Itinerary [1.18]
602.74 - Contents of a Flight Plan or Fight Itinerary [1.18]
602.75 - Filing of a Flight Plan or Fight Itinerary [1.18]
602.76 - Changes in the Flight Plan [1.18]
602.77 - Requirement to File an Arrival Report [1.18]
602.78 - Contents of an Arrival Report [1.18]
602.79 - Overdue Aircraft Reports [1.18]
602.86 - Carry-on Baggage, Equipment and Cargo [1.18]
602.87 - Crew Member Instructions [1.18]
602.88 - Fuel Requirements [1.18]
602.89 - Passenger Briefings [1.18]

OPERATIONS AT OR IN THE VICINITY OF AN AERODROME
602.96 - General [CPL]
602.97 - VFR and IFR Aircraft Operations at Uncontrolled Aerodromes within an MF Area [CPL]
602.98 - General MF Reporting Procedures [CPL]
602.99 - MF Reporting Procedures before Entering Manoeuvring Area [CPL]
602.100 - MF Reporting Procedures on

Departure [CPL]
602.101 - MF Reporting Procedures on Departure [CPL]
602.102 - MF Reporting Procedures when Flying Continuous Circuits [CPL]
602.103 - Reporting Procedures when Flying through an MF Area [CPL]
602.104 - Reporting Procedures for IFR Aircraft when Approaching or Landing at an Uncontrolled Aerodrome [CPL]
602.105 - Noise Operating Criteria
602.106 - Noise Restricted Runways

VISUAL FLIGHT RULES
602.114 - Minimum Visual Meteorological Conditions for VFR Flight in Controlled Airspace [1.14]
602.115 - Minimum Visual Meteorological Conditions for VFR Fight in Uncontrolled Airspace [1.14]
602.116 - VFR Over-the-Top [1.14]
602.117 - Special VFR Flight [1.14]

INSTRUMENT FLIGHT RULES
602.121 - General Requirements [1.19]
602.122 - Alternate Aerodrome Requirements [1.19]
602.123 - Alternate Aerodrome Weather Minima [1.19]
602.124 - Minimum Altitudes to Ensure Obstacle Clearance [1.19]
602.125 - En Route IFR Position Reports [1.19]
602.126 - Take-off Minima [1.19]
602.127 - Instrument Approaches [1.19]
602.128 - Landing Minima [1.19]
602.129 - Approach Ban - General [1.19]
602.130 - Approach Ban - Cat III [1.19]

RADIOCOMMUNICATIONS
602.136 - Continuous Listening Watch [1.19]
602.137 - Two-way Radiocommunication Failure in IFR Flight [1.19]
602.138 - Two-way Radiocommunication Failure in VFR Flight [1.14]

EMERGENCY COMMUNICATIONS AND SECURITY
602.143 - Emergency Radio Frequency Capability [1.36]
602.144 - Interception Signals, Interception of Aircraft and Instructions to Land [1.36]
602.145 - ADIZ [1.30]
602.146 - ESCAT Plan [1.30]

604 - Private Operator Passenger Transportation

GENERAL
604.01 - Application [1.13]

FLIGHT OPERATIONS
604.10 - Checklist [N/A]
604.11 - Operational Flight Data Sheet [N/A]
604.12 - VFR Flight Minimum Flight Visibility -

Uncontrolled Airspace [N/A]
604.13 - No Alternate Aerodrome - IFR Flight [N/A]
604.14 - Take-off Minima [N/A]
604.15 - Instrument Approach Procedures [N/A]
604.16 - Flight Attendant Requirement [N/A]
604.17 - Cabin Safety Procedures [N/A]
604.18 - Briefing of Passengers [N/A]

FLIGHT TIME AND FLIGHT DUTY TIME LIMITATIONS AND REST PERIODS
604.26 - Flight Time Limitations [N/A]
604.27 - Flight Duty time Limitations and Rest Periods [N/A]
604.28 - Split Flight Duty Time [N/A]
604.29 - Extension of Flight Duty Time [N/A]
604.30 - Unforeseen Operational Circumstances [N/A]
604.31 - Delayed Reporting Time [N/A]
604.32 - Requirements for Time Free of Duty [N/A]
604.33 - Flight Crew Positioning [N/A]

EMERGENCY EQUIPMENT
604.38 - Survival Equipment [N/A]
604.39 - First Aid Kits [N/A]
604.40 - Protective Breathing Equipment [N/A]
604.41 - Hand-Held Fire Extinguishers [N/A]

PERSONNEL REQUIREMENTS
604.65 - Designation of Pilot-in-command and Second-in-command [N/A]
604.66 - Crew Member Qualifications [N/A]
604.68 - Validity Period [N/A]
604.73 - Training Program [N/A]

MANUALS
604.80 - Requirements relating to Operations Manual [N/A]
604.81 - Contents of Operations Manual [N/A]
604.82 - Distribution of Operations Manual [N/A]
604.83 - Aircraft Operating Manual [N/A]
604.84 - Standard Operating Procedures [N/A]

605 - Aircraft Requirements

GENERAL
605.03 - Flight Authority [1.28]
605.04 - Availability of Aircraft Flight Manual [1.28]
605.05 - Markings and Placards [1.28]
605.06 - Aircraft Equipment Standards and Serviceability [1.28]
605.07 - Minimum Equipment Lists [1.28]
605.08 - Unserviceable and Removed Equipment - General [1.28]
605.09 - Unserviceable and Removed Equipment - Aircraft with a Minimum Equipment List [1.28]
605.10 - Unserviceable and Removed Equipment - Aircraft without a Minimum Equipment List [1.28]

AIRCRAFT EQUIPMENT REQUIREMENTS
605.14 - Power-Driven Aircraft - Day VFR [1.14]
605.15 - Power-driven Aircraft - VFR OTT [1.14]
605.16 - Power-driven Aircraft - Night VFR [1.14]
605.17 - Use of Position and Anti-collision Lights [CPL]
605.22 - Seat and Safety-Belt Requirements [1.34]
605.23 - Restraint System Requirements [1.34]
605.24 - Shoulder Harness Requirements [1.34]
605.25 - General Use of Safety Belts and Restraint Systems [1.34]
605.26 - Use of Passenger Safety Belts and Restraint Systems [1.34]
605.27 - Use of Crew Member Safety Belts [1.34]
605.28 - Child Restraint System [1.34]
605.29 - Flight Control Locks [CPL]
605.30 - De-Icing and Anti-Icing Equipment [1.41]
605.31 - Oxygen Equipment and Supply [1.35]
605.32 - Use of Oxygen [1.35]
605.33 - Flight Data Recorder and Cockpit Voice Recorder Requirements [1.39]
605.34 - Use of Flight Data Recorders and Cockpit voice Recorders [1.39]
605.35 - Transponder and Automatic Pressure Altitude Reporting Equipment [1.17]
605.36 - Altitude Alerting Systems or Devices [1.40]
605.37 - Ground Proximity Warning Systems [1.40]
605.38 - ELT [1.43]
605.39 - Use of ELTs [1.43]
605.40 - ELT Activation 1.43]
605.41 - Standby Attitude Indicator [1.42]
605.84 - Aircraft Maintenance - General [1.28]
605.85 - Maintenance Release and Elementary Work [CPL]
605.86 - Maintenance Schedule [CPL]
605.87 - Transfer of Aeronautical Products between Maintenance Schedules [N/A]
605.88 - Inspection after Abnormal Occurrences [N/A]

TECHNICAL RECORD
605.93 - Technical Records - General [1.28]
605.94 - Journey Log Requirements [1.28]
605.95 - Journey Log - Carrying on Board [1.28]
605.96 - Requirements for Technical Records other than the Journey Log [1.28]
605.97 - Transfer of Records [1.28]

Miscellaneous
606.01 - Munitions of War [N/A]
606.03 - Synthetic Flight Trainer [N/A]

CARS PART VII - COMMERCIAL AIR SERVICES

700 - Commercial Air Services

GENERAL
700.01 - Definitions [1.08]

FLIGHT TIME AND FLIGHT DUTY TIME LIMITATIONS AND REST PERIODS
700.15 - Flight Time Limitations [1.08]
700.16 - Flight Duty Time Limitations and Rest Periods [1.08]
700.17 - Unforeseen Operational Circumstances [1.08]
700.18 - Delayed Reporting Time [1.08]
700.19 - Time Free From Duty [1.08]
700.20 - Flight Crew Positioning [1.08]
700.21 - Flight Crew Members on Reserve [1.08]
700.22 - Long Range Flights [1.08]
700.23 - Controlled Rest on the Flight Deck [1.08]

702 - Aerial Work

GENERAL
702.01 - Application [1.10]

FLIGHT OPERATIONS
702.13 - Flight Authorization [1.10]
702.14 - Operational Flight Plan [1.10]
702.16 - Carriage of Persons [1.10]
702.17 - VFR Flight Minimum Flight Visibility - Uncontrolled Airspace [1.10]
702.18 - Night, VFR OTT and IFR Operations [1.10]
702.20 - Aircraft Operating over Water [1.10]
702.22 - Built-up Area and Aerial Work Zone [1.10]
702.23 - Briefing of Persons other than Flight Crew Members [1.10]

AIRCRAFT EQUIPMENT REQUIREMENTS
702.42 - Night and IMC Flights [1.10]
702.43 - Additional Equipment for Single-pilot Operations [1.10]
702.44 - Shoulder Harness [1.10]

PERSONNEL REQUIREMENTS
702.64 - Designation of Pilot-in-Command and Second-in-command [1.10]
702.65 - Flight Crew Member Qualifications [1.10]
702.66 - Validity Period [1.10]

MANUALS
702.83 - Distribution of Company Operations Manuals [1.10]
702.84 - Standard Operating Procedures [1.10]

703 - Air Taxi Operations

GENERAL
703.01 - Application [1.09]

FLIGHT OPERATIONS
703.17 - Flight Authorization [1.09]
703.18 - Operational Flight Plan [1.09]
703.20 - Fuel Requirements [1.09]
703.21 - Admission to Pilot's Compartment [1.09]
703.22 - Transport of Passengers in Single-engined Aircraft [1.09]
703.23 - Aircraft Operating over Water [1.09]
703.24 - Number of Passengers in Single-engined Aircraft [1.09]
703.26 - Simulation of Emergency Situations [1.09]
703.27 - VFR Flight Obstacle Clearance Requirements [1.09]
703.28 - VFR Flight Minimum Visibility - Uncontrolled Airspace [1.09]
703.29 - VFR Flight Weather Conditions [1.09]
703.30 - Take-off Minima [1.09]
703.31 - No Alternate Aerodrome - IFR Flight [1.09]
703.32 - Enroute Limitations [1.09]
703.33 - VFR OTT Flight [1.09]
703.34 - Routes in Uncontrolled Airspace [1.09]
703.35 - Instrument Approach Procedures [1.09]
703.37 - Weight and Balance Control [1.09]
703.38 - Passenger and Cabin Safety Procedures [1.09]
703.39 - Briefing of Passengers [1.09]

AIRCRAFT EQUIPMENT REQUIREMENTS
703.64 - General Requirements [1.09]
703.65 - Airborne Thunderstorm Detection and Weather Radar Equipment [1.09]
703.66 - Additional Equipment for Single-pilot Operations [1.09]
703.67 - Protective Breathing Equipment [1.09]
703.68 - First Aid Oxygen [1.09]
703.69 - Shoulder Harness [1.09]

EMERGENCY EQUIPMENT
703.82 - Equipment Standards and Inspection [1.09]

PERSONNEL REQUIREMENTS
703.86 - Minimum Crew [1.09]
703.87 - Designation of Pilot-in-Command and Second-in-Command [1.09]
703.88 - Flight Crew Member Qualifications [1.09]
703.91 - Validity Period [1.09]

MANUALS
703.106 - Distribution of Company Operations Manual [1.09]
703.107 - Standard Operating Procedures [1.09]

704 - Commuter Operations

GENERAL
704.01 - Application [1.11]

FLIGHT OPERATIONS
704.12 - Operating Instructions [1.11]
704.13 - General Operational Information [1.11]
704.16 - Flight Authorization [1.11]
704.17 - Operational Flight Plan [1.11]
704.19 - Checklist [1.11]
704.20 - Fuel Requirements [1.11]
704.22 - Simulation of Emergency Situations [1.11]

CANADIAN MINIMUM NAVIGATION PERFORMANCE SPECIFICATIONS (CMNPS) AIRSPACE
1 General [1.37]
2 Partial or Complete Loss of Navigation Capability [1.37]
3 Position Reporting [1.37]

CANADIAN MINIMUM NAVIGATION PERFORMANCE SPECIFICATIONS (CMNPS) CERTIFICATION
1 General [1.37]
2 Certification [1.37]
3 Navigation System Requirements [1.37]
4 Transition Between CMNPS Airspace and the Canadian Domestic Airway Structure [1.37]
5 Separation Minima [1.37]

ATC SPECIAL PROCEDURES
1 Adherence to Mach Number [1.44]
2 Parallel and Offset Procedures [1.44]
3 Structured Airspace [1.44]
4 Required Navigation Performance Capability Airspace (RNPC) [1.44]
5 Canadian Minimum Navigation Performance Specifications Airspace (CMNPS) [1.44]
6 Canadian Domestic Routes [1.44]
7 Canadian Track Structures [1.44]
8 Traffic Alert and Collision Avoidance Systems (TCAS) [1.44]

NORTH ATLANTIC OPERATIONS
1 General Aviation Aircraft [1.45]
2 North American Routes (NAR) [1.45]
3 NAT Organized Track System [1.45]
4 Flight Rules and Flight Planning Procedures [1.45]
5 Clearances, Position Reports, Communications Failure [1.45]
6 Transponder Operation [1.45]

REDUCED VERTICAL SEPARATION MINIMA (RVSM)
1 General [1.46]
2 RVSM Airspace [1.46]
3 RVSM Transition Airspace [1.46]
4 Air Traffic Control (ATC) Procedures [1.46]
5 Aircraft Requirements [1.46]

2. Airframes, Engines, Propellers and Aircraft Systems

AIRFRAMES
1 Flight Controls [4.01]
2 Flaps [2.01]
3 Slots/Slats [2.01]
4 Spoilers [2.01]
5 Wing Fences [2.01]
6 Winglets [2.01]
7 Canards [2.01]
8 Vortex Generators [2.01, 4.01]
9 Trimming Devices [2.01]

ENGINES
1 Principles of Reciprocating Engines [2.02]
2 Handling Procedures for Reciprocating Engines [2.02]
3 Principles of Turbo-prop Engines [2.03]
4 Handling Procedures for Turbo-prop Engines [2.03]
5 Principles of Turbo-jet Engines [2.03]
6 Handling Procedures for Turbo-jet Engines [2.03]
7 Engine Controls [2.02 - 2.03]

PROPELLERS
1 Propeller Thrust and Torque [2.04]
2 Geometric and Effective Pitch [2.04]
3 Slipstream, Gyroscopic Effect and Asymmetric Thrust [2.04]
4 Controls [2.04]
5 Ground and Flight Range [2.04]
6 Constant Speed [2.04]
7 Feathering [2.04]
8 Reversing [2.04]

AIRCRAFT SYSTEMS
1 Fuel [2.05, 2.06]
2 Oil [2.08]
3 Electrical [2.09]
4 Hydraulic [2.10]
5 Pneumatic [2.15]
6 Warning (E.G. Ice, Fire, GPWS and Altitude Alert) [2.00, 3.00]
7 Fire Protection [2.14]
8 Heating [2.17]
9 De-Icing and Anti-icing [2.13]
10 Oxygen [2.16]
11 Air Conditioning [2.19]
12 Pressurization [2.18]
13 Landing Gear and Brakes [2.12]
14 Autopilot [2.20]
15 Avionics [2.21]
16 Flight Controls [2.22]

3. Meteorology

THE EARTH'S ATMOSPHERE
1 Properties [6.01]
2 Vertical Structure [6.01]
3 ICAO Standard Atmosphere [6.01]

ATMOSPHERIC PRESSURE
1 Pressure Measurements [6.01]
2 Station Pressure [6.01]
3 Mean Seal Level Pressure [6.01]
4 Pressure Systems and Their Variations [6.01]
5 Effects of Temperature [6.01]
6 Horizontal Pressure Differences [6.01]

METEOROLOGICAL ASPECTS OF ALTIMETRY
1 Pressure Altitude [6.02]
2 Density Altitude [6.02]
3 True Altitude [3.05, 5.02]
4 Altimeter Setting [6.02]
5 Effects of both Pressure and Temperature [6.02]

TEMPERATURE
1 Heating and Cooling of the Atmosphere - Convection Advection/Radiation [6.03, 6.04]
2 Horizontal Differences [6.03, 6.04]
3 Temperature Variations with Altitude [6.03, 6.04]
4 Inversions [6.03]
5 Isothermal Layers [6.03]

MOISTURE
1 Relative Humidity/Dewpoint [6.04]
2 Sublimation/Condensation [6.04]
3 Cloud Formation [6.04]
4 Precipitation [6.04]
5 Saturated/Dry Adiabatic Lapse Rates [6.04]

STABILITY AND INSTABILITY
1 Lapse Rate and Stability [6.04, 6.05]
2 Modification of Stability [6.09]
3 Characteristics of Stable/Unstable Air [6.05]
4 Surface Heating and Cooling [6.05]
5 Lifting Process [6.05]
6 Subsidence/Convergence [6.05]

CLOUDS
1 Classification [6.06]
2 Formation [6.04]
3 Types and Recognition [6.06]
4 Associated Precipitation and Turbulence [6.06]

TURBULENCE
1 Convection [6.07]
2 Mechanical [6.07]
3 Orographic [6.07]
4 Clear Air Turbulence [6.07, 6.12]
5 VIRGA - Evaporation Cooling [6.06]
6 Reporting Criteria [6.07]
7 Mountain Waves [6.08]

WIND
1 Pressure Gradient [6.08]
2 Deflection Caused by the Earth's Rotation [6.08]
3 Low Level Winds-Variation in Surface Wind [6.07]
4 Friction [6.07]
5 Centrifugal Force [6.07]
6 Veer and Back [6.07]
7 Squalls and Gusts [6.07]
8 Diurnal Effects [6.07]
9 Land and Sea Breezes [6.07]
10 Katabatic/Anabatic Effects [6.07]
11 Topographical Effects [6.07]
12 Wind Shear, Types and Causes [6.07, 4.16]

JET STREAMS
1 Frontal Jet Streams [6.12]
2 Wind Distribution/Location [6.12]
3 Temperature Distribution [6.12]
4 Seasonal Variations in Latitude and Speed [6.12]
5 Arctic Stratospheric Jets [6.12]
6 Subtropical Jet Streams [6.12]
7 Turbulence [6.12]

AIR MASSES

1 Definition and Characteristics [6.09]
2 Formation [6.09]
3 Classification [6.09]
4 Modification [6.09]
5 Factors that Determine Weather [6.09]
6 Seasonal and Geographic Effects [6.09]
7 Air Masses Affecting North America [6.09]

FRONTS

1 Structure [6.10]
2 Types [6.10]
3 Formation [6.10]
4 Cross-sections [6.10]
5 Discontinuities Across Fronts [6.10]
6 Frontal Waves and Occlusions [6.10]
7 Frontogensis and Frontolysis [6.10]

FRONTAL WEATHER

1 Warm Front [6.10]
2 Cold Front [6.10]
3 Stationary Front [6.10]
4 TROWAL and Upper Fronts [6.10]

AIRCRAFT ICING

1 Formation [6.13]
2 Types of Ice [6.13]
3 Reporting Criteria [6.13]
4 Cloud Types and Icing [6.13]
5 Freezing Rain and Drizzle [6.10, 6.13]
6 Icing in Clear Air (Hoar Frost) [6.13]
7 Collection Efficiency [6.13]
8 Aerodynamic Heating [2.13, 6.13]

THUNDERSTORMS

1 Requirements for Development [6.14]
2 Life Cycle [6.14]
3 Classification - Air Mass, Frontal, Squall Line, Convective [6.14]
4 Tornadoes and Hurricanes [6.15]
5 Hazards - Turbulence, Hail, Rain, Icing, Altimetry, Lightning, Gust Fronts, Downbursts and Microbursts [6.14]

SURFACE BASED LAYERS

1 Fog Formation [6.16]
2 Fog Types [6.16]
3 Haze and Smoke [6.16]
4 Blowing Obstructions to Vision [6.16]

METEOROLOGICAL SERVICES AVAILABLE TO PILOTS

1 Aviation Weather Briefing Service (ABWS) [6.17]
2 Aviation Weather Information Service (AWIS) [6.17]
3 Flight Service Stations (FSS) [6.17]
4 Weather Broadcasts by Flight Service Stations [6.17]
5 Atmospheric Environment Service Weather Briefing [6.17]
6 Transcribed Weather Broadcasts (TWB) [6.17]
7 DUATS - Commercial Weather Service [6.17]
8 Automatic Terminal Information Service (ATIS) [6.17]
9 VOLMET (HF) Broadcast [6.17]
10 Pilots Automatic Telephone Weather Answering Service (PATWAS) [6.17]

AVIATION WEATHER REPORTS

1 Aviation Routine Weather Report (METAR) [6.21]
2 SPECI [6.21]
3 Decoding [6.21]
4 AWOS
5 Pilot Reports (PIREP/AIREP) [6.23]

AVIATION FORECASTS

1 Times Issued/Validity Periods [6.18, 6.19, 6.20]
2 Decoding [6.18, 6.19, 6.20]
3 Graphical Area Forecasts (GFA)/AIRMET [6.18, 6.24]
4 Terminal Area Forecasts (TAF) [6.19]
5 Upper Level Winds and Temperature Forecasts (FD) [6.20]
6 Significant In-flight Weather Warning Messages (SIGMET) [6.22]

WEATHER MAPS AND PROGNOSTIC CHARTS

1 Times Issued/Validity Periods [6.27, 6.28]
2 Symbols/Decoding [6.27, 6.28]
3 Surface Weather Map [6.27]
4 Prognostic Surface Chart [6.27]
5 Upper Level Charts - ANAL (850mb, 700mb, 500mb & 250mb) [6.28]
6 Upper Level Charts - PROG (FL240, FL 340 & FL450) [6.28]
7 Significant Weather Prognostic Chart FL 100 - 250 (700 - 400mb) & FL250 - 600 (400 - 100mb) [6.28]

4. Instruments

FLIGHT INSTRUMENTS - PRINCIPLES AND OPERATIONAL USE

1 Pitot Static System [3.01]
2 Airspeed Indicator [3.02]
3 Machmeter [3.03]
4 Altimeter and Encoding Altimeter [3.03]
5 Radio/Radar Altimeter [3.03]
6 Outside Air Temperature [3.07]
7 Turn-and-Bank Indicator/Turn Co-ordinator [3.11]
8 Vertical Speed Indicator (VSI) [3.04]
9 Heading Indicator [3.10]
10 Attitude Indicator (AI) [3.09]
11 Flight Director [3.12]
12 Radio Magnetic Indicator (RMI) [3.15]
13 Horizontal Situation Indicator (HSI) [3.13]
14 Angle of Attack Indicator [3.16]

FLIGHT MANAGEMENT INSTRUMENTS

1 Flight Management System (FMS) [3.14]
2 Electronic Flight Instrument System (EFIS) [3.14]

ENGINE INSTRUMENTS - PRINCIPLES AND USE

1 Tachometer [3.17]
2 Manifold Pressure [3.17]
3 Oil Pressure [3.17]
4 Oil Temperature [3.17]
5 Exhaust Gas Temperature [3.17]
6 Cylinder Head Temperature [3.17]
7 Carburettor Air Temperature [3.17]
8 Intake Air Temperature [3.17]
9 Fuel Pressure [3.17]
10 Fuel Flow [3.17]
11 Torquemeter [3.17]
12 Engine Pressure Ratio (EPR) [3.17]
13 Turbine Temperature (ITT/TIT) [3.17]

AIRCRAFT COMPASS SYSTEMS

1 Construction [3.06]
2 Use [3.06]
3 Limitations and Faults [3.06]
4 Gyromagnetic Remote Indicating Compass [3.06]

5. Navigation General

NAVIGATION TERMS

1 Air Position [5.01]
2 Great Circle [5.01]
3 Rhumb Line [5.01]
4 Greenwich Hour Angle [5.05]

MAPS AND CHARTS

1 Lambert Conformal [5.06]
2 Transverse Mercator [5.06]
3 Enroute Low and High Altitude Charts [5.06]

TIME AND LONGITUDE

1 Time Zones and Relation to Longitude [5.03]

FLIGHT PLANNING CALCULATIONS

1 Heading and True Airspeed [4.02, 5.02]
2 Wind and Windspeed [5.02]
3 IAS-CAS-EAS-TAS [5.02]
4 Track and Groundspeed [5.01, 5.02]
5 Mach [4.02]
6 Time [5.03]
7 Weight and Balance [4.03]
8 Flight Planned Fuel Requirements [4.03]
9 Fuel Load/Zero Fuel Weight [4.03]
10 Pay Load/Weight Shift [4.03]
11 Critical Point (CP) [5.08]
12 Equal Time Point (ETP) [5.07]
13 Flight Plans [1.18]
14 Flight Itinerary [1.18]

COMPUTERIZED FLIGHT PLANS

1 Decode [1.18]
2 Analysis and Interpolation [1.18]

EN ROUTE NAVIGATION

1 Use of Aeronautical Charts [CPL]
2 Calculation of Heading and G/S [5.02]
3 Use of Radio Aids to Determine Position and

Transferring Position Lines [7.03, 7.04]
4 Gyro Steering Techniques in Areas of Compass Unreliability
5 Maintaining Flight Log (Air Position) [CPL]
6 Determination of Wind Velocity [5.02]

6. Radio Communications and Aids to Navigation

RADIO
1 Elementary Theory [7.01]
2 Wave Length and Frequency [7.01]
3 Frequency Bands Used in Communication and Navigation [7.01]
4 Characteristics of Low, High and Very High Frequency Radio Waves [7.01]
5 Ground Waves and Sky Waves [7.01]
6 Skip Distance [7.01]
7 Reflection and Refraction [7.01]
8 Night Effect [7.01]

AIRCRAFT RADIO TRANSCEIVERS
1 VHF [7.01]
2 HF [7.02]
3 DATALINK [7.01]

SELECTIVE CALL SYSTEM (SELCAL)
1 VHF [7.01/7.02]
2 HF [7.02]

EMERGENCY LOCATOR TRANSMITTER (ELT)
1 Requirements [1.43]
2 Testing [1.43]
3 Flight Planning [1.43]
4 Accidental Transmission [7.15]
5 Pilot Response to Signals [7.15]
6 Downed Aircraft Procedures [7.15]

RADAR
1 Elementary Theory [7.13]
2 Primary Returns [7.13]
3 Secondary Returns [7.13]
4 Weather Radar [7.13]

NAVIGATION SYSTEMS
1 Automatic Direction Finder (ADF) [7.04]
2 VHF Omnidirectional Range (VOR) [7.03]
3 Distance Measuring Equipment (DME) [7.06]
4 Co-located VOR and TACAN (VORTAC) [7.03, 7.05]
5 Long Range Area Navigation (LORAN C) [7.10]
6 Global Navigation Satellite System (GNSS - GPS) [7.13]
7 Very High Frequency Direction Finding (VHF-DF) [7.16]
8 Area Navigation System (RNAV) [7.07]
9 Inertial Navigation System (INS) [7.07]

APPROACH AIDS
1 Instrument Landing system (ILS) [7.08]
2 Global Navigation Satellite System (GNSS-GPS) [7.13]
3 Surveillance Radar (ASR & AASR) [7.12]
4 Precision Approach Radar (PAR) [7.12]
5 Secondary Surveillance Radar (SSR) [7.12]
6 VASIS/PAPI [4.09]

TRANSPONDERS [7.12]

ACAS/TCAS
1 General [1.44]
2 Use of TCAS/ACAS [1.44]
3 Pilot Immunity from Enforcement Action [1.44]
4 Pilot/Controller Actions [1.44]
5 Pilot and Controller Interchange [1.44]

7. Flight Operations

ATMOSPHERIC EFFECTS ON FLIGHT
1 ICAO Standard Atmosphere [6.01]
2 Temperature and Pressure/Air Density [6.02]
3 Humidity/Rain [2.03, 6.02]
4 Cold Temperature Corrections [1.25]

PERFORMANCE
1 Indicated and True Stalling Speeds [CPL]
2 Slow Speed Flight Characteristics: Turbo-prop, Turbo-jet [4.01, 4.02]
3 High Speed Flight Characteristics: Turbo-prop, Turbo-jet [4.01, 4.02]
4 Relationship of Speed to Angle of Attack [4.01]
5 Cruising for Range/Endurance [4.01, 4.07]
6 Flight Performance "V" Speeds - Definition and Use [4.04, 4.05]
7 Weight and Balance - Load Adjustment [4.03]
8 Effect of Changes in Weight and Load Distribution [4.03]
9 Hydroplaning [4.08]
10 Wind Shear - Effects and Avoidance [4.17, 7.08]
11 Landing Techniques [4.17]

CHARTS AND GRAPHS
1 Weight and Balance [4.03]
2 Take-off [4.05]
3 Climb [4.06]
4 Cruise [4.08]
5 Buffet Boundary [4.02]
6 Descent [N/A]
7 Landing [CPL]
8 Crosswind [4.08]
9 Weight, Altitude, Temperature (WAT), Take-off/ Landing Performance Charts [4.06, 4.08]

CRITICAL SURFACE CONTAMINATION
1 Clean Aircraft Concept - Practices and Techniques [4.12]
2 Frozen Contaminants Including Cold-Soaking Phenomenon [2.13, 4.12]
3 De-Icing and Anti-Icing Fluids [2.13, 4.12]
4 De-Icing and Anti-Icing Procedures [2.13, 4.12]
5 Variables that Can Influence Holdover Time [2.13]
6 Critical Surface Inspections [4.12]
7 Pre-take-off Inspection [4.12]
8 Health Affects [2.13]
9 Application Guideline Tables [2.13]

WAKE TURBULENCE
1 Causes and Effects [4.15]
2 Avoidance Procedures [4.15]
3 Separation Criteria and Waiver [1.22]

FLIGHT MANUAL
1 Approved Information

VOLCANIC ASH
1 Hazards [4.18]

AIRMANSHIP/RULES OF THUMB
1 General [Throughout]

8. Theory of Flight

FORCES ACTING ON AN AEROPLANE
1 Load Factor [4.01]
2 Relationship of Weight and Load Factor to Stalling [4.01]
3 Gust Loads [4.01]
4 Stability [4.02]
5 Lift/Weight/Thrust/Drag [CPL]

WING DESIGN
1 Wing Tip Vortices [4.01, 4.15]
2 Sweepback [4.02]
3 Leading and Trailing Edge Flaps [2.01]
4 Winglets [2.01]
5 Canards [2.01]
6 Vortex Generators [2.01, 4.02]
7 Wing Fences [2.01]
8 Spoilers [2.01]

9. Human Factors

AVIATION PHYSIOLOGY
1 Hypoxia/Hyperventilation [4.19]
2 Gas Expansion Effects [4.19]
3 Decompression (including SCUBA diving) [4.19]
4 Vision/Visual Scanning Techniques [4.19]
5 Hearing [4.19]
6 Orientation/Disorientation (including visual and vestibular illusions) [4.19]
7 Positive and Negative "G" [4.19]
8 Circadian Rhythms/Jet Lag [4.19]
9 Sleep/Fatigue [4.19]

THE PILOT AND THE OPERATING ENVIRONMENT
1 Personal Health / Exercise/Fitness
2 Obesity/Diet/Nutrition [4.19]
3 Medications (prescribed and over-the-counter) [4.19]
4 Substance Abuse (alcohol and drugs) [4.19]
5 Pregnancy [4.19]
6 Heat/Cold [4.19]
7 Noise/Vibration [4.19]
8 Effects of Smoking [4.19]
9 Toxic Hazards (including carbon monoxide) [4.19]

AVIATION PSYCHOLOGY

1. The Decision-Making Process [4.20]
2. Factors That Influence Decision-Making [4.20]
3. Situational Awareness [4.20]
4. Stress [4.20]
5. Managing Risk [4.20]
6. Attitudes [4.20]
7. Workload (Attention and Information Processing) [4.20]

PILOT - EQUIPMENT/MATERIALS RELATIONSHIP

1. Controls and Displays: Errors in Interpretation and control; Information Selection (e.g. "glass" cockpits) [4.21]
2. Alerting and Warning Systems: Appropriate Selection and Set Up; False Indications; Distractions and Responses [4.21]
3. Standard Operating Procedures (SOPs) [4.21]
4. Correct Use of Charts, Checklists and Manuals [4.21]
5. Cockpit Visibility and Eye Reference Position/ Seat Position [4.21]

INTERPERSONAL RELATIONS

1. Communications with: Flight and Cabin Crew; Passengers; Company Management; Flight Operations; Maintenance Personnel; Air Traffic Services [4.22]
2. Crew Problem Solving and Decision Making [4.22]
3. Crew Management/Small Group Dynamics [4.22]
4. Operating Pressure: Family; Peer Group; Employer [4.22]

Crew Resource Management (CRM) [4.22]

Controlled Flight Into Terrain (CFIT) [4.22]

Accelerate Your Career With Our Fast Track Exam Prep Seminars!

We offer 2-day accelerated exam prep seminars from our aviation training centre at YVR Airport. Our courses are perfect for those wishing to comprehensively review all required subjects so as to fast track their study and succeed on the latest Transport Canada written exams. For further information on seminar dates or to register, visit our website at AcceleratedAviation.com, drop by our training centre at #130-5980 Miller Road YVR International Airport, Richmond BC (near the main terminal complex, in the vicinity of the Air Canada hangars) or call us at 604-279-0179.

Private/Recreational

2-day exam prep seminar for Transport Canada Private Pilot** (PPAER) and Recreational Pilot (RPPAE) writtens. Topics covered: Air Law and Procedures, Aircraft Engines, Airframes and Aircraft Systems, Theory of Flight, Flight Instruments, Meteorology, Navigation and Navigational Aids, Flight Operations, Human Factors, exam question analysis and problem solving techniques.

Course fee: $375.00*

Commercial Pilot

2-day exam prep seminar for Transport Canada Commercial Pilot** - Aeroplane (CPAER) written exam. Topics covered: Air Law and Procedures, Aircraft Engines, Airframes and Aircraft Systems, Theory of Flight, Flight Instruments, Meteorology, Navigation and Navigational Aids, Flight Operations, Human Factors, exam question analysis and problem solving techniques.

Course fee: $475.00*

ATPL/IATRA

2-day exam prep seminar for Transport Canada Airline Transport Pilot, Aeroplane (SARON and SAMRA) and for Individual Aeroplane Type Rating (IATRA) writtens. Topics covered: Air Law and Procedures, Air Transport Operator Regs, Transport Category Aeroplane Engines and Systems, Theory of Flight and Aerodynamics, Flight Instruments, Advanced Meteorology, Navigation, Flight Planning, Aids to Navigation, Flight Operations, Human Factors, exam question analysis and problem solving techniques.

Course fee: $575.00*

Instrument Rating

2-day exam prep seminar for Transport Canada Instrument Rating (INRAT) written. Topics covered: Air Law, IFR Rules and Procedures, Flight Instruments Instrument Flying Procedures, Meteorology, IFR Charts, IFR Flight Planning, Navigational Aids, Human Factors, exam question analysis and problem solving techinques.

Course fee: $475.00*

Helicopter CPL

2-day exam prep seminar for Transport Canada Commercial Pilot - Helicopter (CPHEL) written**. Topics covered: Air Law and Procedures, Aircraft Engines, Airframes and Systems, Rotary Wing Theory of Flight and Aerodynamics, Flight Instruments, Meteorology, Navigation and Navigational Aids, Helicopter Flight Operations, Human Factors, exam question analysis and problem solving techniques.

Course fee: $575.00*

Helicopter ATPL

2-day exam prep seminar for Transport Canada Helicopter Airline Transport writtens (HARON, HAMRA). Topics covered: Air Law, Helicopter Air Operator Regs, Large Helicopter Engines, Airframes and Systems, Advanced Rotary Wing Theory of Flight Flight Instruments, Meteorology, Navigation and Navigational Aids, Helicopter Flight Operations, Human Factors, exam question analysis and problem solving techniques.

Course fee: $675.00*

Flight Dispatcher

2-day exam prep seminar for Transport Canada Flight Dispatcher Generic writtens. Topics covered: Air Law and Procedures, Air Operator Regs, Transport Category Aeroplane Engines and Systems, Theory of Flight and Aerodynamics, Flight Instruments, Basic and Advanced Meteorology, Navigation and Flight Planning, Navigational Aids, Flight Operations, Human Factors, exam question analysis and problem solving techinques.

Course fee: $675.00*

AME Regs

2-day exam prep seminar for Transport Canada Aircraft Maintenance Engineer Regulatory Requirements written exam. Topics covered: detailed review of all applicable CARs, CARs standards and related airworthiness and aircraft maintenance enactments, exam question analysis.

Course fee: $475.00*

AcceleratedAviation.com

**We are not a registered flying school: our courses serve as supplemental training/tutoring services only and cannot be used to replace or supplant formal Transport Canada ground school attendance requirements. PPL/CPL students are responsible to obtain necessary certifications i.e. letter of recommendation for written etc. from their primary flight school: we have no authority to provide any such certifications or letters of recommendation.

*Prices shown are in $CDN; training texts, supplies and applicable taxes are extra. Prices subject to change.

FOR EXAM PURPOSES ONLY - NOT FOR NAVIGATION

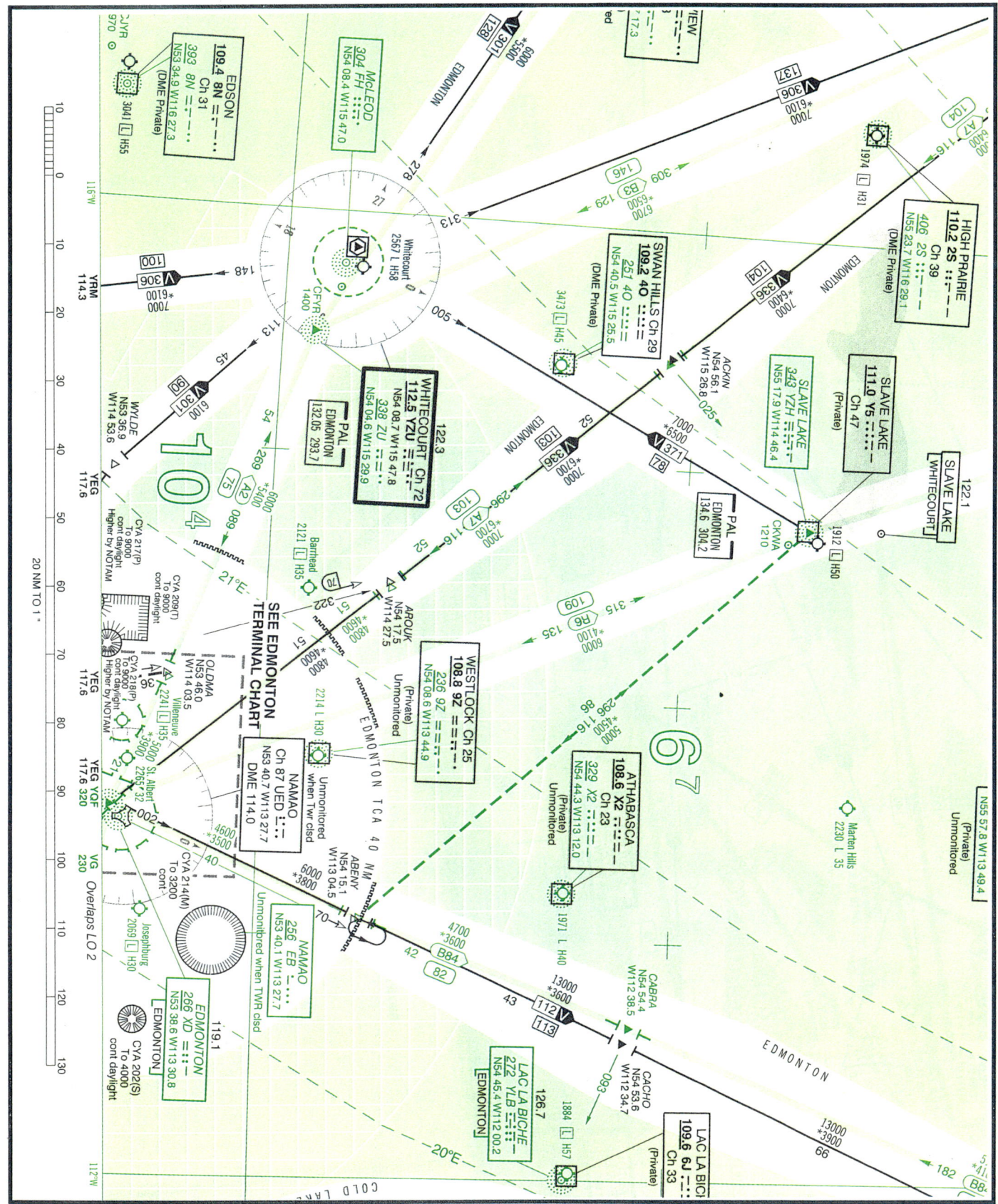

Fig. A-1: LO Chart Extract.

FOR EXAM PURPOSES ONLY - NOT FOR NAVIGATION

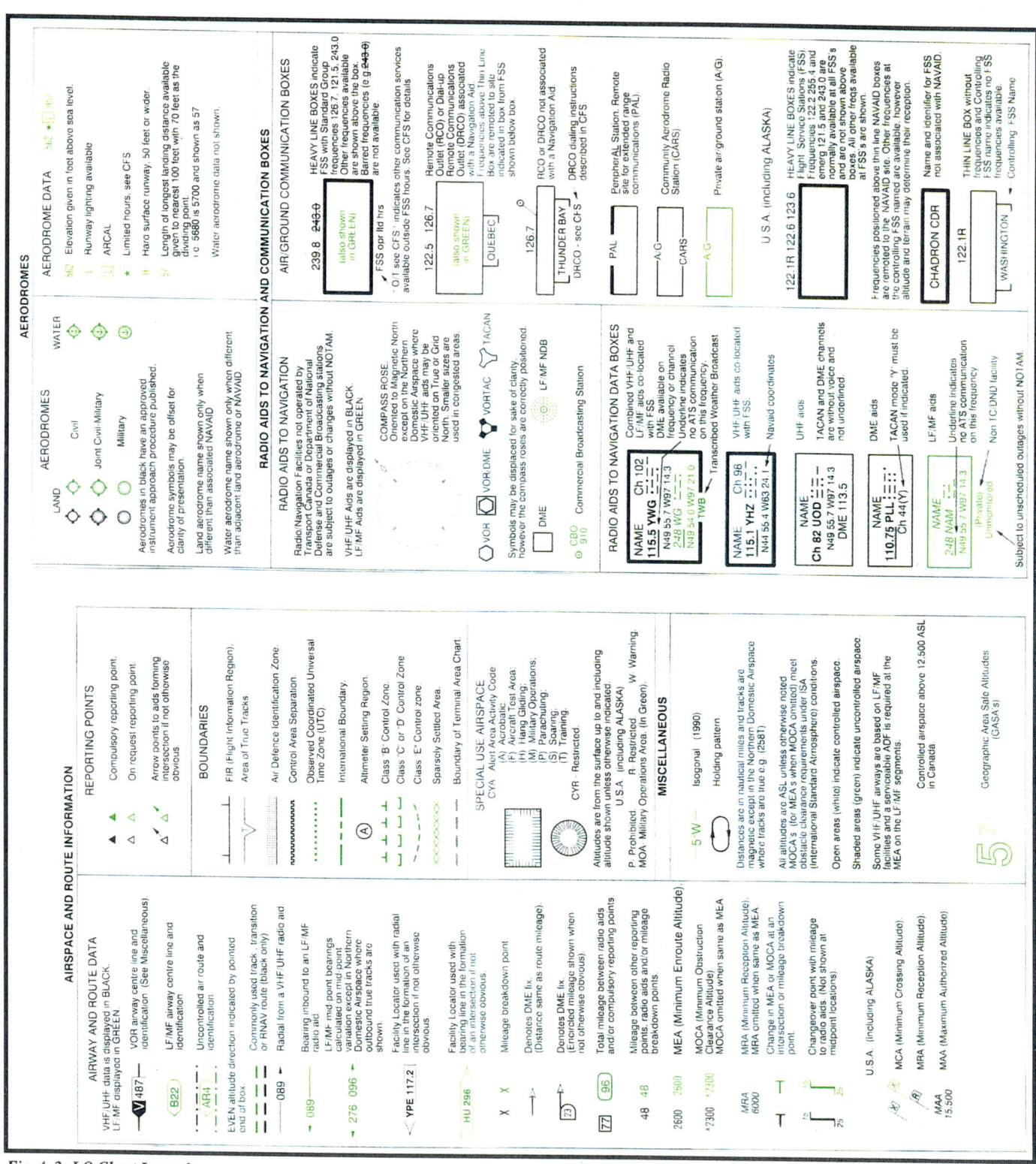

Fig. A-2: LO Chart Legend.